DEVELOPING MODELS IN SCIENCE EDUCATION

Developing Models in Science Education

Edited by

John K. Gilbert

and

Carolyn J. Boulter

School of Education,
The University of Reading,
Reading, U.K.

KLUWER ACADEMIC PUBLISHERS
DORDRECHT / BOSTON / LONDON

A C.I.P. Catalogue record for this book is available from the Library of Congress.

ISBN 0-7923-6652-2

Published by Kluwer Academic Publishers,
P.O. Box 17, 3300 AA Dordrecht, The Netherlands.

Sold and distributed in North, Central and South America
by Kluwer Academic Publishers,
101 Philip Drive, Norwell, MA 02061, U.S.A.

In all other countries, sold and distributed
by Kluwer Academic Publishers,
P.O. Box 322, 3300 AH Dordrecht, The Netherlands.

Printed on acid-free paper

Printed in the Netherlands.

CONTENTS

Preface

This book arises from the collaborative work of a group of international researchers who are members of the Centre for Models in Science and Technology: Research in Education (CMISTRE). Based at The University of Reading in the UK, the Centre has a widely scattered membership, i.e. currently also in Australia, Brazil, Israel, New Zealand, Netherlands. Information about its present work can be accessed via The University of Reading's web pages on http://www.rdg.ac.uk/~ems97pc/MISTRE.

Formed in 1995, the first years of the group were spent in formulating a common language with which to talk about models and modelling and in negotiating the boundaries of the areas to be investigated. In this period of time the main themes which are addressed in this book started to be formulated, based on the interests and experiences of the collaborating members. All this was fuelled by academic visits by members to each others, by regular seminars where new papers and ideas were discussed, in conference symposia, both national and international, where these ideas were subjected to a wider audience and, more lately, by publications in journals. In most senses the Centre is typical of a research group in any field: a commitment by a group of academics to enquiry in a theme held in common. It might differ from many in two ways. First, it draws on the insights of a number of established disciplines: philosophy of science, historical studies of the development of science, the sociology and language of science, the psychology of the teaching and learning of science. Second, it has entailed a greater commitment to collaborative ways of working and to a reflection on the contextual nature of the understandings that are forged. Within the Centre, smaller sets of members often collaborate in particular areas of interest and expertise. This has given rise to the three main areas of interest that are reflected in the Sections of this book.

Although the common language is presented in detail in Chapter 1, it may be helpful to readers if the components of the framework and the agenda are summarised here. A model has been taken to be 'a representation of an idea, object, event, process, or system'. Mental modelling is defined as an activity undertaken by individuals, whether alone or within a group. The results of that activity can be expressed in the public domain through action, speech, writing or other symbolic form. Those expressed models, as we term them, which gain social acceptance following testing by the community of professional scientists play a central role in the conduct of both research and development, becoming consensus models. Whilst those consensus models which are currently in use at the frontiers of science may be termed scientific models, those produced in specific historical contexts may be called historical models. Curricular models are those versions of consensus models which are included in science curricula. Teaching models are those developed to assist in the understanding of curricular models and hence the phenomena that they represent. Hybrid models are those formed for teaching purposes by merging the characteristics of several distinct consensus models in a field of enquiry. A model of pedagogy is that used by a teacher in the planning and provision of science education in classrooms and laboratories.

Mental, expressed, and consensus models play key roles in the conduct and dissemination of the outcomes of science and technology. Together with curricular, hybrid, and teaching models, they play key roles in the teaching and learning of science and technology. CMISTRE is thus concerned with a broad question:

> What parts do models play in the production, dissemination, understanding, and use of knowledge in science and technology?

This question is being addressed by exploring:

1. The ways in which individuals construct and use mental models.

2. The ways in which these models are presented as expressed models.

3. The processes by which expressed models gain social acceptance to become consensus models.

4. The relationships between the historical models in an area of enquiry.

5. The processes by which teaching models are developed and used to facilitate the understanding of consensus models.

6. The uses made of models of all types both in science and technology and in science education and technology education.

7. Models of the curriculum in science and technology education.

8. The development and use of models of pedagogy by teachers.

The components of this common language and agenda of enquiry are returned to throughout the book, notably in the Preface to each Section and in Chapter 18. The work on the nature of models and the roles that they play in science and technology and in science and technology education is represented in Section One, 'On the Nature and Significance of Models'. The important differences between mental models in the private domain and expressed models in the public domain determined the area of research collected together in Section Two, on 'The Development of Mental Models'. From the development of the theoretical base represented in these two Sections, the work has progressed into the practical task of investigating models and modelling in settings where teaching and learning are the focus. This forms the content of Section Three, 'Teaching and Learning Consensus Models'.

Chapter 18, the last in the book, looks at the challenges of the position that has now been reached and at the various practical projects which are in operation using the theoretical framework developed. It then looks into the future to describe the areas for possible future research both in the theory of models and in their practical expression in situations of teaching and learning.

Acknowledgements

The writing and production of this book have involved many people who have given their time and expertise generously. As the Editors, we would like to thank all who have contributed to the text and to its proof reading. Above all, we would like to thank Mrs. Helen Apted who has put this book together, struggling with templates and pagination to produce the text that you see. She has done this with great calmness and presence of mind for which all the contributing authors are most grateful.

John K. Gilbert and Carolyn J. Boulter
School of Education
The University of Reading
UK

Section One: On the Nature and Significance of Models

Preface

This Section is concerned with the importance of models in both science and technology and in science and technology education. In the first Chapter, the place of modelling in the process of scientific investigation and in the production of technological artefacts and processes is discussed, together with how they relate to an understanding of authentic education in these disciplines. The terminology that has developed within the group of researchers represented in this book is described, providing a framework for subsequent Chapters. The second Chapter analyses the ways in which three prominent philosophers (Kuhn, Nersessian, Bunge) have used modelling to explore the relationships between models, theories and their understandings of the nature of the world-as-experienced. The case is made for the key role that models have in forging links between reality as perceived and reality as idealised. These links have implications for how constructivism can be interpreted in science and technology education. The representation of models expressed in classroom settings, an important component of constructivism, forms the basis for the third Chapter, which puts forward a typology for these expressed models. The range of possible models is defined through their 'aspects' and 'modes' of representation. This typology opens the door to future avenues for research into teaching and learning with and about models in classrooms. The final Chapter of the Section takes up the idea of mode of representation and connects this to the enactive, iconic and symbolic modes of Bruner, concentrating upon mathematical models. This Chapter defines the special way in which mathematical models represent real and theoretical objects. It shows how the rules, which can be applied to a mathematical model, facilitate the production of particularly important predictions and hence form a key link between experimentation and the making of theory. A strong case for the development and teaching of mathematical models as a core component of scientific understanding closes the Section.

Chapter 1

Positioning Models in Science Education and in Design and Technology Education

John K. Gilbert[1], Carolyn J. Boulter[1], Roger Elmer[2]
[1]*The University of Reading, UK;* [2]*King Alfred's College of Higher Education, Winchester, UK*

INTRODUCTION

The purpose of this Chapter is to establish the place of modelling and models in science education and in technology education (the U.K. terminology of 'design and technology education' is introduced and used during the Chapter). It is argued that both the processes and outcomes of science and of technology *per se* have a great deal in common. 'Authentic' educations in science and in technology must reflect the natures of the parent disciplines as far as is practicable. Modelling and models are common to both, thus providing a potential bridge between science education and technology education. The basic terminology of modelling and models used throughout this book is presented.

THE ROLE OF MODELLING IN SCIENTIFIC ENQUIRY

The central roles that modelling plays in the processes of scientific enquiry and that models play as the outcomes of that enquiry are well established (e.g. Giere, 1988). As a consequence, modelling and models should make major contributions to 'authentic' (Roth, 1995) science education. This book is, primarily, an exploration of that potential contribution, for it is not yet fully realised in the classroom and laboratory. However, there is a secondary purpose. Barnes (1982) has argued that there are considerable similarities between the processes and outcomes of science and of technology. This suggests that some commonalities ought to exist between science education and technology education. Modelling and models should be capable of

J.K. Gilbert and C.J. Boulter (eds.), Developing Models in Science Education, 3–17.

forming a bridge between the two. This book is a first step in constructing such a bridge. Whilst the emphasis is on the role of modelling and models in science education, because much of the relevant research and development work so far has been done there, this Chapter makes the case for such a bridge whilst Chapters 7,14, and 18 explore some of the issues involved.

The essence of much of the thinking that underlies this book is reflected in the report *Beyond 2000: Science Education for the Future* (Millar and Osborne, 1999). A recommendation made is that:

> The science curriculum from 5 to 16 (years) should be seen primarily as a course to enhance general 'scientific literacy'. (para. 4.2)

It is suggested that one structural element of such a curriculum should be 'explanatory stories', which are:

> The heart of the cultural contribution of science ... a set of major ideas about the material world and how it behaves ... (presented in) one of the world's most powerful and persuasive ways of communicating ideas ... narrative form. It is these accounts ... which interest and engage pupils. (Para. 5 2.1)

It is also proposed that

> Work should be undertaken to explore how aspects of technology and the applications of science currently omitted could be incorporated within a science curriculum designed to enhance 'scientific literacy. (para. 5.2.3)

We intend to establish that the theme of 'modelling and models' is both a highly suitable basis for the construction for many 'explanatory stories' and that it can provide a valuable link between science and technology in education.

THE CONDUCT OF SCIENCE AND OF TECHNOLOGY

Educational provision under the labels of 'science' and 'technology' should be as 'authentic' as possible (Roth, 1995), that is they should be as faithful to the intellectual structures of the parent disciplines as possible. Syllabuses should reflect three things. First, the processes by which science and technology are conducted (their epistemologies). Second, the value systems underlying such activities, the situations in which and the purposes for which they take place (their contexts). Third, the entities with which they deal and which are their outcomes (their ontologies).

This reflection of epistemologies, contexts, and outcomes should be as accurate as is possible under the circumstances within which education is conducted. For 'authenticity' to be possible, there must be a reasonable prior understanding of them by both practitioners and educators. The natures of these processes and outcomes are discussed in the sections below. These are complicated matters: for example, only simplified versions of 'processes' (given below) are even partially acceptable to their practitioners.

The Nature of Technology as Process and as Outcome

Pacey (1983) has defined 'technology-practice' as:

> ... the application of scientific and other knowledge to practical tasks by ordered systems that involve people and organisations, living things and machines...

The 'practical tasks' most commonly addressed focus on the improvement of the physical conditions of human life (UNESCO,1983). Pacey's (1983) 'technology-practice' consists of three, simultaneously operational, elements: the technical aspect, the organisational aspect, and the cultural aspect. The technical aspect consists of:

> ... knowledge, skill and technique; tools, machines, chemicals, liveware; resources, products, wastes...

In short, it is the aggregate of human resources brought to bear on these 'practical tasks', the means by which these are deployed, and the material focus and outcomes of this deployment. The organisational aspect is:

> economic and industrial activity, professional activity, users and consumers, trade unions.

These are the social organisations in which technology as an activity takes place, together with those which support, in one way or another, the conduct of that activity. The cultural aspect consists of relevant:

> ... goals, values and ethical codes, belief in progress, awareness and creativity.

within which solutions to practical problems are both framed and evaluated. In respect of the core idea of 'values', Pacey (1983) notes that:

> ... the culture of technology comprehends at least two overlapping sets of values, the ones based on rational, materialistic, and economic goals, and the other concerned with the adventure of exploiting the frontiers of capability and pursuing virtuosity for its own sake (p.89)

Striking a balance between the influence of these two sets of values in technology education is very difficult. It will be manifest in the outcomes of technological activity, the technologies that are produced, the solutions to the practical tasks arrived at: objects (products, e.g. cars, clothes) and systems (processes, e.g. ways of making cars, clothes). What emerges from Pacey's (1983) ideas is that technological process consists of thoughtful actions by individuals taken within social contexts to produce solutions to problems which it is intended will be valued.

The Nature of Science as Process

Science is about finding explanations for natural phenomena in the world-as-experienced. The document *Science for All Americans* (Rutherford & Ahlgren, 1990) states that:

> Science presumes that the things and events in the universe occur in consistent patterns that are comprehensible through careful, systematic, study. (p.3)

Matthews (1994) has identified ten philosophical theses which inform the view of science-as-a-process in *Science for All Americans*. These may be summarised as follows:

(1) Realism. The material world exists independent of human experience and knowledge.
(2) Fallibilism. Although human knowledge of the world is imperfect, it is possible to make reliable comparisons between competing theories about the nature of the world.
(3) Durability. Science modifies the ideas that are produced about the world, rather than abandoning them if they are found to be inadequate.
(4) Rationalism. The validity of scientific arguments is tested, sooner or later, against the criteria of inference, demonstration, and common sense.

(5) Antimethodism. There is no fixed set of steps in a scientific enquiry, for knowledge involves an element of human creativity rather than emerging directly from experiment.
(6) Demarcation. Although there is no fixed method for scientific enquiry, it does involve a series of features which enable it to be distinguished from other, non-scientific, endeavours.
(7) Predictability. Successful science predicts observations which are then made.
(8) Objectivity. Although science is a human activity, it attempts to rise above subjective interests in the pursuit of truth.
(9) Moderate Externalism. The direction of scientific research is influenced by prevailing views on what questions are worth addressing, and what methods will prove productive.
(10) Ethics. Ethical considerations determine what topics are researched and arise in the actual conduct of research.

The outcomes of science are the broadly-conceived notion of 'scientific methodology', together with descriptions of how the material world behaves, ideas about the entities of which the world is believed to consist or with which it can be reliably analysed (concepts), proposals for how these entities are physically and temporarily related to each other in the material world (models), and general sets of reasons why these behaviours, concepts, and models can be thought to occur (theories). Science then consists of thoughtful actions by individuals within social contexts producing explanations of the natural world which it is hoped will be valued. The similarity of these overviews of science and technology suggests that there is a relationship between them.

The Relationship Between Science and Technology

The ways that science and technology relate, which cover both the processes involved and the outcomes achieved are undoubtedly complicated. It is possible to argue that the processes of technology first provide solutions to problems. Science afterwards explains the reasons for the success of these solutions. For example, steel was initially developed empirically as a way of producing harder iron, whilst the consequences for the structure of iron of the addition of small amounts of other elements, e.g. cobalt, were only explained long afterwards. It is possible also to argue that science precedes technology in time, such that technology is the application of science. For example, that enquiries into the sequences of amino acids within genetic material are leading to the rapid development of the industry of biotechnology.

A third interpretation is that the two are bound together in a synergistic relationship, as Barnes (1982) argues, a view which justifies the similarities between their definitions. Both involve invention, being creative, constructive activities conducted within social contexts which draw extensively on prior achievements and which are subject to no one major constraint on their success. People constitute the link between the two, with an individual often moving between scientific and technological activity. There is a traffic in knowledge and skills between the two, whilst they are both concerned to achieve definite outcomes. In short, science and technology are interdependent. We argue that this interdependency should inform 'authentic' science education and technology education. At the moment they generally do not, hence the call by Millar and Osborne (1998).

THE NATURE OF EDUCATION IN SCIENCE AND IN TECHNOLOGY

In most countries the curricula in science and in technology are currently organisationally separate.

The Nature of Science Education

Science has, in one form or another, been a 'subject' at school level in many countries for well over a century. As a consequence, there has been an extensive sharing of experience, so that the structure and substance of that provision has become fairly homogeneous at world level. Beyond transitory fashions (e.g. 'learning by discovery') the differences, such as they are, tend to be couched in terms of the 'applications' of science to 'practical tasks' of local (often national) importance, e.g. water purification in rural areas, wind-powered electrical generation. In general terms, there are three major components to all curricula which are drawn from diluted versions of the academic subjects of biology, chemistry, and physics. These are either taught independently of each other, or with the curricula co-ordinated so as to avoid repetition and the teaching of different interpretations of the same ideas, or in an integrated fashion, with the material built around some common theme or topic.

Hodson (1993) has identified three purposes for science education which cut across the structure and content of whatever provision is made. One major purpose is to 'learn science', that is, to come to understand the major achievements of science, the concepts, the models, the theories. A second major purpose is to 'learn about science', that is, to develop an understanding of the nature and methods of science, how it is conducted.

The third purpose is to 'learn to do science', that is to become able to engage in and develop expertise in the practice of scientific enquiry.

The U.K. National Curriculum for Science (DfEE, 1995a), for example, is organised into four threads ('Attainment Targets') which run across the full age spread of compulsory science education in state schools (5 to 16 years). These are concerned with the processes of science ('Experimental and Investigative Science'), biology ('Life Processes and Living Things'), chemistry ('Materials and their Properties'), and physics ('Physical Processes'). The emphasis is on 'learning science' and to some extent on 'learning how to do science', with a relative neglect of 'learning about science'.

The Nature of Technology Education

In many countries, 'technology education' is evolving, at an uneven pace, from 'craft education' (McCormick, 1991). Craft education, the physical making of things, was traditionally reserved for students of lower academic achievement, e.g. in Canada (Ontario Ministry of Education and Training, 1993), and of lower social status, e.g. in South Africa (Department of National Education, 1991). There are global trends towards the automation of industrial production and increased competition in the innovation of products. These are leading to an increased instrumental valuation of the design element of technology education. At the same time, the intrinsic valuation of this element is also increasing as the contribution that it can make to the development of creativity (see Chapter 7) is appreciated. The design element is concerned with deciding on the optimum fit (a value-laden term) between the problem for which a solution is sought, the structure of that solution, and the materials from which it is to be made.

The trend towards the emphasis on the design element is reflected, for example, in the U.K. National Curriculum for the significantly named school subject of 'Design and Technology' (DfEE, 1995b). This consists of the two threads ('Attainment Targets') of 'Designing' and 'Making'. This book will adopt the U.K. nomenclature of 'Design and Technology' (hereafter D&T) for the school subject of 'technology education'. This is done for two reasons. First, to differentiate it from the U.S.A. use of the word 'technology' as meaning 'anything to do with computers'. Second, as a reminder of the curricular tension between 'making' and 'design' which is reflected in the titles of the two Attainment Targets.

EXPLORING LINKS

There are a number of strong reasons why an exploration of possible links between science education and D&T education is desirable. Living in the everyday world entails solving a continuous series of practical problems. Decisions have to be taken about personal matters, for example, what type of diet to follow in order to remain healthy. Decisions have to be taken socially about matters of collective importance, for example, whether a waste incinerator should be built in the neighbourhood. Decisions have to be taken about economic matters, for example, whether an individual should seek work in an emerging field of employment. Taking informed decisions about an ever-increasing number of examples of all three of these types requires a substantial knowledge and understanding of the processes, contexts, and outcomes of both science and of technology. The perspective brought to bear on such problems often involves an integration, or at least co-ordination, of the insights drawn from the two fields of endeavour.

It was argued above that science and technology, as spheres of intellectual activity and as practical pursuits, are interdependent. However, a review of the curricular interactions between these two domains, within timetabled 'science', shows a wide and fragmentary range of provision (Gilbert, 1992): minimal reference to technological outcomes during science lessons; the teaching of science followed by a discussion of its 'application' in technology; teaching which starts from science and which then leads on to technological outcomes; teaching which starts from technological outcomes and which leads on to the associated science; and, of course, the use of technological outcomes in the teaching of science. To these should be added the design and making of technological outcomes within 'design and technology education' with very varying degrees of reference to the processes and ideas of science. 'Authentic' education in both, whether as a preparation for everyday decision-taking or for disciplinary activity *per se*, must draw them closer together by focusing attention on those aspects which they share or where considerable similarity exists between them. If students are to learn about these shared aspects, it would seem more desirable to use them as a partial basis for structuring the overall curriculum pursued rather than persisting with the present *ad hoc* arrangements.

The theme of 'modelling and models' can be manifest in three ways in the school subjects of science and of D&T. First, modelling as a process and the models that are an outcome of that process can be part of the substance of curricula. The issue is one of the values of modelling and models *in* curricula. Second, science education and D&T education can be represented for analysis with the aid of models of the curriculum. The issue is one of the value of modelling and models of curricula. Third, modelling and models in and of science curricula and D&T education can form an important way of exploring the relationship between them. The emphasis in this book is, because of the circumstances created by the emphasis of research to date, on the first of these three, and on science education in particular. However, we attempt to establish the conditions of a future treatment of the other two possibilities.

ON MODELLING AND MODELS

Modelling and Models in Science Education

A model in science is a representation of a phenomenon initially produced for a specific purpose. As a 'phenomenon' is any intellectually interesting way of segregating a part of the world-as-experienced for further study, models are ubiquitous. The specific purpose for which any model is originally produced in science (or in scientific research, to be precise) is as a simplification of the phenomenon to be used in enquiries to develop explanations of it (see Chapter 10). Many models are composed of entities which are concrete, objects viewed as if they have a separate existence (e.g. a wheel) or as if they are part of a system (e.g. a wheel on a car). A model of an object can be either smaller than the phenomenon which it represents (e.g. of a train), or the same size as it (e.g. of the human body), or bigger than it (e.g. of a virus). Other models are composed of abstractions, entities which are treated as if they are objects, e.g. forces, energy. A model can thus be of an idea. A model can consist of a mixture of entities which are concrete (e.g. masses) and of entities which are treated as if they are concrete (e.g. forces acting on masses). A model can be of a system, a series of entities in a fixed relation to each other (e.g. of the stations and the connections between them in a metro railway). A model can be of an event, a time-limited segment of the behaviour of one or more entities in a system (e.g. a model of an athletics race). A model can be of a process, one or more events within a system which have a distinctive outcome (e.g. of the Bosch-Haber method of making ammonia from nitrogen and hydrogen). A Thought Experiment (see Chapter 8) is a model of that group of processes known as a 'scientific experiment' carried out entirely within the mind as an idea, a mental model.

A classification of the ontological status of models is possible:

- A *mental model* is private and personal cognitive representation. It is formed by an individual either on their own or whilst within a group (see Chapter 5).
- An *expressed model* is placed in the public domain by an individual or group, usually for others to interact with, through the use of one or more modes of representation (see next paragraph). The relation between any one mental model and the apparently corresponding expressed model is complex. Any reflective person who has set out to express a mental model will be aware that the act of expression has an effect on a mental model: expressing it changes it.
- Different social groups, after discussion and experimentation, can come to an agreement that an expressed model is of value, thereby producing a *consensus model*. In particular, scientists produce a wealth of expressed models of the phenomena which they are investigating. An expressed model which has gained acceptance by a community of scientists following formal experimental testing, as manifest by its publication in a refereed journal, becomes a *scientific model*. It then plays a central role in the conduct of scientific research for a length of time which is governed by its utility in producing predictions which are empirically supported.
- Those consensus models produced in specific historical contexts and later superseded for many research purposes are known as *historical models* (see Chapter 11).
- That version of an historical or scientific model which is included in a formal curriculum, often after some further simplification, is a *curricular model*.
- As the understanding of consensus, historical, and curricular models (as well as the phenomena that they represent) is often difficult, *teaching models* are developed to assist in that process. Teaching models can be developed by either a teacher or by a student (see Chapter 10).
- An *hybrid model* is formed by merging some characteristics of each of several distinct scientific, historical, or curricular models in a field of enquiry. It is used for curriculum and classroom teaching purposes as if it were a coherent whole (see Chapter 11).
- A *model of pedagogy* is used by teachers during the planning, practical management, and reflection on, classroom activity and is concerned with the nature of science, the nature of science teaching, and the nature of science learning (see Chapter 17).

One or more of five *modes of representation* are significant in expressed models of any phenomenon and can be used to construct typologies (see Chapter 3):

- The *concrete mode* consists of the use of materials, e.g. a metal model of a railway engine, a polystyrene model of a molecule.
- The *verbal mode* consists of the use of metaphors and analogies in speech, for example when talking about teaching models (see Chapter 10) and in written form, for example in textbook descriptions.
- The *mathematical mode* consists of mathematical expressions, including equations, e.g. the universal gas equation.
- The *visual mode* makes use of graphical pictorial forms in graphs and in diagrams.
- The phrase '*symbolic model*' includes the visual, verbal, and mathematical modes.
- The *gestural mode* consists of actions, e.g. movements of the hand.

Modelling and models make three major contributions to science education. First, it is believed that the formation of mental models and the public presentation of expressed models are central to the development of an understanding of any phenomenon or body of information (see Chapter 5). Mental modelling is thus as important in achieving all three of Hodson's (1993) purposes for science education as it is in the learning of any other subject. Second, the production and experimental testing of expressed models plays a central role in the processes of science (see points 2,3,5,6,7 in Matthews {1994} list, given above). Hodson's (1993) purposes of 'learning about science' and 'learning to do science' thus involve modelling and model testing. Third, historical and scientific models are major outcomes of science (see Chapter 2 and points 1,2,3,4,8,9 in Matthew's {1993} list). Hodson's (1993) purpose for science education of 'learning science' must involve the development of an understanding of major historical and scientific models, if only through curriculum models.

Modelling and Models in Design and Technology Education

Modelling and models are also used for specific purposes in D&T education. Harrison (1992) puts it thus:

> The critical question to ask about a model is … for what purpose is it intended? Indeed this intention will determine the nature of the model, against which the usefulness will be evaluated. (p.32)

Several systems for classifying these purposes have been put forward. Harrison (1992) divides them into: helping with thinking; communicating form or detail; evaluating a design or features of it. Whilst others, e.g. Liddament (1993), have longer lists, a telling division is into 'communicating with oneself' and 'communicating with others' put forward by Kelly et al. (1987). This latter division reflects the situation in design *per se*, where the designer uses modelling to reflexively develop personal ideas, to work with other members of the design team by facilitating communication, as well as in communicating intentions to and in negotiation with clients (Baynes, 1992).

Design and technology education, in contrast to science education, is often substantially 'authentic' in nature, i.e. it involves a student acting as a 'designer' and then going on to be a 'maker'. However, there are three major problems inherent with such an expectation. First, the educational system is badly capitalised, so that the range of resources available to the student with which to construct concrete models is bound to be poorer than that available in an industrial context. Second, although industry has to produce solutions to problems within a given period, the time allocated to a task is usually fairly negotiable at the outset. Schools, on the other hand, divide time into blocks of rigid size, perhaps leading to conservative designs and poor 'making'. Indeed, the act of producing a concrete model may act as a substitute for full product realisation. Third, the matter of assessment becomes complicated, for:

> There is some confusion between industrial and educational perspectives on the activity (of D&T Education). In education, the concern is to expose pupils to designing technological experiences in order that they may develop understanding and capability. In industry, that design and technological capability is directed towards a manufacture of a product. (Kelly et al. 1987, p.7)

Within the educational perspective on student work, what Downey and Kelly (1986) call the 'intrinsic' aim, the teacher-as-assessor has to evaluate the quality of the processes undertaken during the activity and their personal significance for the student, including the capability to work appropriately with others. This latter tension will be manifest in terms of the models presented: good work within the industrial perspective, for example within the visual mode, may lead students to use professional means of presentation, e.g. computer graphics, whilst good work within the educational perspective may lead students to use more personally expressive means of presentation, e.g. a series of rough sketches on paper. Liddament

(1993) has explored the conflict between models used to teach concepts of design and technology and models used to assist and advance design activity *per se*. They include issues of the 'ethics of representation' (Baynes, 1992): modelling-for-the-public-domain will be in tension with modelling-for-the-private-domain.

Modelling and models is not a well-developed theme in the literature of design and technology education. As is typical of any discipline in what Kuhn (1970a) calls a pre-paradigmatic phase, a wide range of words is used to cover ill-defined, perhaps similar, meanings. For 'modelling', these include 'imaging', 'cognitive modelling', 'concrete modelling', 'making'. For 'model', these include 'mock-up', 'lash-up', 'prototype'. One aspect of an exploration of the scope of modelling and models in forming links between science education and design and technology education must be the extent to which a common terminology can be used in the two fields.

That which goes on in a designer's head has been described by Kimbell et al. (1996, p.114) as 'creative concrete thinking'. However, the processes involved and the consequences achieved seem to be very close to what were called 'mental models' earlier in this Chapter (see also Chapter 5). The next step in the design process is the generation of an expressed model in an appropriate mode of representation. A visual mode, for example a sketch, or a concrete mode, for example the use of modelling clay, are commonly used. This is then subjected to a cycle of development, testing, further development, and so on, until the designer is convinced that the outcome can be presented to the client (or, in an educational context, the teacher-as-surrogate client) in the form of a *prototype*. This prototype will be subsequently altered in response to the client's reaction and, perhaps more significantly, in the light of the materials used in fabrication when the product is manufactured.

This developmental process parallels, in many ways, the changes that take place as a consensus model is produced in a D&T classroom. Common to both is:

- The notion of a developmental cycle, with changes taking place to the nature of the testing imposed and to the model itself, leading to an outcome through a rolling programme.
- The notion of 'fitness for purpose' in respect of a 'design specification' being used as a judgmental criterion at the end of the developmental programme.

- The notion of that evaluation being conducted on behalf of both the immediate social group (the class) and an external reference agency (the client).

However, perhaps there is greater variability in the design and technology education context, as compared with a science education context, in respect of:

- The modes of representation used. Thus, whilst both make use of visual modes, e.g. diagrams, D&T education makes more use of the concrete mode.
- The range of materials used within the concrete mode in D&T education. Card, plastic, modelling clay, are all commonly used. At each stage in the developmental process the 'convenience of use' of a material is balanced against its 'analogical capability' (the range of ideas that it can express).
- The range of perspective adopted. Whilst models are all too often only presented in 2D in science education, not only is much greater use of 3D made in D&T education but the perspective adopted can vary during the developmental cycle.
- The range of scales used. Whilst modelling in the science education context often sticks to one scale (that which can be contained within one page of a school exercise book!), scale in D&T education can either remain small (where the product would be large, e.g. a bridge) or be gradually increased to that of the prototype (where the product would be of 'human scale', e.g. a kettle).

The notion of 'teaching model' has some relevance to D&T education. This is where an 'exemplary solution to a problem' is used to teach the principles of D&T. There is no direct equivalence to the idea of a scientific or historical model, in that there are, axiomatically, no generic solutions to design problems. However, the existence of a 'school' of design, where general principles of colour, form, and composition are employed. For example, the Bauhaus (Pevsner 1960, Wingler 1969) comes moderately close through the notion that a recognisable approach to problem solution is widely adopted.

MODELLING AND MODELS AS A BRIDGE

It would seem that modelling and models do constitute a possible bridge between science education and D&T education. It is a defining feature in the conduct and outcomes of both science and D&T. Although Erika

Zimmermann, in Chapter 17, having researched science education, restricts herself to that field, it will be possible to speculate in the future on the existence of possible parallel models of pedagogy for D&T education.

The purpose of modelling in both fields is to facilitate communication through a visualisation of the relation between the intention and the outcome of the activity. In the case of science education, the intention is to provide an explanation, which can be defined as an answer to a question about the nature of the world-as-experienced (see Chapter 10). The quality of an explanation produced can be evaluated by consideration of the predictive value of the model produced. In the case of D&T education it is usual to define the problem with some precision before the design process begins. A model enables the suitability of any proposed solution(s) (the fitness-for-purpose) to be evaluated against that problem before fabrication of the final outcome takes place.

The first stage of creative activity in both fields is the construction of a mental model. In both cases this mental model is subjected to a developmental process, through the medium of a series of expressed models, towards a version which is socially accepted (a scientific or historical or curriculum model and prototype respectively). The range of modes of representation used in both fields is somewhat similar, although the incidence of their use probably varies between the two. Whilst a consensus model, a scientific/historical/curriculum model, is the 'final' outcome of science education, the prototype model in D&T education should be followed by another stage (the manufacture of the product outcome) but this happens all too infrequently. Exemplary designs in D&T education have a role similar to that of historical models in science education: they represent solutions to problems which, whilst they have now been overtaken by events, were valued in their day.

This Chapter has outlined the basic terminology which has developed and through looking at the nature of authentic education in science and D&T has suggested that modelling and models should be taught across both fields as a way of linking them. Further chapters in this book take up these ideas and develop them further within a wide variety of situations and from perspectives which deal with the processes, contexts, and ontologies of modelling.

Chapter 2

Science and Education: Notions of Reality, Theory and Model

John K. Gilbert[1], Mauricio Pietrocola[2], Arden Zylbersztajn[2], Creso Franco[3]
[1]The University of Reading, UK; [2]Federal University of Santa Catarina, Brazil; [3]Catholic University of Rio de Janeiro, Brazil

INTRODUCTION

It was argued in Chapter 1 that science education and technology education should both be as 'authentic' as possible and that modelling and models, for which a typology was proposed, can form a bridge between the two. However, modelling and models must be seen within a broader context, that of the relationship between notions of 'reality', 'theory' and 'model', for two reasons. First, science education, which aspires to be authentic, must be based on an historically and philosophically valid view of the nature of science, in which these three notions play important parts. Second, it can be argued that perhaps, to some extent and in some way, the development of ideas by an individual parallels (or can be seen as a metaphor for) the development of ideas in science. The treatment of the reality/theory/model relationship given in this Chapter, which is of importance in its own right, is set within the second of these two reasons because it subsumes the first.

SCIENCE, SCIENCE EDUCATION AND CONSTRUCTIVISM

In the last twenty years or so, a very large body of research data has been accumulated into the nature of students' understanding of specific elements of the content and processes with which science is concerned (Pfund and Duit, 1988). This output, known as 'alternative conceptions' or 'alternative frameworks' or 'naive understandings' or 'children's science', may be summarised as follows:

J.K. Gilbert and C.J. Boulter (eds.), Developing Models in Science Education, 19–40.

1. From an early age, and prior to any formal teaching and learning of science, children develop both meanings for many words used in science teaching and views of the world.
2. Children's ideas are often strongly held and significantly different from the current views of scientists.
3. Children make sense of many new experiences by constructing meanings based on their existing ideas.
4. Students retain, modify, or change their existing understandings when they are taught, as well as acquiring ideas.

That spectrum of ideas known as 'constructivism', broadly definable as 'using existing ideas to construct meaning from new experiences whilst using acquired experience for producing new ideas', has achieved wide support as being the best available explanatory psychological framework within which to set these conclusions. It has also been seen by many as providing the best approximation available to the conditions for teaching and learning needed to achieve 'authenticity' (Tobin, et al. 1994).

Inevitably, the constructivist 'movement' has acquired its critics, the strongest of whom is Matthews (1994, 1998). As he puts it:

> For many, constructivism has ceased being just a learning theory, or even an educational theory, but rather it constitutes a worldview or *Weltanschuung*... constructivism is committed to certain epistemological positions that are very contentious and, given the widespread educational influence of the doctrine, deserve close scrutiny. (Matthews, 1994, p.139)

Osborne (1996) has analysed the achievements, strengths, and weaknesses, of constructivism, with an emphasis on the epistemological assumptions which underpin the pedagogical actions that are most commonly taken by teachers in 'the constructivist classroom'. He points out that an identification of the difficulties that students experience in learning has led to practical measures to help them become more aware of their own understandings, to the rejection of the *tabula rasa* assumption in teaching and to improved teacher skills of formative assessment. However, he suggests that constructivism misrepresents the nature of science by failing to accept the notion of realism which underpins scientific practice and confuses the contexts of knowledge making and knowledge learning. The consequences of these weaknesses are, in his view, that it offers no guidance on how students might adjudicate between competing theories, offers no guidance to teachers on the selection and sequencing of content in curriculum, and rejects didacticism as an approach to teaching. The major criticism he levels against the 'radical constructivist' perspective is based on its alleged vehement rejection of realism. The criticism he levels against the 'social constructivist' perspective is the failure of schools to create the social context for such scientific learning to take place, whilst he sees 'personal construct psychology' to have failed to produce testable predictions about learning.

Just the existence of these criticisms suggests that those who wish to base the teaching of science on constructivist principles must do so with approaches which are both well-founded epistemologically and well-developed pedagogically. The assertion underlying this book is that modelling and models should and can have a major impact on the curriculum, teaching, and learning of science in schools in the movement for authenticity. As this impact is likely to be made within constructivist assumptions, it follows that there must be the greatest possible ontological and epistemological clarity over what might be done, why, and how. A number of key questions must in addressed in an attempt to obtain such clarity. Is 'realism' an acceptable epistemological and ontological basis for science education? What meanings and roles have modelling and models within realism? How do notions of model and theory, which are intertwined within most discussions of science and science education, relate to each other within an acceptable realist assumption? How can modelling and models contribute to a well-developed pedagogy for science education? We start with the notion of realism.

NOTIONS OF THE WORLD-AS-EXPERIENCED

The first key element of a realist view is the assertion that the world-as-experienced actually exists independently of humanity, being composed of entities of a fixed nature. Ogborn (1995), concerned with the relation between science and science education, believes that it is the only sure foundation on which science, and hence authentic science education, can be based:

> … knowledge in the natural sciences is made by human beings, is never, because nothing could ever be, absolutely certain, and yet provides solidly reliable accounts of the material world, upon which we can certainly act. (p.6)

The second key element of a realist view is the assertion that science can gradually approach a complete knowledge of reality. This has been explored by Bhaskar (1978). He argues that, although knowledge is produced by the application of a social product (scientific methodology) leading to the modification of other antecedent social products (theories and models), a realist view requires the assumption of two 'dimensions' of understanding and two kinds of 'objects' of knowledge:

a. The 'intransitive dimension' in which the object is the real structure or mechanism that exists and acts quite independent of human beings and the conditions which allow them access to it.
b. The 'transitive dimension' of ideas about the nature of the entities of which the world is thought to exist. These are produced, communicated, and changed, are historically situated and contingent, and are thus a human achievement.

Using this terminology, practising scientists can be said to adopt a policy of viewing the ideas within the transitive dimension as provisionally real (intransitive), such that suitably informed individuals anywhere can use them to act upon the world (Harré, 1986). Action in the form of experiments, using these ideas as tools for enquiry, is needed to test the validity of the assumption of intransivity. If such action is always successful and the ideas do not infer facts which are not found in the world then they gradually come to be viewed provisionally as true, as factual, as permanently part of the assumed intransitive dimension. However, if the world actually exists independently of what we know of it, such ideas always remain fallible and open to modification, even to refutation.

Every type of entity of which the world-as-experienced is intransitively composed, in the realist view, has a distinctive nature, can do only specific things, and can only have certain things done to it. An entity shows all of its range of behaviour in 'open systems' (those unaltered by human action), although systematic, scientific, enquiry usually involves the construction of limited, hence artificial, conditions, or 'closed systems', to prevent other entities intruding into the behaviour of those on which enquiry is focused (Bhaskar, 1978). The gap between the nature of a closed system in which a candidate entity is explored and an open system in which it usually exists in the natural world places an inherent limitation on a readiness to accept it as 'real'. The most complete acceptance of an entity as real, as intransitive, comes about when it is very successful in providing explanations of open systems.

There is, as one would expect, an anti-realist view of the world-as-experienced. This asserts that it is never possible to conclude that the world-as-experienced is actually composed of particular types of entities if they are not directly observable. Such entities must, on this argument, remain entirely the products of the human imagination. For some anti-realists of positivist inclination, theories are just useful summaries of data, which are collected by experiment, with better theories incorporating greater quantities of more accurate data.

Which is the more suitable basis for science education: realism or anti-realism? As has already been said, practising scientists tend to adopt a provisionally realist stance: so it is, perhaps, up to anti-realists to justify their views. They do so by attacking realism. How is this done and to what effect? Ogborn (1995) argues that anti-realists start from the assumption that science is conducted in accordance with fixed rules of rationally using a fixed empirical methodology based on a realist (intransitive) view of entities. Anti-realists, according to Ogborn (1995), then show that the actual practice of science deviates from this representation of it. They then conclude *both* that scientific knowledge is socially constructed (because scientific methodology is not context-independent) *and* that the assumption of realism is unfounded. The anti-realist argument thus denies the differentiation between the intransitive and the transitive dimensions, believing the world-as-experienced to be entirely transitive. The weakness in the argument, for Ogborn (1995), is that, although science-as-practised is demonstrably not conducted on strictly rationalist lines and by the mechanical application of an algorithmically-applied empiricist methodology, it cannot be inferred from this that reality does not exist. As the assumptions of the anti-realists are, for Ogborn (1995), false, so must be the anti-realist view itself. What emerges from these arguments is that it would be unsafe to base an authentic science education on an anti-realist view of science.

On what interpretation of the realist view should science education then be based? Should a 'strong' interpretation be used, where it is assumed both that the world-as-experienced exists and that science can gradually determine the true nature of the entities of which it is composed? Or should a 'weak' interpretation be used, where only the realism of the world-as-experienced is assumed, with the question of whether science can progressively get closer to an understanding of the true nature of its elements being left to one side. Both would allow for the assumption of the transitive dimension.

In the next three sections we set out the ideas on the nature of 'theory' and 'model' expounded by Thomas Kuhn, Nancy Nersessian, and Mario

Bunge. These three are all realists, but they differ in the way that their philosophical positions would seem to relate to the 'strong' and 'weak' interpretations of realism. Kuhn, we will argue, takes a 'weak' view, Bunge a 'strong' view, with a reading of Nersessian, who is mainly concerned with processes of model change in science, being capable of supporting either view. In the last section of this Chapter, we look at the implications of the 'weak' and 'strong' views of realism for Osborne's (1996) criticisms of constructivism.

THOMAS KUHN ON REALITY, THEORY AND MODEL

Kuhn (1970a) was concerned with the representation of change at the macro level in science. He introduced the notion of 'paradigm', in which the set of problems to be addressed in a field of enquiry, the theories and models adopted, the experimental techniques used, the criteria applied in the evaluation of results obtained, are fixed. Work in a new field of scientific enquiry shows no clear agreement between participating scientists on such matters: it is 'pre-paradigmatic'. This is then followed by a period of 'normal science', in which the operating paradigm seems to be agreed by scientists and can be identified by an observer. Dissatisfaction with some aspect of the explanations produced during a normal science period leads to a chaotic 'revolutionary science' period, as new problems, theories and models, methodologies etc., are tried out. This settles down into a different paradigm in a new period of normal science.

Kuhn on the Nature of Theories

In his work, Kuhn says very little about theories as such. His representation of science does not include an explicitly developed theory of theories (Giere, 1988, pp.35-36), a remark that can also be made about his treatment of models (Abrantes, 1998). The reason for this is that he was concerned with the processes by which scientific knowledge changes at the macro level, rather than with the logical structure of the detailed products of research (Kuhn, 1970b, p.1). In 'Postscript-1969', a section added in the second edition (Kuhn, 1970a, pp.174-210) of his most famous book, he points out the main source of confusion which arose from his original treatment of his ideas (Kuhn, 1962). It is that the concept of paradigm was used in Kuhn (1962) both in a general and in a restricted sense. In the general sense, 'paradigm' was employed to mean the entire constellation of group commitments shared by the members of a scientific community. It was to denote this meaning more clearly that he later suggested the expression 'disciplinary matrix' in Kuhn (1970a):

> Scientists themselves would say that they share a theory or a set of theories, and I shall be glad if the term can be ultimately recaptured for this use. As currently used in philosophy of science, however, 'theory' connotes a structure far more limited in nature and scope than the one required here. Until the term can be freed from its current implications, it would avoid confusion to adopt another. For present purposes I suggest 'disciplinary matrix': 'disciplinary' because it refers to common possession of the practitioners of a particular discipline; 'matrix' because it is composed of ordered elements of various sorts, each requiring further specification. (p.182).

The main components of the disciplinary matrix were identified by Kuhn (1970a) as being: symbolic generalisations (expressions, either in mathematical or verbal form, deployed without question or dissent by group members), models, shared values, and the exemplars (concrete solutions to problems, that serve as models for the solution to similar problems). The last component represented the restricted sense in which the word paradigm was originally used by Kuhn (1962), and one which he considered to be of the uttermost importance both for the education and practice of members of a scientific community.

Kuhn on the Nature of Models

Two distinct senses in which the notion of model is used can be found in Kuhn's treatment of the notion of 'disciplinary matrix'. One sense has to do with the role played by the 'exemplars'. This is concerned with the processes of learning to become a scientist and later of actually doing science as an independent scientist. In both cases what is involved are problem-solving activities modelled on solutions already accepted within the paradigm:

> As the student proceeds from his freshman course to and through his doctoral dissertation, the problems assigned to him become more complex and less completely precedented. But they continue to be closely modelled on previous achievements as are the problems that normally occupy him during his subsequently independent career. (Kuhn, 1970a, p.47).

These exemplary problem solutions were regarded by Kuhn as one of the essential vehicles for learning the cognitive content of a theory, which he saw as consisting, among other things, of verbal and symbolic generalisations together with examples of their function in use (Kuhn, 1977a, p.501). For him, 'normal science' research is mostly guided by a direct modelling of these exemplary problem solutions, as opposed to the application of abstracted rules (Kuhn, 1970a, p.47).

The second sense in which the notion of model is used is concerned with beliefs in particular types of models. In the 'Postscript-1969' (Kuhn, 1970a), he refers to a spectrum of types ranging from 'ontological' to 'heuristic' models. Ontological models were regarded by Kuhn as objects of metaphysical commitment, deeply held by scientists, about what actually exists in the universe and about what their main features are. In this category Kuhn (1970a) included beliefs such as 'heat is a constituent property of bodies' and 'all perceptible phenomena are due to qualitatively neutral atoms in a vacuum, or alternatively, to the interaction of matter and force, or to fields'.

Heuristic models were seen as analogies, which enable an object of study to be fruitfully considered as if it was like another, more extensively understood, object even though the latter is known to be completely different in nature. Scientists are not committed to them in any permanent way as objects of belief and they are viewed and used pragmatically and instrumentally. Examples of this variety that he gives are 'an electric circuit may be regarded as a steady-state hydrodynamic system' and 'a gas behaves like tiny elastic billiard balls in random motion'.

In spite of the difference in commitment to the two varieties of models by scientists, in 'Postcript-1969' (Kuhn, 1970a) stressed the similar functions that they serve for a group, a community, of scientists:

> Though the strength of such commitments varies, with non-trivial consequences, along the spectrum from heuristic to ontological models, all models have similar functions. Among other things they supply the group with preferred or permissible analogies and metaphors. By doing so they help to determine what will be accepted as an explanation and as a puzzle-solution; conversely, they assist in the determination of the roster of unsolved puzzles and in the evaluation of the importance of each. (Kuhn, 1970a, p.184).

Models, for Kuhn, perform these functions by virtue of being a source of similarity relations which can be either external (between essentially different objects and situations) as in the case of heuristic models, or internal (between objects and situations essentially of the same type) as in the case of ontological models (Hoyningen-Huene, 1993).

The status of heuristic models is quite clear in Kuhn's work. They are used to demonstrate a formal similarity between laws and theories in different domains (Abrantes, 1998), so that those from one can help to explain or to investigate the other. The case is not so clear, however, when ontological models are considered. Kuhn used expressions like 'metaphysical models' (Kuhn, 1970c, p.271), 'objects of metaphysical commitment' (Kuhn, 1977b, p.463) and 'ontological models' (Kuhn, 1970a, p.184) interchangeably. In doing so, he introduced a confusion, by equating 'ontological' models with 'metaphysical' models. Since ontology refers to assertions about the nature of reality and metaphysics does not, what did Kuhn means by 'metaphysical'? His work does not (at our reading) include a definition, so we must assume that he was using 'metaphysical' as a synonym for 'philosophical', as opposed to 'scientific' in the realist sense of the latter term. This interpretation is supported by his statement:

> And as the problems change, so often does the standard that distinguishes a real scientific solution from a mere metaphysical speculation, word game, or mathematical play. (Kuhn, 1970a, p.103)

Kuhn on the Nature of Reality

Why did he refer to scientists' commitments to ontological models as being 'metaphysical'? This can be answered by considering Kuhn's views of the relation between theory and reality as given in 'Postscript-1969' (Kuhn, 1970a). For him a theory was better than its predecessors only in the sense of being a better instrument for discovering and solving puzzles. He was not a 'strong' realist, in the sense defined earlier in this Chapter, despite realism being the prevalent perspective on nature adopted by both philosophers of science and lay people at the time of his major publication (1962):

> There is, I think, no theory-independent way to reconstruct phrases like 'really there'; the notion of a match between the ontology of a theory and its 'real' counterpart in nature now seems to me illusive in principle. Besides, as a historian, I am impressed with the implausibility of the view. I do not doubt, for example, that Newton's mechanics improves on Aristotle's and that Einstein's improves on Newton's as instruments for problem solving. But I can see in their succession no coherent direction of ontological

> development. On the contrary, in some important respects, though by no means in all, Einstein's general theory of relativity is closer to Aristotle's than either of them is to Newton's. Though the temptation to describe that position as relativistic is understandable, the description seems to me wrong. Conversely, if the position be relativism, I cannot see that the relativist loses anything needed to account for the nature and the development of the sciences. (Kuhn, 1970a, p.206).

In not accepting an ontological approximation to reality in the historical development of theories, it was only natural for Kuhn to consider the beliefs held by scientists about what exists in nature as metaphysical. He thus saw no harm in broadening the usage of the word 'model' to include objects of belief such as atoms, fields, or forces acting at a distance (Kuhn, 1977b, p.463, Note 9). For him, assertions about what exists in nature were to be seen as being a model of it and never as a claim about what is really there. For him, a scientist may fully believe that there is a match between theoretical entities and their real counterparts, but those ontological beliefs are, at the end of the day, no closer to reality than the electric circuit is to the steady-state hydrodynamic system, or a molecule is to a billiard ball.

On our analysis, then, Kuhn was a 'weak' realist. He was prepared to accept that the world-of-experience actually existed, if only for the sake of argument. His use of 'model' to refer to the 'model solutions to problems within a paradigm' is relevant to the concerns of this Chapter and needed to be clarified. However, his recognition of heuristic models as helpful analogies and his ambivalence over the status of ontological models suggests that, whilst he was willing to accept what Bhaskar (1978) subsequently called the transitive dimension, he was not willing to accept that the progress of science shifts entities in the transitive dimension to the intransitive dimension. Scientific change is not necessarily moving closer to an understanding of the intransitive, but rather to a different presentation of the transitive, made for different purposes.

Nancy Nersessian on Reality, Theory, Model

Nersessian (1992a) emphasises the importance of overcoming the traditional separation of the analysis of the context of discovery from that of the context of justification in philosophy of science if we are to address the status of Bhaskar's (1978) intransitive/transitive distinction. She pointed out that Kuhn did not fully clarify the discovery/justification issue within his notion of 'paradigms of science'. In her own words (Nersessian, 1992a, p.7) she says that:

> [Kuhn] identifies conceptual change as the 'last act' (in paradigm change), when 'the pieces fall together'. Thus portrayed, conceptual change appears to be something that happens to scientists, rather than the outcome of an extended period of construction by scientists. A change of 'gestalt' may be an apt way of characterising this last point in the process, but focusing exclusively on this last point has contrary to Kuhn's aim provided a misleading portrayal of conceptual change; has reinforced the widespread view that the processes of change are mysterious and unanalysable; and has blocked the very possibility of investigating how precisely the new gestalt is related to its predecessors.

In order to grasp the process of conceptual change in science, Nersessian (1992a) developed a system of cognitive-historical analysis, which sought to investigate the context of development in which

> A vague speculation gets articulated into a new scientific theory, gets communicated to other scientists, and comes to replace existing representations of a domain (op. cit., p 6).

In the cognitive historical perspective, developing scientific theories is a problem-solving process that consists of modelling activities which involve generating new conceptual representations from existing ones (Nersessian, 1992a, p.12). The modelling capabilities of the mind are exercised through a set of abstracting techniques, which include imagistic reasoning, analogical reasoning (see Chapter 5), thought experiments (see Chapter 8) and limiting case analysis.

In her 1992(a) study Nersessian analyses the specific mechanisms by which scientific theories are developed. Her thorough examination of Maxwell's studies of the electromagnetic field, based on Faraday's ideas, shows how the use of analogical and imagistic reasoning supported Maxwell's development of a new theory. Her sketch analysis of Galileo's and Einstein's studies focuses on thought experiments and limiting case analysis as tools for modelling new theories. In the case of Galileo, for instance, she analyses the establishment of the law of falling bodies by means of considering the fall of a body in a medium and exploring the consequences of reifying the medium down to the limit of a vacuum. In her analysis, Nersessian emphasised some features of Galileo' s approach. These are: (i) the assumption, for the first time, of an Archimedian model for framing the issue of falling bodies, an approach which was already used in hydrostatics; (ii) modelling the consequences of the initial assumptions via thought experiments and limiting case analysis, thus redeveloping for the context of falling bodies some features which were already known in the context of hydrostatics.

An important feature of Nersessian's work on modelling is her use of theoretical tools, e.g. abstraction techniques, which both deal with the particularities of the specific scientific domain under investigation and also which represent some degree of generalisability in approach. In other words, she bridged the gap between context independent and context dependent approaches accounts of the processes of knowledge building.

By comparing the work of Kuhn and Nersessian, it is possible to establish two distinct but complementary patterns for the relationship between theories and models. According to Kuhn, models are a constitutive component of an already established disciplinary matrix. Such models, which are constrained by the intransitive nature of the phenomena studied within an existing paradigm, offer analogies on the basis of which the phenomena might be conceived within that paradigm. Complementarily, Nersessian points out that models are important as a starting point for the development of theories. She also stresses that modelling activities are carried out by scientists using abstracting techniques. The creation of new ideas leads the scientific enterprise to results that go beyond the model which was the starting point.

Nersessian says nothing which leads to the conclusion that she doubts the existence of a human-independent reality. However, she seems ambivalent over the intransitive/transitive issue. She has clarified the nature, status, and mode of operation of models as a key element in the transitive domain. However, her position is not so clear on whether or not the change between paradigms, produced by modelling in the transitive dimension, does or does not lead scientists nearer to a complete understanding of reality. Her position could be said to be capable of supporting both a ‘strong’ and a ‘weak’ view of realism.

Mario Bunge on Reality, Theory and Model

As we have seen, Kuhn was primarily concerned with periods of ‘normal science’, in which theories and models play a stable but ill-defined yet mutually supportive role, interspersed with periods of ‘revolutionary science’. As we have also seen, Nersessian is concerned both with the role that models have within paradigms and in the formation of new paradigms. Bunge is concerned with the relationship between theory and model at any time in the development of a field of scientific enquiry. He views the development of theoretical knowledge as the main purpose of science. Thus:

> To convert concrete things into richer and deeper conceptual images and to expand them into progressively complex theoretical models, increasingly faithful to the facts, is the only effective method of apprehending reality by thought. (Bunge, 1974,p.12)

He is a realist (Cupani, 1991) who both accepts the existence of the intransitive dimension and sees science as capable of providing, eventually, a full understanding it. This, as we shall see, involves him in using modelling and models within the transitive dimension. These views qualify him as a 'strong' realist.

His scheme of analysis has three components:

- *Generic Theories*. These are abstractions produced by reason and intuition, which are potentially capable of applying to any part of reality.
- *Model-Objects*. These represent the common properties of a group of real objects.
- *Theoretical Models* (otherwise called *Specific Theories*). These, what in this book we would call 'models' (whether expressed, scientific, or historical see Chapter 1), are produced by applying a generic theory to a model object, interpreting the latter in terms of the former.

He summarises the relationship as follows:

> When suppositions and special data referring to a particular body (a model-object) are associated with classical mechanics and classical gravitation theory (generic theories), a specific theory is produced (a theoretical model) about that body. In this way we have Lunar theories, theories about Mars, theories about Venus, and so on. (Bunge, 1973,p.54)

The Table below contains the outlines of a number of Bunge's actual examples (Bunge, 1973,p.53). Thus:

SYSTEM	MODEL OBJECT	THEORETICAL MODEL	GENERIC THEORY
Moon	Spherical solid rotating about its axis, in rotation about a fixed point, etc.	Lunar Theory	Classical Mechanics and gravitation theory
Moonlight	Plane polarised electromagnetic wave	Maxwell equations for the void	Classical electromagnetism
Piece of ice	Linear casual chain of beads	Statistical mechanics of casual chains	Statistical Mechanics
Crystal	Grid plus cloud of electrons	Bloch's Theory	Quantum Mechanics

He has most to say about model-objects and theoretical models.

A model-object is an idealisation, a generalised object produced by the simplification of a number of real objects so as to emphasise their commonalties. It is an arbitrary idealisation, being the product of what Bunge (1974,p.16) refers to as 'fictional materialism', which must be evaluated in terms of 'fitness for purpose' rather then in terms of being right or wrong. Such objects are treated temporarily as if they were the reality from which they were abstracted: Harré's (1986) 'policy realism'. Doing so enables scientists to focus on specific aspects of a complex reality. Bunge (1977) believes it to be unimportant if the model-object is constructed by the use of analogy: the issue is the quality of the insight gained when it is combined with a generic theory. For example, in the early years of the study of heat and electricity, model-objects based on the idea of an 'incompressible fluid', derived by analogy from the well-developed science of fluid mechanics, enabled considerable progress to be made in those fields of enquiry.

A theoretical model (in this book, a 'model' whether expressed, scientific, or historical) occupies a scientifically vital intermediary position between a model-object, which being an idealised empirical object cannot yield knowledge by the direct application of logic, and a generic theory, which being entirely the product of imagination cannot be directly applied to reality. A theoretical model includes a representation of the properties and behaviour of the model-object and of the entities of which it is constructed. This enables the application of hypothetico-deductive reasoning to produce predictions, which can be subsequently tested. The main attribute of any theoretical model is that it can represent a domain of reality. Indeed, for

Bunge (1974,p.22) it can simulate the real, thus enabling the internal mechanisms which support the relationships between the entities of which it is thought to be composed to be defined. He differentiates between theoretical models, in which internal mechanisms are postulated and 'black box' models where they are not. These mechanisms, within a realistic perspective, are not accessible to perception, but are merely inferred. He calls these 'hidden mechanisms'

> A hypothesis of hidden mechanisms can only be considered as confirmed if it satisfies the following conditions: to explain observed operations; to foresee new facts other than the ones foreseen by black-box models; and to be compatible with known laws. (Bunge, 1974, p.22)

The significance of a generic theory is evaluated by considering its success, when used to interpret a model-object so as to yield a theoretical model, in leading to predictions which are empirically confirmed (Bunge, 1974,p.19).

The relationship between these three ideas can be shown through an example given in the Table included above. The phenomenon which is called *moonlight* was simplified and abstracted into the model-object of *a plane-polarised electromagnetic wave* and interpreted through the generic theory of *classical electromagnetism* to yield the theoretical model known as *Maxwell's Equations*. The latter enable predictions to be made and tested, e.g. the effect of a polarise on the brightness of moonlight, the effect on the plane of polarisation of a magnetic field. Confirmation of the anticipated outcomes validated the worth of the model-object, the theoretical model, and most importantly of all, the generic model.

Bunge's contribution, within the concerns of this book, has been to show the role of models in forging a link between reality-as-perceived and reality-as-idealised. He is a 'strong' realist, subscribing not only to the notion of reality but also to the view that science can, in due course, provide a full understanding of that reality.

MODELS, THEORIES AND REALITY

In this section we will attempt to draw together what has been said and to focus it on the central concern of this book: models.

Models are important in respect of the second element of realism: whether or not science is capable of eventually providing a full understanding of the world-as-experienced. If, like Bunge, one adopts the 'strong' view of realism, then models can acquire one of two statuses. Those that are believed to fully represent the world-as-experienced become incorporated in Bhaskar's (1978) intransitive dimension: they are thought to be the truth. However, refutation is still possible and the fact that a model has gone unchallenged may just be due to inertia on the part of the science community. One has only to look at the apparently unassailable position of the 'inert gas configuration model' in the late 1950s and that of the 'only two allotropes of carbon model' until the mid 1990s to appreciate that 'the hubris of models' is always possible. The other, much more common, status of models is as part of Bhaskar's (1978) transitive dimension. They are overly constructed by analogy, initially for some specific purpose, and survive in active scientific enquiry just so long as they are useful. Thereafter they become 'historical models' (see Chapter 1) and are condemned to be used only for routine enquiries and to that graveyard of all science, the school (and university?) curriculum. In the 'weak' view of realism, on the other hand, all models forever remain part of the transitive dimension.

It would have been very satisfying to have produced a definitive relationship between theories and models, and of both with reality, but we have not. After all, philosophers have kept this ball in play for some hundreds of years: fame has eluded us (for the moment). What we have done is to bring the theory/model/reality relation in the work of our chosen three philosophers to the fore. Kuhn, as we have seen, has little to say in detail about either theory or model. They are lumped together in the notion of 'paradigm'. He can be considered a 'weak realist' who believed in the existence of an external reality, which constrains the construction of scientific theories, but he did not expand on the details of this relationship. Nersessian sees the formation of models by analogy as a key element in the formation of theories: the fog between the two is somewhat dispelled. Bunge offers a route from phenomena to (in his terms) theoretical models through the construction of 'model objects'. However, this allows us to see how theories may be applied to phenomena through the medium of theoretical models, but has little to say about how theories themselves are constructed. The general view of the relationship between a theory and a model may be summarised in the following way: a model is a readily perceptible entity by means of which the abstractions of a theory may be brought to bear on some aspect of the world-as-experienced in an attempt to understand it.

NATURE OF SCIENCE, INDIVIDUAL'S LEARNING AND CONSTRUCTIVISM

The Nature of Science

Authentic science education must be based, as far as possible, on an acceptable view of the nature of science, i.e. one which is received as being reasonably valid by historians and philosophers of science and also by practising scientists. So how do the chosen three philosophers stand up to this test? Such a decision is especially important if it is decided both to base the science curriculum on the view of science presented by one of them and to make students overtly aware of that basis.

When the ideas of Kuhn first appeared in 1962, they were a radical alternative to the logical positivist approach and to Popper's ideas which had dominated thinking for many years beforehand. In his own terms, they represented a new paradigm and fostered extensive work to exploit its potential. Inevitably, with its application to the history of specific fields, the cracks began to appear in the system. It seemed that change was not so coherent as Kuhn's system suggested. Moreover, as Nersessian (1992a) points out, the context of the justification of scientific knowledge was treated adequately by Kuhn but the context of the discovery of scientific knowledge was not, although the notion of 'normal science' does provide a framework within which the two can reside. Kuhn's scheme was succeeded by those proposed by Lakatos (1974) (see Chapter 11) and by Laudan (1977). Nevertheless, although the explanation for change in science given by Kuhn is rather course-grained, it could be a valuable first step into the field of philosophy of science at school level. Kuhn's idea of 'exemplars' as 'models *of* problems' in science is a novel way of looking at the idea of models, although his treatment of models *in* science is neither well-developed nor clearly distinguished from that of theories, they being conflated together under the heading of 'disciplinary matrix'.

The ideas of Nersessian (1984, 1987, 1989, 1992a,b) are too recent to have been explored to the same extent as those of Kuhn. Her own application of the cognitive-historical approach to the ideas of Faraday, Maxwell, Lorentz, and Einstein have provided valuable insights into the 'context of discovery' of scientific knowledge. However, she pays little attention to the context of justification. A practising scientist, upon reading her case studies, should come to the conclusion that they are a valid reconstruction of scientific discovery/invention. Able students should be able to follow the cognitive processes involved and all students will empathise with the demonstration of scientific thinking as an example of human creativity. However, the approach is limited to cases where detailed documentary evidence is available. Although she does not seem (at our reading) to take position on the issue of 'realism', she does propose that models are used in the development of new theories.

The work of Bunge (1973,1974,1977) is very helpful in that it deals with the relationship between the notions of 'model' and 'theory' in some detail. The scheme would seem to be applicable to scientific enquiry at any stage in the process of change, from the situation (in Kuhn's terms) of 'normal science' to that of 'revolutionary science'. With suitable examples, it should be intelligible to students.

The Nature of Science and of Learning by an Individual

The question to be addressed here is: to what extent is the work of the three philosophers discussed above an adequate basis on which to view the learning of science by an individual? This is a complex question, to which only a preliminary treatment can be given in the space available.

There is no general agreement about the existence of a relationship between change in science and change in the cognition of an individual: see Schwitzgebel (1999) and Gilbert (1999) for recent discussion of the issues. Researchers have taken widely differing positions on the matter. Piaget and Garcia (1989) saw the processes to be identical, with the mechanisms of equilibration, assimilation and accommodation being at the heart of both. Whilst Nersessian (1992a) sees a strong analogy between the two, other researchers have just pointed out the parallels between them whilst maintaining that the social psychological circumstances of science and of science education are very different (e.g. McClelland, 1984; Lythcott, 1985).

Ten years after the initial publication of their ideas (Posner, et al. 1982) in which they developed a strong analogy between Thomas Kuhn's representation change in science and change in the cognition of an individual, Strike and Posner (1992) revised those ideas. In the intervening years, numerous studies had taken place to test their basic premises: that conceptual change should take place if an individual is dissatisfied with a current conception and if an alternative is both intelligible, plausible, and fruitful. Although the results of those studies had been very mixed, Strike and Posner (1992) maintained the credibility of their scheme, only seeing the need to pay closer attention to the learner's 'conceptual ecology'. The pattern of results may be taken to indicate that the original analogy is not, in fact, a strong one. However, it may just reflect the weaknesses in Kuhn's scheme, not least the relative ambiguity of the roles of 'model' and 'theory' in it. It may be that indicators of having changed a conception bear little relationship to the processes involved in undertaking that change.

If this last point is of any merit, then it might point to greater success if Nersessian's ideas are used as the basis of the analogy. She is herself cautiously optimistic about the possibilities:

> Conceptual change as it has occurred in the history of science provides a valuable resource for gaining an understanding of the general issues of restructuring and, in some cases, may even aid the formation of hypotheses about the dimensions along which to probe students representations. (Nersessian, 1989, p.164)

Although little work seems to have been done to test these possibilities, the clear and central role for models in her scheme leads us to expect success, if only because model formation and use is a key element in the development of understanding (Johnson-Laird, 1983). Similar arguments also apply to Bunge's scheme, although it has not yet been even suggested that it be used as the basis for an analogy to an individual's cognition.

The Nature of Science and an Acceptable Version of Constructivism

One of the central questions addressed in this Chapter has been: to what extent are the views of the nature of science of the three philosophers, particularly in respect of the reality/theory/model relationship, an epistemologically and ontologically adequate basis for an acceptable pedagogy based on constructivist principles? Providing a direct answer might provide support for the fairly common practice of using these principles as a template for the design and conduct of classroom teaching and learning. On the basis of evidence of a lack of widespread success of 'the constructivist classroom', Tobin and Tippins (1993) cast doubt on the value of these principles as a template. Nevertheless, the key issues of adequacy remain, for Tobin and Tippins (1993) see a well-founded view of constructivism as a valuable critical referent against which to evaluate a wide variety of classroom practices, a variety far wider than that normally encompassed within 'the constructivist classroom'.

Whether used for planning classroom activity or as a critical referential scheme, constructivism in science education must be responsive to the five issues raised by Osborne (1996):

1. The need for a basis in realism:

All three philosophers accept the first element of a realist view: the world-as-experienced exists independently of humanity and constrains the theories that are acceptable by scientific communities. However, they differ over the second element: whether or not science can eventually discover its true constitution. As we have shown, Kuhn may be termed a 'weak' realist, accepting the first element but not the second. Bunge is a 'hard' realist, accepting both elements, whilst Nersessian seems somewhat ambivalent on the issue. If one wanted to portray science as leading to a true picture of the world, as do many senior academic scientists, then one would consider Bunge's ideas. Those wishing to present a view of science that emphasises a gap between the world-as-experienced and the possibility of science ever fully describing this world would choose a framework influenced by Kuhn's ideas.

There is one set of issues which is being consciously addressed in nations which are more alert to the ethnic diversity of the people of which they are constituted. These issues concern potential tension or even conflict between the cultural base of formal science, which may be termed 'White, Western. and Male', and that of other communities. Whether viewed as matters of cultural hegemony (Cobern, 1998), or more pragmatically as the problems that ethnic minorities have in believing some or all of the conclusions of WWM science (Aikenhead, 1996), the question of 'whose reality?' arises. A sensitivity to diverse 'voices', amongst which must be those of women of all ethnic groupings, is called for (Gilbert, et al. 1998b). Nersessian's approach, with its emphasis on a recognition of how individuals think, seems important in this context.

2. The need for an effective treatment of the contexts of discovery and of justification:

Kuhn deals well with the context of justification (the operation of normal science), but is not so successful with the context of discovery (seen as an undifferentiated element in an inchoate period of 'revolutionary science'). Bunge is also effective in dealing with justification (the production of 'theoretical models' from theories prior to experimental testing) but less successful over discovery (he has apparently little to say about how new theories are produced). Nersessian, on the other hand, is very convincing over the context of discovery, but does not address the context of justification.

3. The provision of guidelines for theory adjudication:

Kuhn gives a clear indication of where theory adjudication has taken place, at least in the case where it is failing/has failed, heralding a period of revolutionary science. However, he gives little treatment of how this takes place psychologically, concentrating instead on its sociological manifestation. Bunge is the opposite: he shows how theoretical models are produced and tested as a theory is to be evaluated, but says little of the sociology involved. Nersessian is effectively silent on the subject.

4. The provision of clear guidelines for the selection and sequencing of content:

A consideration of the ideas of all three philosophers suggests that students might be introduced to the evolving theories and models in a given area of enquiry in the order of their historical sequence. However, if this is to be done, then close attention must be paid to providing a historically valid representation. This is one in which the circumstances of change, the manner of change, and the consequences of change, are discussed not only from the vantage point of the present day but, much more importantly, also as these processes were seen as they actually took place in the past. Nersessian has a lot of invaluable detailed methodology to contribute to this approach, perhaps viewed (at least to a first approximation) within the framework provided by Kuhn. The treatment of this theme through the medium of Bunge's ideas has apparently not yet taken place.

5. The placing of a suitable value on didactic approaches to teaching:

This is only an issue if constructivism is seen as the direct basis for classroom activity. If, as is suggested by Tobin and Tippins (1993), it is just seen as the basis for the critical review of pedagogic practice, then the issue is not significant. Didactic approaches to teaching can then have their place, e.g. in defining the curriculum and in the teaching of ideas which students are unlikely to have come across in everyday life.

Looking back over the discussion of the above issues identified by Osborne (1996), it does seem that all three philosophers have something to contribute to several of those issues. More might have been said if closer consideration had already been given to the educational implications of the more recent philosophers, i.e. Nersessian, Bunge. It may be tempting to educationalists to pick individual aspects of the models of science presented by several philosophers and to combine them into a model constructed especially for pedagogic purposes. As has been shown elsewhere (Justi, this volume; Justi and Gilbert, 1998a), whilst such hybrid models can be useful in solving particular educational problems, they are not open to rational (as opposed to expedient) replacement as they have no origin and hence no status in the philosophy of science.

To close this Chapter, we would observe that constructivism, if broadly defined as at the begnning of our discussion (i.e.'using existing ideas to construct meaning from new experiences whilst using acquired experience for producing new ideas'), is not necessarily incompatible with realism, since no one would deny that scientific theories are social constructions. The extent to which one believes either that they advance in the direction of a true picture of that reality or that they are only constrained by that reality positions one in the divide between weak and strong realists. We would argue that an acceptable version of constructivism for science education should entertain, at least, a weak version of realism. To fall short of this could lead to idealism by rejecting the existence of an independent external reality. It could also lead to extreme 'post-modern' versions of relativism, in which the scientific enterprise is denied its status as the best available way of understanding the world-as-experienced. To do the latter would be to forfeit the support of the academic science community. That support is, of itself, the cornerstone of success for the enterprise of science education.

Chapter 3

Constructing a Typology of Models for Science Education

Carolyn J. Boulter, Barbara C. Buckley
The University of Reading, UK

INTRODUCTION

Representations and expressed models abound in science classrooms and vary widely on multiple dimensions. In order to encourage systematic research and principled curriculum development, we have developed a typology for categorising diverse kinds of representations and models. This chapter articulates an operational typology of models based on the attributes and modes of representations employed. It emerged from analysis of a range of models of the heart and the lunar eclipse. We conclude with a discussion of the utility of this typology for supporting research in model-based teaching and learning and its link to the study of the parts of models found in Chapter 6.

THE NEED FOR A TYPOLOGY OF MODELS

There are a plethora of models in use in science classrooms. They have arisen in a range of contexts (in history, within science, by teachers) and play diverse roles in learning. These models vary in the relationship to the phenomenon they represent, their perceived utility and function, and how they are used by teachers and learners. In the classroom, students seldom understand that they are building and using models to explain a phenomenon. Rather they encounter representations that they know they must learn as part of the science curriculum, but which are presented to them without any explicit discussion of their nature and functioning as models. (Gilbert, 1991; Grosslight et al. 1991) Even when historical scientific models

J.K. Gilbert and C.J. Boulter (eds.), Developing Models in Science Education, 41–57.

(see Chapter 11) are presented, these may not be seen as forming a link between theory and phenomena (Leatherdale, 1997). Pupils therefore often confuse the simplified, incomplete, and decontextualised models presented with the phenomena themselves. Furthermore, they are not encouraged to think about the different ways in which one phenomenon can be represented by different models. This means that, unless they also encounter sufficient complementary information and experiences, it is not only difficult for them to perceive the importance of models in explaining phenomena, it is also difficult for them to construct the robust coherent mental models needed to develop understanding (Johnson-Laird, 1983). These mental models are used in the discourse of the classroom to construct expressed models within speech, writing and action (see Chapter 15). Into this already obscure situation, teachers often introduce their own teaching models which they invent and use to provide a bridge to understanding phenomena (Brown, 1994). Both teachers and pupils lack a critical appreciation of the discrepancies and usefulness of these transient and pragmatic teaching models (Treagust et al., 1992) and how they relate to other models of the same phenomenon. They do not understand the nature of the relationship between phenomena and their representations in models. We believe the articulation of model-based teaching and learning (Chapter 6) will help address these problems. However, the idiosyncratic nature of the representations and models created for diverse purposes presents difficulties in understanding how expressed models function in science learning. A necessary part of the articulation is understanding the range of models for specific phenomena and how the different kinds of representation compare. Our first step is to produce a classification based on observable characteristics. We need categories and labels that enable us to look at the particulars and compare them in different learning contexts. What do different kinds of models of the same phenomenon have in common that enables them to facilitate or hinder learning?

CLASSIFYING MODELS IN SCIENCE EDUCATION

Categorisations of all kinds are based upon the elucidation of the properties of sets of things. This enables groupings according to their similarities, and, by default, their differences. Classifications in social science are usually constructed to highlight these similarities and differences between types, but also to facilitate description and to reduce complexity. ***Typology*** is just another term for a particular sort of classification in which the cells are constructed by combining (generally) two dimensions. The resultant display is easy to read, to make comparisons within, and to see which cells have been used and which have not. A good typology should show an exhaustive set of types and the dimensions upon which they are based. Typologies allow for types to be located quickly, discussed and compared across different areas of the typology, and for relationships among types to be explored (Bailey, 1994).

We hope that the production of such typologies may help us to:

- Chart the range of models of a phenomenon.
- Compare the range used in different learning contexts.
- Facilitate teacher reflection on the models they use and create for their students.
- Facilitate pupils' understandings of the usefulness and limitations of models within a possible range for a phenomenon.
- Assist in understanding the developmental patterns in pupils' expressed and mental models.

Categorisation has synthetic power; it should allow us to begin to structure and give coherence to the world of models and to organise the diverse range of models into a usable form. Categorisation has predictive power; it may enable us to predict patterns as we seek to fit new models into the categories. It may also enable us to ask useful questions about the progression of models in the learning process and within the development of science. As we begin to get clues as to how to answer them, it will enable us to build new theory. This is a lofty agenda. The process described in this Chapter is but the start of a research programme that we believe can be extended to a wide range of other phenomena.

CATEGORIES AND PARTICULARITIES – WAYS OF PERCEIVING THE WORLD

Categorisation or classification is a crucial personal process in making sense of the world. The human mind is set upon making sense of the enormous range and complexity of the sense impressions that we are able to experience (Bailey, 1994; Billig, 1987; Tversky, 1989). From an early age a person is able to perceive individual objects as separate from each other and to distinguish things about their individuality. The features of individual objects are paid close attention to as infants learn to identify particular physical and living forms. They use speech to name them, the parts of which they are composed and their properties. Curricula for very young children recognise the importance of this exploration, the naming of parts, and the properties of objects and events (SCAA, 1996). This vital process of attention to the particular continues throughout life. Alongside this identification process lies the companion process of building groups of similar objects. In this categorisation process objects are sorted into groups according to their similarities and differences on certain criteria, often their properties. This process too is highlighted in curricula for infants (Hohmann et al., 1979).

The interaction between categorisation and particularisation is seen in the research of Tversky (1989) who was concerned with the way her subjects viewed types of objects. When asked to list subdivisions of a common concept such as 'fish', half of her subjects produced subdivisions based upon kinds of fish, such as trout or perch, and half on parts of fish, such as fins or tail. Tversky considers "fish" to be an example of the basic level of categorisation. It is the most abstract level for which an image could be formed, at which behaviour patterns were similar, and the level named earliest by children. Her findings illustrate two ways of organising knowledge into kinds and into parts. In this chapter we begin to organise the models of science education into a typology, a categorisation based on kinds. Tversky considers partonomies, categorisations based on parts, to be a necessary and useful complement to taxonomies, empirical categorisations based on kinds. In Chapter 6 we deal with the categorisation based on parts.

CLASSIFICATION, CATEGORIES AND TYPOLOGIES

To aid us in constructing our classification scheme, we consulted Bailey (1994), who distinguishes three levels as the focus of classification conceptual, empirical, and operational, the levels with which he believes social science is concerned. Typologies can be classifications of conceptual entities. In this classification the categories are deduced from cases but no empirical cases are shown. The conceptual entities in the field of models would be the kinds of expressed models, such as, diagrams, verbal analogies, orreries, or space-filling molecular models. Empirical typologies, on the other hand, represent the opposite approach and are formed by clustering actual cases from empirical data. The empirical cases in the field of models are the models actually used in the classroom, such as a particular teaching model to explain the refraction of light, a particular plastic model of the eye, or a drawing of the digestive system. Sharing attributes of both of these approaches, the operational typology may be constructed by two routes; either deductively by creating a conceptual typology then identifying cases for the cells, or inductively, by creating conceptual labels for cases clustered in the cells of a taxonomy. Thus operational typologies represent a mapping from both the conceptual and the empirical levels.

In producing an operational typology of models one approach would involve the production of a typology of kinds of expressed model followed by the production of examples for each cell; this might be called the deductive 'classical strategy' (Bailey, 1994). Another approach would entail grouping collected models from the classroom and the creation by induction of conceptual labels for the groups; a 'grounded theory' approach (Glaser and Strauss, 1967).

Given these different approaches we could begin with a conceptual framework and, through top-down analysis, create a conceptual typology based on theory, irrespective of whether a model exists for any given cell. Or we could create an empirical taxonomy using a bottom-up approach from a mass of examples. Or thirdly, we could create an operational typology using both processes iteratively. Because our work is concerned with the production of theory, the collection of data in the classroom, and the interaction between these two activities, we chose to create an operational typology that allows us to use both our developing theoretical frameworks and our classroom data. We began deductively by creating an array of conceptual types of models. We then tested it by populating the array with particular models of two phenomena: the heart within the circulatory system and the lunar eclipse within the solar system.

The phenomena selected come from work on model-based learning or model-building which refers to the construction of mental models of phenomena through a recursive process of model formation, testing, and rejection or revision. The array of particular models of the circulatory system arose during classroom-based research on the use of an interactive multimedia resource for learning about the circulatory system (Buckley, 1992; 1995). This cognitive case study documented a case of intentional model building in a high school biology classroom. The array of particular models of the lunar eclipse draws on a study of learning through collaborative discourse to answer questions about the eclipse in a primary classroom. (Boulter, 1992; Boulter et al., 1998). In this study various models of the eclipse were used and built as the children discussed their understanding of what happens during a lunar eclipse. (See also Chapter 15.)

In attempting a similar task, Mirham (1989) proposed criteria that attempted to be natural (for the purpose of explanation) and generic (as fundamental as possible). He chose the extent to which the models are material or symbolic, whether they are static or dynamic, and whether they have a defined and known outcome (deterministic) or an outcome based on probabilities (stochastic). He combined these three criteria into a two-dimensional array. We have found these criteria and dimensions a useful starting point in examining models of phenomena that are themselves static or dynamic with deterministic or stochastic outcomes. We have adopted the static/dynamic/deterministic/stochastic dimension as one dimension of our typology which we call 'attributes of representation'. In light of the extensive research and theorising in cognitive psychology about visual, spatial, and verbal processes in human information processing, we felt the material/symbolic dimension needed further elaboration before it would be useful as the 'mode of representation' dimension.

MODES OF REPRESENTATION

Mode of representation describes the medium in which the model is rendered. Expressed models may employ different modes of representation, what Twyman refers to as modes of symbolisation (Twyman, 1985). His schema for the analysis of graphic language uses four modes of symbolisation: verbal/numerical, pictorial, pictorial plus verbal/numerical, and schematic. We collapsed his pictorial and schematic modes into one category (visual), and then separated his verbal/numerical into two categories (verbal, mathematical). We added a category for concrete mode from Mirham's analysis, because this non-typographical mode is often found in science classrooms. In the verbal category we included the spoken as well as the written mode. We also included gestural mode which is known to be an important aspect of teaching models (Crowder, 1996) and a significant aspect of conveying meaning (Iverson, 1998). Expressed models often require multiple modes to convey information about the phenomenon, such as diagrams for structural aspects, plus verbal description of behaviour and/or explanation of the causal mechanism; or animation of structures to convey behaviour plus narration to explain the causal mechanism. Therefore, we included mixed modes for all categories. The following list contains the modes of representation we currently feel are most salient and which therefore form one axis of our typology:

- Concrete: 3D material models; e.g. a plastic heart.
- Verbal: models that are heard or read, of description, explanation, narrative, argument, analogy, and metaphor; e.g. 'The heart is a pump.'

- Visual: models that are seen, such as diagrams, animations, some simulations, video; e.g. circle and line drawing of eclipse.
- Mathematical: models that are formulae, equations, and some simulations; e.g. equations of planetary motion.
- Gestural: models that are movements of the body or its parts; e.g. a solar system made of pupils moving around each other.
- Concrete mixed: concrete models with visual, verbal, and/or numerical components; e.g. orrery with an explanatory label.
- Verbal mixed: text with visual or numerical components added; e.g. a text explanation of the structure of the heart with a related diagram.
- Visual mixed: visual models with verbal and/or numerical components; e.g. an annotated diagram of the structure of the heart.
- Mathematical mixed: equations and formulae with verbal explanations; e.g. a boxed planetary motion formula with text under it.
- Gestural mixed: acted out representations with verbal explanations; e.g. pupils talking about their movements as the movements of the earth and moon.

ATTRIBUTES OF REPRESENTATION

The other axis categorises the attributes of representations based on whether the models are static or dynamic and if they have a predictable outcome. When considering the empirical cases it became clear that there was an additional criterion needed to distinguish between models that bear a quantitative relationship to the phenomenon from those bearing a qualitative relationship. The following list contains the attributes of the representation we currently feel are most salient for the other axis of our typology:

The first divides the axis by the relationship to quantification.
Quantitative vs. Qualitative: Is the representation precise as in scale drawings or equations or is it qualitative?

The second divides both groups by their behaviour through time.
Static vs. Dynamic: Is the representation a static one such as a diagram or a dynamic one such as an animation?

The third divides the dynamic groups according to the reproducibility of the behaviour of the representation.
Deterministic vs. Stochastic: If the representation is dynamic, is the behaviour of the representation always the same

(deterministic) or is its behaviour based on probabilities (stochastic) and therefore variable?

Together these two dimensions, the attributes of the representation and the modes of representation, form the dimensions for the cells for our typology of expressed models used in science classrooms.

Paralleling the process of producing an operational typology we shall start by presenting a typology of the types of representations and move on to typologies of examples of the eclipse and of the heart.

Each axis is divided in half with categories arrayed symmetrically on each side in order to display the complexity of the four criteria. The Attributes of Representation axis has a qualitative half and a quantitative half with static, deterministic and stochastic arrayed on each half. The Modes of Representation axis has single and mixed mode halves with concrete, visual, verbal, mathematical and gestural arrayed on each side.

CATEGORISATION OF TYPES OF MODELS

How do the generic kinds of models encountered in science classrooms fit in these dimensions? Among the kinds of expressed models we find used in science education are: 3D and scale models; diagrams; verbal models within descriptions, explanation, narrative, argument, analogy, metaphor; animations; simulations; video of phenomena; mathematical formulae and equations. These kinds of representation are abstract although it is generally possible to form an image of each of them by way of an example of that kind. They are at a basic level (Tversky, 1989) at which the parts and the functions often coincide.

The typology of these generic expressed models shown in Figure 3.1.

Modes of Representation

Single mode | Mixed mode

Attributes of representation		*Concrete Material*	*Visual Pictorial*	*Verbal Written/oral*	*Mathematical Formulaic*	*Gestural Bodily*	*Gestural*	*Mathematical*	*Verbal*	*Visual*	*Concrete*
Qualitative	*Static*	3D model	Diagram Drawing	Analogy Description Metaphor		Showing positions	Showing positions with talk		Analogy with drawing	Diagram with labels	3D model with labels
Qualitative	*Dynamic: Deterministic*	3D models that move	Sequenced diagrams Animations			Acting out set movements	Acting out with talk			Animation with verbal	
Qualitative	*Dynamic: Stochastic*	Physical simulations				Hand gestures	Hand gestures with talk				Physical simulation /labels
Quantitative	*Dynamic: Stochastic*		Graphical displays		Formulae					Graphical Display	
Quantitative	*Dynamic: Deterministic*	Working scale replicas	Video of live phenomena		Formulae Computer simulations	Gesturing relative behaviours	Gesturing behaviour with quantities described	Computer simulations		Video with verbal	Working scale replica with verbals
Quantitative	*Static*	Scale models	Photographs	Description with size or distance	Equations Chemical Formulae	Showing size	Showing size with talk	Equations with diagram	Description with size and gesture	Photos with labels	Being an object with verbals

Figure 3.1. Typology of expressed models

CATEGORISATION OF PARTICULAR MODELS USED IN SCIENCE CLASSROOMS

Turning to the examples of the selected phenomena, we begin with the heart within the circulatory system. We conclude that a plastic or plaster model of the human heart is a concrete model that is qualitative and static. The heart diagram is also qualitative and static, but it is visual. The addition of labels or description makes either of these mixed mode models. They are qualitative if neither the model nor the diagram is to scale, and more generally, because of the natural variability of organs and organisms. Phrases such as 'the heart is two pumps' are verbal, qualitative and static, while an animation of the heart cycle is visual, dynamic, deterministic in that the portrayal is consistent, and again qualitative. Video of a dead heart is visual and largely static despite changes in view, while video of a beating heart is dynamic. Both are quantitative; if the video includes an indication of scale and the time rate is known, the size of the heart, its visible parts, the pulse rate, etc. could be determined. In contrast, a computer simulation of the heart cycle could combine visual and mathematical modes, be dynamic, and either deterministic or stochastic depending on the programming underlying the simulation. It could be considered qualitative due to the simplifying assumptions made during the construction of the simulation, or quantitative because the behaviour of the model is mathematically prescribed. Gestural models using the fist to model the relative size and position of the heart and its contraction could show either qualitative or quantitative features and be static or dynamic. The presence of verbal elements, either in text or audio, qualify any of these models for mixed mode. The array of models of the heart shows a wide range of model types only some of which were used in the high school biology classroom studied by Buckley (1992). The array includes a substantial number of gestural, visual, verbal and concrete examples within all the categories. The regularity of a healthy heart and the knowledge level of the students means that most stochastic phenomena would not be encountered until university or medical school level physiology courses. Therefore most stochastic representations would not be needed either, although simple fist clenching to represent heart contractions has a stochastic outcome. The types of representations used in the biology classroom included video of live phenomena, animations, simulations, photographs and diagrams, all with associated text. Gestural models were also observed during interviews. The analysis is given in Figure 3.2.

Modes of Representation

Attributes of representation		Single Mode: Concrete	Single Mode: Visual	Single Mode: Verbal	Single Mode: Mathematical	Single Mode: Gestural	Mixed Mode: Gestural	Mixed Mode: Mathematical	Mixed Mode: Verbal	Mixed Mode: Visual	Mixed Mode: Concrete
Qualitative	*Static*	Plastic heart	Heart diagram.	'heart is two pumps' Describing action of two pumps and blood flow		Showing the position of heart in chest	Showing position and describing		Heart as two pumps with sketch	Heart diagram with labels	Plaster heart with labels
Qualitative	*Dynamic: Deterministic*	Artificial heart	Animation of heart cycle, blood flow							Heart cycle animation with verbal explanation	
Qualitative	*Dynamic Stochastic*					Clenching fist for heart contraction	Clenching fists and describing				
Quantitative	*Dynamic: Stochastic*		Graphical display of heart beating EKG trace		Heart monitor display	Showing relative contractions		Heart monitor display verbally explained		Graphical display of heart beat with labels	
Quantitative	*Dynamic: Deterministic*	Heart lung machine	Video of beating heart showing rate	Describing heart action in terms of timings	Heart beat cycle display		Showing relative contractions with quantities	Heart beat cycle display sketched in diagram		Computer simulation of heart cycle	Artificial heart with verbal description
Quantitative	*Static*	Precise scale model of heart	Photo of heart Video of dead heart	Describing heart size or volume	Flow and pressure equations	Showing size by clenched fist	Showing size with size description	Equation for pressures put on diagram	Describing volume with diagram	Photo with labels	Exact scale model with labels

Figure 3.2. Models of the heart placed in the conceptual space defined by the attributes and modes of representation

Turning to the lunar eclipse, there are a number of used in the classroom. The concrete models include orreries. The simple non-mechanical orrery with a light source, a globe or large sports ball and a smaller ball, is concrete and qualitative, and generally single without other modes of representation such as labels. It is not easily moved and is thus static. Often schools possess motor-driven orreries where the earth and moon orbit on circular orbits at speeds relative to each other which have a quantitative nature. These models are dynamic with relative distances and sizes. To enhance pupils' understandings of the scale of the solar system pupils are sometimes helped to make a playground model with other pupils representing the sun, earth and moon standing at scaled distances, or they place objects to represent relative sizes. Both these playground models are concrete, static models with different quantitative features and are generally single mode. Children may construct mobiles of the sun, planets and moon and hang them to move in the air in the classroom. These often use pictures or photos of the planets and moon mounted on card and cut to shape: they are visual. They move relative to each other in a stochastic manner and seldom have any quantitative features. The visual models also include diagrams which may show perspective 3D shading and may be mixed mode with labels to show orbits and direction of spin. There may be sequenced diagrams of the earth, sun and moon through time, giving a greater insight into the dynamic nature of the phenomenon. These diagrams are qualitative and static in themselves. Photographs of the eclipse are visual and static. Quantitative features may be added by textual labelling of size in mixed modes. Videos of the eclipse are visual and dynamic, quantitative features may be added by a verbal commentary in a mixed mode. Computer simulations of the orbits of the earth and moon which are beginning to be used in classrooms are mixed mode, dynamic, deterministic and include mathematical modes of representation that may be hidden behind the visual model. In any classroom dealing with explanations of the lunar eclipse, common verbal models will be used about the solidity of the earth, its shadow, and the nature of light travelling and shining. Gestural acting out of the positions of the earth, the sun and the moon may be used with verbal explanations. Many of these models were used in the whole class discussions analysed (Boulter, 1992). (See Figure 3.3.)

Attributes of representation		Single mode					Mixed mode				
		Concrete	*Visual*	*Verbal*	*Mathematical*	*Gestural*	*Gestural*	*Mathematical*	*Verbal*	*Visual*	*Concrete*
Qualitative	*Static*	Fixed orreries	Circle and line drawing Perspective drawing	'Shining and shadow' 'The moon is a moving screen'		Showing position with hands of earth and moon	Showing position with verbal description		'The moon's shadow' with a drawing	Diagram of eclipse with labels	Non-mechanical orrerry with description
Qualitative	*Dynamic: Deterministic*		Sequenced diagram of Earth and moon.			Showing movement of light	Showing movement and describing			Sequenced diagram with text	
Qualitative	*Dynamic: Stochastic*	Torch and balls simulation	Solar system mobiles using card			Acting out the planetary movements	Acting out with verbals			Solar system card mobiles with description	Torch and balls with description
Quantitative	*Dynamic: Stochastic*										
Quantitative	*Dynamic: Deterministic*	Mechanical orreries	Video of eclipse		Computer simulation of orbits		Gestures with description of time taken	Computer simulation of orbits with talk		Video of eclipse with verbals	Mechanical orrery with description
Quantitative	*Static*	Playground models using vegetables and seeds	Photos of eclipse. Remote sensing	'Moon is 40th distance Sun – Earth' 'It takes 6 mins for light to reach us from the sun'	Equations of planetary motion	Playground models using pupils	Playground models with talk	Equations of orbits with diagrams	Describing distance with gestures	Photographs of eclipse with labels	Playground model with description

Figure 3.3. Models of the lunar eclipse placed in the conceptual space defined by the attributes and modes of representation

DISCUSSION

The chapter has described the operational typology that we produced by arraying types of expressed models. We then arrayed models of the heart and the sun, earth and moon in a lunar eclipse as they were manifest in two classroom studies. In this exercise we paid close attention to the modes of representation and the qualitative, dynamic and deterministic attributes of the models. We noticed some difficulties in this process. It is difficult in the abstract to place models in the conceptual space. Expressed models are human artefacts and are therefore immensely and arbitrarily variable. Whether a model is single vs. mixed mode, stochastic vs. deterministic, or qualitative vs. quantitative is not necessarily a function of the type of model but of the decisions made in constructing it. Placing types of models in the conceptual space requires making assumptions about the characteristics of the model types. It forces the making of distinctions and differentiations that are significant in defining the relationship between the model and the phenomenon. Placement of mixed modes such as simulations is particularly problematic and may not be possible in the abstract.

Furthermore, in classrooms, models of the kinds we have illustrated may be initially presented to students through four main avenues: (a) through objects - concrete models that can be handled and visualised, (b) by discourse - verbal textual models that can be spoken and listened to, (c) by print-based means that occur in books and diagrams and writing on the board, and (d) by screen-based means in electronic information resources and computer simulations. Although each of these has its own strengths and limitations, they share many features. The screen has print-based features, printed text has discourse-based features, and features of object-based learning may be present in all as well. We believe the dimensions of our typology capture these shared features and facilitate comparisons across different modes and avenues of representation, however idiosyncratic.

We set out to create an operational typology that enables comparisons across representations of a phenomenon and facilitates exploration of relationships and the recognition of patterns of use and development. Using the typologies that we have produced it is straightforward to compare the different modes of representation and the different attributes of the representation that focus on their quantification, dynamic behaviour through time and predictability of outcome. It is easy to see which cells are empty and then try to fill them or to speculate about the reasons for absence. This could be done either on practical grounds that none were seen in the classroom or on theoretical grounds that such models may never exist. With these arrays the range of models for these phenomena can be seen at a glance as well as the characteristics of the representations and how they might relate to each other. For instance, the range of concrete models for the eclipse used in the classroom ranged from the simple balls on fixed wires (non-mechanical orreries) through ball and torch simulations and clockwork or other motorised orreries where the behaviour of the rotations is determined by the mechanism, to the playground models where cabbages, grapes and poppy-seeds represent the relative sizes of the planets and the distances are to scale. The empty cell of dynamic, deterministic, qualitative models might be filled by an orrery where the planets are represented by balls of one size suspended by wires at even distances from each other around a central column, and which can be moved around their prescribed orbits in an qualitative fashion. The empty cell of dynamic, stochastic, quantitative models might be filled by a hanging mobile of the planets of scaled size in relative order and which move at random on the breeze in the classroom.

Our typologies facilitate reflection on the relationship between the model and the phenomenon represented by making explicit the attributes of the representation which can then be compared with the attributes of the phenomenon. Each mode of representation imposes constraints on how a phenomenon is represented. Some modes work better for certain aspects than others, i.e. structural aspects of the phenomenon are easily represented with concrete models and diagrams, while dynamic aspects are more easily portrayed in animations, simulations, and video. Explanations of structure and dynamic processes are likely to require mixed modes of symbolisation with structural and dynamic attributes carried in one mode and the causal explanation carried in verbal mode. However, the distinctions among attributes of different representations and phenomena aren't always clear cut. Video of a dead heart is the informational equivalent of a photograph; static. The stochastic behaviour of a phenomenon might be deterministically represented in a video. Deterministic behaviour might be represented by gestures that are anything but predictable. The potential mismatches between the attributes of representations and that of phenomena no doubt give rise to misinterpretations of the representation and misunderstanding of the phenomenon.

Our typology also enables us to explore relationships and patterns among the models of a particular phenomenon. Developmental sequences in model building and learning are one example of the relationships and patterns sought. Piaget (1969) describes stages of cognitive development from the concrete to the formal. These stages relate to the mode of representation dimension of our typology and may suggest a sequence of models that facilitates learning appropriate to the learner's cognitive development. Bruner (1983) suggests that, for a learner at any stage of cognitive development, entering a new domain may progress along an enactive, iconic, symbolic path. This too parallels the mode of representation dimension. The concrete and gestural modes of representation map to the concrete stage of cognitive development and the enactive phase of learning while the visual mode relates to the iconic phase, and the verbal and mathematical modes to the formal stage and the symbolic phase.

Thus, the typology we have presented is likely to prove useful in the agenda that we set at the start of this chapter. Indeed it has already been used in the categorisation and analysis of models of air resistance in texts for teachers and in given lessons (Simpson, 1999). It allowed the charting of the range of models and the comparison between these two contexts. Such work facilitated teacher reflection upon the models she used and those she omitted in her lesson. It may help to provide pupils with overviews of the scope of models of a particular phenomenon if the teacher chooses to work explicitly with model-based teaching. Given a means of assessing the range of models that pupils use to explain phenomena, the developmental changes could be charted.

Despite our agenda being addressed by the typologies there are still limitations to its construction and use. Although we have tried to make the typology exhaustive, it was created using examples from just two phenomena of science. The two phenomena are respectively of human scale and of much larger scale. Use of the typology with models of phenomena at much smaller and less accessible scale, such as the phenomena of chemistry, or those that take place over long time spans, such as the geological and biological evolution of earth, may result in elaboration or revision of the categories and criteria we have used. Although we have constructed a typology that enables comparisons and facilitates the finding of patterns it cannot help in investigating how these representations function in model-based teaching and learning. For that we need to return to the analysis of the parts of models and their relationship to the aspects of the phenomenon and how they function in learning from representations. This is the topic of Chapter 6.

CONCLUSION

We have produced an operational typology of expressed models in science education through an iterative process of categorisation and particularisation of conceptual and empirical entities. The typology begins to give structure and coherence to the field of models in science education, giving us a tool to look for patterns of models used in different situations. It will help us develop a more differentiated explanation of how models operate effectively in practice. We believe that the distinctions we have drawn will be useful in research that analyses the development and progression of scientific models in the history of science, of curricular models in classrooms, and in the development of individual's expressed and mental models. Such research will in turn enable the principled selection and construction of curricular models during the development of instructional materials, multimedia information resources and courseware, whether on CD-ROM or the Internet. We believe it will also be useful in providing teachers with a framework for creating and reflecting on the teaching models they use in their classrooms. The dimensions of the proposed typology structure the types of models encountered in classrooms, provide an easily understood overview, and can promote questions about the distribution of models and their relationships to each other. All of these are the foundations for building coherent and effective learning experiences for students in science classrooms.

Chapter 4

Mathematical Models in Science

David Malvern
The University of Reading, UK

INTRODUCTION

In ending his charming and insightful essay *The Unreasonable Effectiveness of Mathematics in the Natural Sciences*, by observing that

> The miracle of the appropriateness of the language of mathematics for the formulation of the laws of physics is a wonderful gift which we neither understand nor deserve

the Nobel Laureate Eugene Wigner (1960) summarises the difficulty in understanding why mathematical models seem to be essential and not merely useful to science. Although there may be a straightforward distinction between mathematics and science, the truths of mathematics are based on deriving consequences from axioms while those of science rest on empirical evidence, so intertwined are they that it may not be so easy to determine whether a given page of text belongs to one or the other, nor to say who is a mathematician and who a scientist. Thales (624-546 BCE) was generally considered the founder of both Greek science and mathematics and perhaps the most popularly known mathematician. Pythagoras and Euclid also worked, apparently seamlessly, in the sciences. Euclid systematised geometry axiomatically but also treated optics as part of geometry. Pythagoras' analysis of sound, which remains unaltered today, was at one with his theories of numbers and of astronomy.

> According to Pythagoras, numbers *were the ultimate essence of reality.* ... Harmony, expressed in mathematical ratios and means, was the controlling force of the cosmos. ...

J.K. Gilbert and C.J. Boulter (eds.), Developing Models in Science Education, 59–90.

> These interlocking disciplines provided a means of comprehending the true nature of the universe.
> (Gouk, 1988)

MATHEMATICS AND SCIENCE

From their very beginnings, mathematical and scientific thought have been braids on the same plait. Sharper and more modern distinctions between their epistemologies neither deny their indissoluble histories nor disguise the extent to which the one informs the other. Feynman (1965), in an echo of Pythagoras, finds it natural that mathematics would be needed to handle the large numbers that arise in complicated situations, citing genetics as an example requiring averages and ratios. The 'strange thing', as Feynman puts it, however, is that mathematics is needed not just for the many bodied and large scale applications of science but for the fundamental laws. In a comment reprising a thought originally made by Galileo, Feynman is emphatic about this:

> If you want to learn about nature, to appreciate nature, it is necessary to understand the language she speaks in. She offers her information in only one form [mathematics]. ... All the intellectual arguments in the world will not convey an understanding of nature to those of 'the other culture'[who do not know mathematics].

As Franco and Colinvaux discuss in Chapter 5, Galileo had observed how different masses fell under gravity and it is possible to describe the result in words: *bodies fall faster and faster towards the earth's surface under the influence of gravity and they fall the same way whatever their mass.* He produced a delightful verbal argument to counter the prevailing dogma that more massive bodies fall the faster. Consider a composite body, he reasoned, made up of two bodies, one more and one less massive, tied together. If more massive bodies fall faster then two equally logical but contradictory predictions emerge. On the one hand, being more massive than either of its parts, the composite body should fall even faster than either. On the other, being tied together the more massive one will pull the less massive on and the less massive one will drag the more massive back and the result will be an intermediate motion somewhere between the behaviour of each independently. A theory which predicts such contradiction cannot be right (see Chapter 8).

Such reasoning assumes that the universe does not contradict itself, that it is consistent, and its consistency is open to logical analysis. Of itself, however, it neither proves nor disproves the hypothesis. At best it shows that the Aristotlean law is incomplete, needing a rule for composite bodies. Without empirical observations to show that bodies fall consistently and how they fall, it cannot be rejected. Galileo made such investigations with experimental results of the following kind:

Time taken (t):	0	1	2	3	4	5	(time units)
Distance fallen (s):	0	5	20	45	80	125	(distance units).

It is the precision of measurement that makes it possible to represent the quantities and what happens to them by numbers. What the mathematical model does is then to apply theories which were developed for numbers, to the numbers which represent the quantities. In this case, for example, it is possible to show just from the pattern in the numbers that it is the acceleration that is constant and to derive an equation relating distance fallen and time taken from the pattern of numbers:

t	0		1		2		3		4		5
s	0		5		20		45		80		125
1st difference (velocity)		5		15		25		35		45	
2nd difference (acceleration)			10		10		10		10		

Using the mathematical operation of taking first, second (and so on) differences the precise form of regularity is exposed, hence

$$s = \tfrac{1}{2}10\, t^2$$

and the result can be refined to state that the acceleration at the earth's surface due to gravity is constant regardless of the mass of the falling body, when timing begins, from where the body falls or is observed etc. It is perhaps not surprising, then, that Galileo wrote in *The Assayer*:

> But it [the universe] cannot be understood unless one first learns to comprehend the language and interpret the characters in which it is written. It is written in the language of mathematics ... without which it is humanly impossible to understand a single word of it.

This is an example of the regularities which are the stuff of science and the one which caused Wigner (1960) to comment that

> Galileo's regularity is a prototype of a large class of regularities. ... classical mechanics ... gives the second derivatives of the positional coordinates of all bodies, on the basis of the knowledge of the positions, etc, of these bodies. It gives no information on the existence, the present position, or velocities of these bodies.

The point is that the universal regularity expressed in a natural law resides not in what is observed directly but in what is mathematically derived from these observations.

Building on Galileo's work, Newton developed a theory of universal gravity, applicable to the heavenly bodies as well as on the Earth. It can be stated as an equation giving the gravitational force (F) in terms of the masses (m_1 and m_2) of two bodies a distance apart (r):

$$F = G\frac{m_1 m_2}{r^2}$$

This is much more unequivocal than anything like 'the bigger the masses the bigger the force'. It is by specifying the law with mathematics that permits it to predict states which can be described with sufficient precision for an experimental test to distinguish its efficacy and to continue to test it beyond the known observation which led to its formulation. The known observations which led Newton to propose his law of gravity could only support it to about 4% accuracy but it has since been tested to less than 0.0001% and it only becomes questionable at even lower levels of accuracy. Anything less than the formulation of a mathematical model would fail to provide sufficient precision in the predicted states to allow such limitations to be questioned.

Moreover, the mathematical model is the only description Newton saw fit to give of gravity. It gives little insight into the physical process of gravity. He claimed *'hypothes non fingo'* (I feign no hypothesis) and this law provides no mechanism. Further, the evidence is that he did not think it was necessary to provide one, that the equation is all there is, that in some sense the mathematics is not a model of the physics but is the physics. Manuel (1968) cites Herivel (1965) as noting that in one of the manuscripts of De Motu Corporum, Newton wrote:

> ...that force, whatsoever it may be, by which the Planets are held in their orbits and perpetually restrained from flying off at a tangent ... [I] use the words *attraction*, *impulse*, or *propensity* of any sort towards the centre indifferently and interchangeably one for the other, considering these forces not in the *physical* sense but only in the mathematical sense. Hence let the reader beware lest he think that by words of this kind I define a type or mode of action or cause or physical reason of any kind.

The full title of his great book, after all, is *Philosophiae Naturalis Principia Mathematica*.

Hackfort (1988) has brought considerable scholarship to bear on Newton's ideas on the mathematical nature of science particularly through the work on optics. As well as the book, *Opticks*, he draws on its precursors - articles and letters in the Royal Society's Philosophical Transactions - and on Newton's lectures published after his death (*Optical Lectures read at the Publick Schools of the University of Cambridge*). The usual emphasis given to the exchanges between Huygens and Newton is that of two champions: the one for a wave theory of light, the other for a corpuscular theory. Rutherford considers the historical development of these two ideas in Chapter 13. Hackfort, however, concentrates on what Newton was trying to do and the important light it sheds on Newton's view of science. Newton essentially separated the experimental-mathematical description from a philosophical explanation of the nature of light. His wish was to develop a rigorous mathematical theory of colour:

> ... [a] naturalist would scearce expect to see ye science of those [colours] become mathematical, & yet I dare affirm that there is as much certainty in it as in other part of Opticks.

and in the *Optical Lectures*, Hackfort notes, Newton said:

> It is affirmed that these propositions [on colour] are to be treated not hypothetically and probably, but by experiment or demonstratively ... though colours may belong to physics, the science of them must nevertheless be considered mathematical ... I hope to show - as it were by my example - how valuable mathematics is in natural philosophy.

It is ironic, then, that Maxwell achieved Newton's aim by arriving at a mathematical synthesis of light and electromagnetics by way of a most elaborate hypothesised mechanism of imagined vortices in 'the ether'. Science has forgotten the mechanism and in a second irony remembers especially the mathematics, as his laws of light are generally referred to as 'Maxwell's Equations'[1]. It is, however, an example of how mathematical modelling is not always an abstract manipulation of symbols but is often accompanied by iconic and enactive thought. Fischbein (1987) asserts that mathematics, what he calls a world of mental constructs,

> seems to mirror all the features which enable the real world to function

and that

> the human mind seems to have learned from the basic general properties of empirical reality how to build an imaginary, structured world, similarly governed by rules and similarly capable of consistency and credibility.

He cites Hilbert as championing a concrete realisation being parallel to axiomatic mathematics:

> Who does not always use, along with the double inequality a>b>c, the picture of three points following one another on a straight line as the geometrical picture of the idea 'between'? ...

It is for Fischbein (1987) a psychological point. For Wigner (1960) it is an illustration of what he calls the 'empirical law of epistemology' and for others an article of faith of the theoretical physicist. It should be noted, however, that it is perhaps not so surprising that mathematics, a product of our imagination, turns out to be so applicable to the real world, if our methods of modelling express psychological preference and given that we, along with our imagination and our psychology, are a product of that real world.

Nonetheless, mathematical things like the hexagons in honey combs or in the close-packing atomic structure in some metals, do not come about because atoms or bees measure out hexagons. Electrons follow the paths they do without the capacity to appreciate calculus and so on. The universe does what it does without solving our equations. Bees operate within space, however, and the outcome of their behaviour in the geometry of the circumstances is open to geometric description. Electrons follow a locus in time and space and the theories of mechanics are applicable to the locus. It can be argued that the ways in which we can describe the contents of the universe are constrained by the geometries of space-time-geodesics, light cones and event horizons, and so on: to paraphrase Einstein, 'nature integrates empirically'.

Our theories about the universe may have to be mathematical, then, not because Nature 'speaks' that way but because this is the way which over a long period of time we have found it best to 'listen'. Our best knowledge of the nature of what scientific theories are about mass, time and space etc. looked at in a restricted, selective way is mathematical, because mathematical modelling is the most powerful way we have developed to predict the unknown from the known with the sort of precision required to expose it to experimental test. Moreover, ever since Galileo and Newton, mathematical modelling as a way of theory making in science has been so successful and powerful, it has become *the* archetype of scientific theory, the paradigm towards which the various branches of science aspire. Layton (1973) would date the point when this was cemented into science and science education as 1833 when J.D. Forbes was appointed to the prestigious chair of natural philosophy at Edinburgh in preference to Sir David Brewster. While to Brewster there were real dangers in subordinating physics to mathematics 'making facts of nature "mere pegs on which to suspend festoons of algebraic drapery"', Forbes championed 'the continuing subjection of physics to mathematical treatment' and, supported by referees from Cambridge, he won the chair. Mathematical methods in science became, to borrow Lakatos' terms, the framework theory of framework theories.

THE NATURE OF MATHEMATICAL MODELS

In part, the usefulness of mathematics lies in its economy of expression. Mathematics puts elaborate statements into a shorter symbolic form making it possible to contemplate complexity in an apparently brief formulation. Moreover, it serves to endow such statements with a sort of universality in that the mathematics can then be read to mean the same thing in different mother tongues. Adler (1972) comments '... the language of mathematics is so natural and so simple in comparison to the spoken languages ...' and Gasking (1960) argues that there is a 'public' or 'over-individual' character to mathematics which gives an incorrigible meaning to mathematical statements.

> A mathematical proposition such as '7+5=12' … is incorrigible, because no future happenings whatsoever would ever prove the proposition false, or cause anyone to withdraw it. You can imagine any sort of fantastic chain of events you like, but nothing you can think of would ever, if it happened, disprove '7+5=12'.

The first property of a mathematical model, then, is that it provides a succinct, incorrigible description. For example, the floral formula is a symbolic description of specific parts and structures of a plant. The formula for a buttercup would be written as: $\oplus K5C5A\infty\underline{G\infty}$ indicating it is actinomorphic, has five free sepals (Calyx K5) and petals (Corolla C5) which are clearly distinguishable while there are a large number (greater than twelve) of stamens (Androecium $A\infty$) and carpels in a superior ovary (Gynaecium $\underline{G}\infty$). On the other hand, $\cdot|\cdot K(5)C(5)A4\underline{G(2)}$ represents the zygomorphic foxglove which differs in having its staples and petals joined, 4 epipetalous stamens and only two carpels (also joined) in its superior ovary. Clearly the floral formula is a model of a plant in the same sense as a floral diagram is, namely a schematic diagram of selected features. It does give the number of petals, distinguishes whether or not the gynaecium is above or below the rest of the flower and so on. It does not represent shape, colour or scent; in short it excludes the features by which we might normally recognise a flower and how they 'paint the meadows with delight'. It does, however, allow plants to be classified in ways which acknowledge their biology rather than their appearance.

Similar formulae appear in chemistry and physics and again portray selected properties of the entities described. In an interesting parallel to floral formulae, Feynman et al. (1970) described writing chemical formulae as drawing in two dimensions. H_20 models water in that it describes its chemical composition; it tells us little of the 'old grey widowmaker' as the Viking poet saw the sea. The nuclide formulae $_{92}U^{235}$ and $_{92}U^{238}$ characterise two isotopes of uranium, indicating the numbers of protons and total number of nucleons in the two nucleuses but not that Oppenheimer described the effects of one with words from the Bhagavad Gita 'I am become Death, the shatterer of worlds'.

The economy and universality of expression in a mathematical description, then, appears to extract a price. It abstracts some features and not others. It shares this with most if not all types of models. Some would say this is a form of reductionism which replaces the real and rich with the pale and limited. As it happens, nothing could be further from the truth.

Firstly because such a 'reduction' is essential, without it there would be no science. Wigner (1960) is emphatic:

> ... It is true that if there were no phenomena which are independent of all but a manageably small set of conditions, physics would be impossible. ... The point which is most significant in the present context is that all these laws of nature contain, in even their remotest consequences, only a small part of our knowledge of the inanimate world. All laws of nature are conditional statements which permit a prediction of some future events on the basis of the knowledge of the present, except that some aspects of the present state of the world, in practice the overwhelming majority of the determinants of the present state of the world, are irrelevant from the point of view of the prediction.

Importantly in being a selective representation, not the full, real thing, a mathematical description is deliberate in selecting aspects which have a required 'over-individuality' in displaying the regularities consistently observed in nature and ignoring the subjective or unpredictable connotations which words may bring. Describing something mathematically, then, is to represent chosen and well defined aspects in a symbolic form of language.

As such a mathematical *description* is a very limited model. Floral formulae, for example, achieve little beyond a parsimony of space and a restriction of the attention to the chosen features. Both of these features have their uses: the first allows ready comparison of many cases and the second prevents distraction. But if mathematical models were no more than a brief form of a verbal description in a restricted range of the language they would not be distinguishable from other abbreviated representations such as scale models or circuit diagrams, useful in themselves but hardly essential for the laws of nature.

What provides mathematical expressions with power is that they can be built into mathematical statements which can then be manipulated under rules. Chemistry provides an obvious example. Chemical formulae can be combined with other symbols into chemical equations, e.g.:

$$CuO + H_2SO_4 \Rightarrow CuSO_4 + H_2O$$

The symbols and their arrangements do not have their 'ordinary' mathematical meaning, that is CuO is not a multiplication, but there are mathematical rules which have to be obeyed for example, the same elements must appear on either side of the equation and the sum of the subscripts for each one on the left hand side must equal that on the right. Nuclear equations are similar in physics:

$${}_{17}Cl^{37} + \upsilon \rightarrow {}_{18}A^{37} + \beta^{-}$$

although the counting rules are obviously different.

This is like moving from words to sentences. To do that successfully requires a form of generalisation in that some understanding of the generality of what can form subjects, objects, verbs (etc.) and permitted word orders is needed. Here the properties of the symbols for elements which permit them to be placed in chemical equations and the addition rules for balance have to be seen, as Harré (1967) points out, as the 'class intension' or the properties that all chemical equations have in common to be chemical equations. He goes on to say: 'They [generalisations] play an important part in reasoning for our rules of inference are derived from them'. Putting mathematical descriptions into sentences, then, invokes rules for inference some of which at least must be mathematical, or as Feynman (1965) puts it '... mathematics is not just another language. Mathematics is a language plus reasoning.'

Simple chemical equations, as an example of mathematical models, provide further insights. For simplicity, let us take as the rules of chemical equations that the same and only the same element symbols appear on each side of the equation and that the sums of subscript for each element (including the unstated 1s) must be equal on both sides of the equation. In an unspecified equation such as:

$$\Delta_a\Theta_b + \Pi\Phi_c\Psi_d \Rightarrow \Pi\Theta + \Delta_e\Phi_f\Psi_g + \Phi_h\Theta_i + \Delta\Psi_j$$

We have $a = e+1$; $b = 1+i$; c $= f+h$; $d = g+j$. Firstly, it should be noted that the algebra is designed to suit so that the mathematical rules model the chemical notions that in such reactions combinations of elements are rearranged but no atom is destroyed, transmuted or otherwise lost. In models of other physical phenomena, the rules can be different; e.g. in the nuclear physics equation above it was not a rule that the same and only the same elements can appear on either side of the equation. Secondly, although it is necessary for these rules to be obeyed for a statement to be true it is not a sufficient condition. That a chemical equation balances does not guarantee that the reaction it models occurs; for that the scientific truth test of empiricism is required.

This apparent limitation on the model is one of its strengths. The rules themselves are chosen to accord with the regularities in previous empirical observation, but the mathematics of chemical equations makes available a generalisation which can be used to search for potential new members of the class of chemical reactions. The mathematical representation makes it possible to contemplate hitherto unobserved reactions, which are permissible by the known mathematical rules, to explore new possibilities consistent with past observations. Ormell et al. (1979) see this as a consequence of a Peircian view of mathematics which characterises its axiomatic form as '*if p then q*'. In science the *p*'s derive from empiric observation and theory is built from modelling them in mathematical form. The model then exploits mathematics' capacity to explore the potential consequences, the *q*'s, through the predictions made by its compendium of 'if ... then ...' statements.

Mathematical operations can be carried out on the original description of known natural phenomena to end up predicting something quite different to that originally observed. This phase of mathematical modelling, the working model in action as it were, can be called the development and to be acceptable a mathematical model must be correct in both its description and its development. Some of the possible consequences may turn out to be in accord with (or at least not contradicted by) experiment, some not. In the above example of a generalised chemical equation some of the equations permitted by the given two rules predict reactions which are found to occur and others do not. When exploring a model's predictions, what turns out to be consistent with experimental evidence extends our understanding of the range of usefulness of the model, and what does not, causes modification of the rules to refine the model or provides evidence to reject it in favour of another.

Mathematical models are not simply a form of describing the real world, they are working models in a theoretical world. In this theoretical world, they first substitute, or model, objects, events or other observable phenomena in the real world with mathematical entities. In a sense they populate the theoretical world with mathematical correlates of what occurs in the real world; for example, Newton's theories are populated with point masses and loci where planets and movement are found in the universe. It is the mathematical model as *description* and its initial state is that of the observed conditions or a particular arrangement which, as Wigner puts it, happens to interest us. Next, in order to work, mathematical rules must be chosen to operate on this population in order to alter its initial state to another. These rules are selected to model in the theoretical world a transition in the real one. Applying the rules is the development which transforms the original state of the given population into a new state in the theoretical world which predicts an equivalent state in the real world. There are, then, three parts to a mathematical model: the 'population model' which defines how the model represents the real objects, events or phenomena and the starting conditions under consideration; the 'transitional model' which consists of the rules which can be applied in the model; and the 'predicted state' which describes what has happened to the population model after the rules of the transitional model have been applied.

Each of these parts will have what Gilbert and Boulter call in Chapter 1 'modes of representation'. Here it will be useful to use terms adapted from Bruner's strategies for problem solving (enactive, iconic and symbolic) (Bruner, 1966) for generalisation of these modes. Firstly, there is a 'physical' or concrete mode in a population model which uses materials of one kind or another. If the transitional model physically manipulates this material (moving, bending, stretching etc) it is like employing Bruner's enactive strategy for problem solving – i.e. it is acting out what happens under change – a transitional model which works in this way, therefore can be said to operate in 'enactive' mode. Secondly, an 'iconic' population model consists of pictorial representations, such as blueprints, diagrams, graphs and so on, and an 'iconic' transitional model acts by altering the original form of such pictures (extrapolating, for example). The 'symbolic' mode is when a population model employs symbols, for example algebraic variables or Cartesian coordinates, to represent real world things like mass, velocity, position etc. and when the rules of the transitional model are couched in terms of how such symbols can be manipulated (the rules of algebraic addition, for instance, or geometric theorems and so on).

The parts of a model can be illustrated in Figure 4.1 where the three dimensions of symbolic, iconic and physical/enactive are shown as orthogonal axes enclosing a space defining the character of the modes in these terms. The three components transitional model, population model and predicted state of any particular model will each occupy a position within this 'character' space. This diagram displays many features of models in general, and a mechanical example can serve to illustrate it. One way of experimenting on a proposed bridge suspension cable (say) would be to use a scale model of the cable (a physical population model) in a machine which subjects it to stretching, shock loading etc. (an enactive transitional model) to destruction (a physical predicted state).

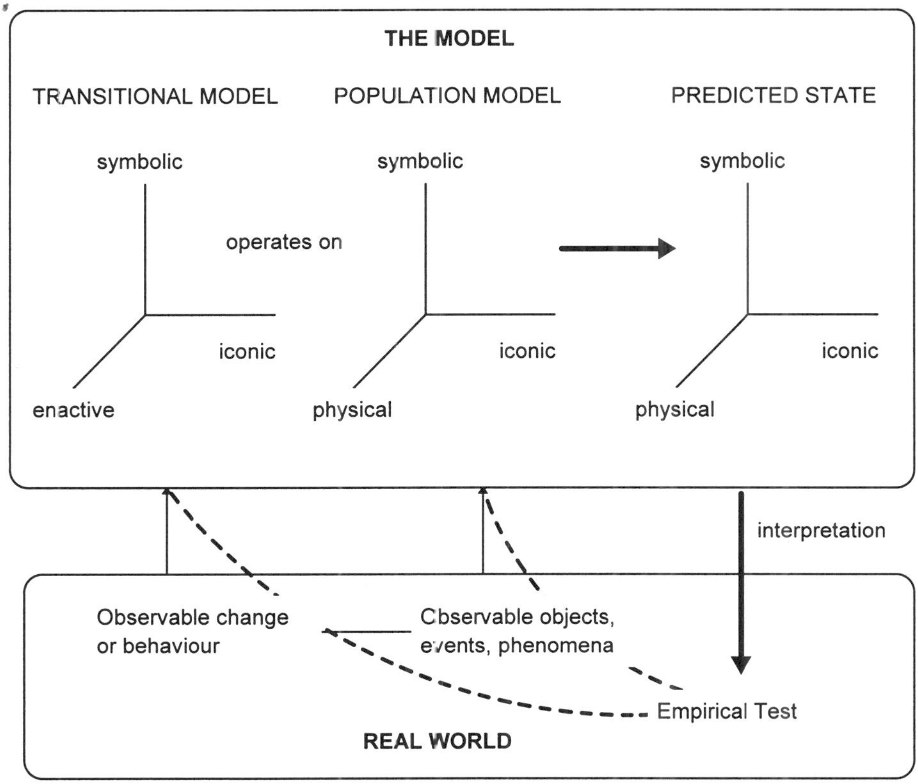

Figure 4.1. The structure of the model

In a mathematical model the usual positions would be (for example):

- a symbolic population model (let m_i = the mass of the ith planet)
- with a symbolic transitional model

$$\left(G\frac{Mm_i}{r_i^2} = \frac{m_i v_i^2}{r_i} K \right)$$

- and symbolic predicted state ($v_i = \ldots$).

The outcome, however, may be given in iconic form and mixed modes are possible, models which combine icons and symbols are particularly common (graphical solutions or a Feynman diagram, for example). In a machine such as an orrery mixed symbolic, enactive/physical models occur, and interestingly computer models frequently combine all three.

Two kinds of transitional rules are applied in the development of a mathematical model in science. One legitimises an equality by appeal to a scientific law or to experimental evidence. There is nothing in mathematics *per se* which states that always $F = ma$. It is a corrigible identity, because only if F is a force, m a mass and a an acceleration might it be true and then only in circumstances sanctioned by Newton's second law. In other words, sometimes an apparently mathematical move made in the development is a scientific act. The other kind of rule is purely mathematical and incorrigible when one identity implies another ultimately under the sanction of a mathematical axiom or theorem. The example given as Figure 4.2 paraphrases a mathematical model in physical chemistry and illustrates both kinds of transitional rules.

In the example in Figure 4.2, which is a model with both symbolic and iconic components, the extrapolation of the graph to the point of zero pressure took the model into an 'ideal' not a real world, illustrating the problems of meaning in Brewster's 'algebraic drapery' (quoted in Layton, 1973). Manipulating symbols under solely mathematical rules, in the abstract as it were, produces mathematically unproblematic entities whose physical identity can be questioned. For example, the mathematical model of a simple pendulum predicts the periodic time as

$$T = 2\Pi\sqrt{\frac{l}{g}}.$$

The right-hand side of this equation contains the square root of the reciprocal of the local gravitation acceleration. It is very hard to say what physical meaning that may have.

A Mathematical Model in Action

The General Gas Equation states that:

$$PV = \frac{m}{M}RT$$

or $$M = \frac{m}{V}\frac{RT}{P}$$

and since $\frac{m}{v} = d$ (the density of the gas)

$$M = \left(\frac{d}{P}\right)RT$$

Hence if the temprature of the gas is constant

$\left(\frac{d}{P}\right)$ is constant for any given gas.

For example, for oxygen at 0° C and 1 atmosphere

$$d = 1.43\ \mathrm{g}\ \mathrm{l}^{-1}$$

and $$\left(\frac{d}{p}\right) = 1.43\ g\ \mathrm{l}^{-1}\ \mathrm{atm}^{-1}$$

At 0° C and 0.5 atmospheres, by Boyle's Law

$$d = \frac{1}{2} \times 1.43\ g\ \mathrm{l}^{-1}$$

and still $\left(\frac{d}{P}\right) = 1.43\ g\ \mathrm{l}^{-1}$

However, if very accurate measures are made

the following figures for oxygen at 0° C are found.

Commentary

This equality is justified by the science. It is an outcome of a symbolic transitional model operating on algebraic variables representing properties of a population model of a gas as consisting of spherical molecules of equal mass and so on.

$m/v = d$ is a definition.

The algebraic substitutions and conclusion (*d/P* is constant at constant temperature) are mathematical moves.

The illustration of a concrete value is science from experiment.

And the next equality (at 0.5 atmospheres) is scientifically justified by Boyle's Law, with its evaluation being mathematical.

P (atms)	d (g l^{-1})	$\left(\frac{d}{P}\right)$ (g l^{-1} atms $^{-1}$)
1	1.428 9	1.428 9
0.666 67	0.952 4	1.428 6
0.5	0.714 15	1.428 3
0.333 33	0.476 0	1.428 0

This is reporting some experimental results and is scientific.

It is setting out the initial state of the iconic population model to follow.

For most gases the relation between P and d/P is linear. Thus the value of d/P can be found as P approaches zero by extrapolation of the graph of d/p against P.

d/P against *P* for oxygen

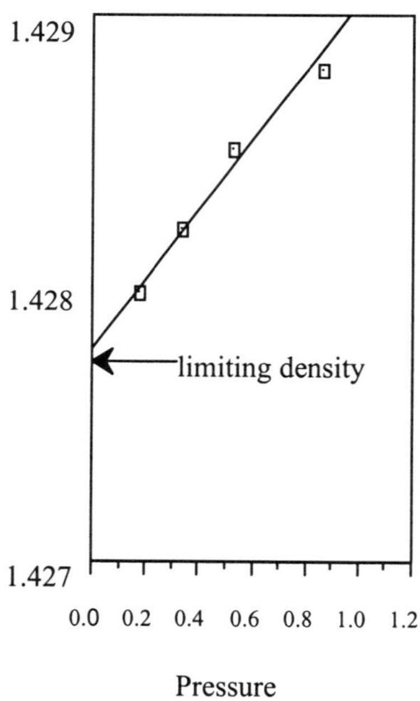

This is an observation based on a pattern of numbers, but it is then converted into an iconic population model (points on a graph) and the manipulation of the graph is entirely mathematical – first calculating the best fit straight line and then extrapolating it to a mathematical point which represents conditions unrealisable in the real world (a gas being a gas at 0^0C but at zero pressure and having a measurable ratio of density to pressure).

This part sees the transitional model (fitting a best straight line and extrapolating it to zero) operating on the initial state of the population model (the points placed on the graph from the experimental results in the table above) to produce the predicted state $(d/P)_0$.

At the hypothetical condition of zero pressure we can assume that gases behave ideally and the limiting density of oxygen from extrapolation is:

$$\left(\frac{d}{P}\right)_O = 1.4276g^{-1}$$

A bold claim for the model! Note particularly that real gases do not behave ideally – this is a statement about the mathematical model of a gas and the limiting density exists and can be evaluated only in the model.

Recalling that $M = \left(\frac{d}{P}\right) RT$

We can use this limiting or ideal value for d/P. Measured at 0°C or 273.16K to evaluate M, The molecular mass of oxygen

$$M = 1.4276 \times 0.082056 \times 273.16 = 32.00$$

Values for R and T in appropriate units are obtained scientifically by experiment, but the substitution and calculation are mathematical – note this produces a predicted state (the predicted molecular mass of oxygen) which is empirically testable in the real world

'This agreement of experiment with theory lends support to the assumptions of the theory. ... Such methods gave the most accurate measurement of atomic masses of gaseous elements before the advent of the mass spectrometer.'

This is a quotation which follows this analysis in the textbook on *The Elements of Physical Chemistry.*

Figure 4.2. A Mathematical Model in Physical Chemistry

Nonetheless, applying the gas laws did lead to a prediction about the real world, and an 'agreement of experiment with theory' and the model of the simple pendulum predicts periodic times which can be measured in reality. What legitimises mathematical models is that, however abstract they become during their development, they result in predicted states which are at least in principle open to refutation or otherwise by empirical test. This phase of the model, the interpretation, is essential. Feynman (1985) makes the point with reference to what he calls the 'most shocking' problem of the meaning of the amplitudes in quantum electronic dynamics by appeal to the theory's ability to predict experimental results accurately:

> ... all the new particles and new phenomena that we are able to observe fit perfectly with everything that can be deduced from such a framework of amplitudes. ... So this framework of amplitudes has *no experimental doubt* about it: you can have all the philosophical worries you want about as to what the amplitudes mean (if, indeed, they mean anything

> at all), but because physics is an experimental science and the framework agrees with experiment, it's good enough for us so far.

Not only is the interpretation phase a necessary qualification for the activity to remain science, it releases the method from the constraint that in at all stages every part has realisable physical meaning what matters is that it results in something that does.

Mathematical models, then:

- use science to select significant aspects from which to build succinct descriptions in terms of mathematical entities- the population model;

- exploit mathematics as 'language plus rules for reasoning' - the transitional model;

- are 'working models' which proceed through development using scientific rules (behavioural observations, laws etc.) and mathematical rules (theorems etc) in the transitional model to operate on mathematical entities in the population model to explore possibilities (if ... then ...) which lead to a predicted state;

- need physical interpretation (which has to take account of the original selection and the way science is used in the development); not necessarily of all the mathematical entities involved, but certainly in respect of the description agreeing with experimental evidence and the predicted state having to be at least in principle open to experimental test.

MATHEMATICS IN SCHOOL SCIENCE

The necessity for and intimate relationship between mathematical models and experimental science described here leads to the expectation of mathematics being deeply embedded in the education of scientists. In particular this is true of the education of physicists. The following review explores the extent and nature of the connection between mathematical and scientific education in the light of the importance of mathematical models developed earlier. Forbe's appointment in 1833 gave the impetus to mathematical methods in the higher education of physicists and by this century they have become commonplace. At Bristol University a General Paper is set in physics finals, and a published collection of the problems set for examinations or preparatory tutorials over a quarter of a century is aptly called *Thinking Like a Physicist* (Thompson, 1987). Although in the introduction, Thompson admits it is not a balanced selection, he does claim that '... more discursive types have been included'. Nonetheless, every one of the answers to the 137 problems in the book contains some sort of mathematics. Even using a simple a criterion for a content analysis shows that 129 of the 137 contain an explicit mathematical equation and all of the remaining eight include something else mathematical (asymptote, tangential, parabola, ratio etc.), or refer by result to necessary calculations not made explicit. Thinking like a physicist has come to mean thinking in mathematical models.

It is not surprising to find that mathematical models are part and parcel of school physics too. A joint Royal Society/Institute of Physics report (1973) in Britain produced two inventories of the mathematics used in school physics: a one of minimum needs and one of very desirable abilities. It also recommended that each examination board should produce an explicit statement of the mathematics that would be required for its A level physics examination. One which did was the London Board, by the simple expedient of copying the report's minimum needs list. Over the twenty five years which followed, this has remained the authoritative list of the most advanced use of mathematics in school science[2]. Before exploring the extent to which it defines the mathematical models in school science, it is worth recalling that both school physics and mathematics are in the larger part service subjects for students planning to proceed in other directions: other sciences, medicine, engineering etc. While some 85% or more of those taking A level physics will go on to higher education only at most 10-15% will do so in physics and in some years it falls below 7%. The figures for mathematics are comparable, with the modal destination of about 55% taking it to proceed to study in the physical sciences or engineering courses. Not all of those taking physics also enter mathematics in the same year. Those who do take both and those who do not provide a striking lesson. Figures for combined performance are available for Anglophone Cameroon, where the examination system is based on that of the London Board.

In one institution between 1990 and 1992, 614 students studied the same physics course with 277 (45%) also taking A level mathematics. Of the mathematics group just under one third (32.9%) achieved high grades (A-C) compared to just less than a quarter (24.3%) of the non-mathematics group, 30% of whom achieved an F compared to only 14% of the mathematics group. Using Rutter's (1994) scoring method (A=7, B=6, and so on), the mean points score for the mathematics group is half a grade higher than that of the non-mathematics group, compared to a difference of 1.2 grades reported among English students.

For the whole group of schools in the Bamenda district only 1.3% of those who took both subjects managed a high grade in physics having achieved a low grade (D-F) in mathematics. It is clearly extremely difficult to do well in physics without mathematical ability. The inverse is not true, 27.4% of those who enter both obtain a high grade in mathematics but only a low grade in physics. A more detailed look makes the point even sharper. Figure 4.3 is a table of the entrants in both physics and mathematics from one school over a period of five years, illustrating the outcome from candidates who were all taught by the same teachers.

These results reflect those for the whole educational district, for example only 1.8% manage a high grade in physics with a low grade in mathematics (and all of them achieve no more than a C). Less than 1% achieves an A in physics without an A in mathematics, and none at all with less than a B. The most striking result, however, is that the modal diagonal is not as might be expected the leading diagonal (AA to FF) which it would be if all this is just a consequence of students' performing equally well in both according to their ability. It is actually offset to the right, showing that the modal pattern is for candidates to have a grade higher in mathematics than in physics.

Physics > ***GRADES*** *Mathematics*	***A***	***B***	***C***	***D***	***E***	***F***
A	13.8	**15.6**	5.5	0.9	0.9	0
B	0.9	4.6	**11.9**	2.8	0.9	0
C	0	3.7	8.3	**11.0**	2.8	0.9
D	0	0	0.9	5.5	**2.8**	0
E	0	0	0	0.9	1.8	**0.9**
F	0	0	0.9	0	0.9	0.9

Figure 4.3. Table of the percentage of candidates from one school achieving combinations of grades in physics and mathematics A level

Again it can be seen that without high mathematical ability candidates do not get high physics grades, while it is possible to do well in mathematics without doing well in physics. Taken together these figures provide convincing evidence that mathematical ability is necessary but of itself not sufficient for success in physics. It would seem that what is missing from those who do well in mathematics and not in physics is a capacity to apply their mathematics, ability in mathematical modelling, as well as a more generalised scientific competence.

To explore how appropriate or not is the minimum required list of mathematics for A level physics, the questions set under the auspices of the London Board in England and Cameroon were analysed for the mathematics required to answer the theory papers of physics examinations over a total period of ten years. The mathematics list was divided into 35 separable topics grouped together under seven topic areas (not of equal size see Appendix 1). The questions were then answered and analysed for which mathematical topics were required. The frequency of use of a topic was recorded as the number of questions requiring it in the answer (regardless of how often it may be repeated in the same answer). Altogether the questions required a total of 1420 uses of mathematics on the list. The first thing to note is that no mathematics topic not on the list was required. Secondly, only one topic on the list was not used at all. This is Topic 24. *Sketch graphs of harmonically varying quantities*. With this single exception, then, the list does encompass the minimum mathematics which is needed for A level physics.

Some topics were needed much more frequently than others, however, and Figure 4.4 shows a bar chart of the relative frequencies of use for the seven topic areas.

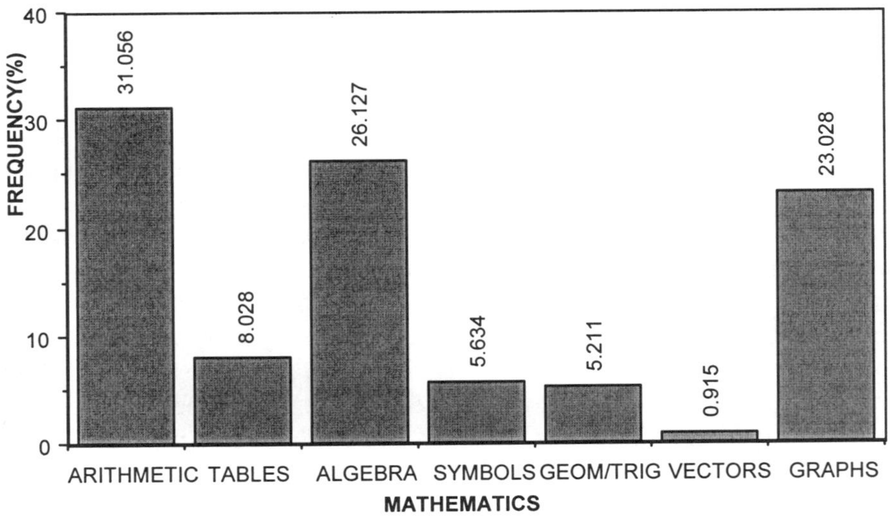

Figure 4.4. Percentage of total use in A level physics answers for topic areas in the minimum required mathematics list

In terms of specific topics, as might be expected by far the most frequent was Topic 1 *Make calculations involving addition, subtraction, multiplication and division of quantities expressed in decimal notation*. The second was Topic 27 *Translate information between graphical, numerical, algebraic and verbal forms,* an essential skill in setting up and interpreting a mathematical model. As too are the third and fourth most frequently used topics, which are even more explicitly central to mathematical modelling: Topic 9 *Substitute physical quantities into physical equations using consistent units so as to calculate one quantity. Check the dimensional consistency of such equations*; and Topic 11 *Formulate simple algebraic equations as mathematical models of physical situations and identify failures of such models*. The evidence is that a facility with the topics in this list and the capacity to deploy them in mathematical models is essential for success in physics at the highest level of schooling.

For school students, the base of their science education is the 16+ courses in biology, chemistry and physics or an integrated course in all three. A more general picture of the mathematics needed for modelling in all the major sciences in schools emerges from an analysis of the 16+ courses. Appendix 2 presents the core topics of typical mathematics courses studied at this level and, from an analysis of a selection of text books, indicates which appear in each of the three sciences (a ‘x’ is placed in the science column opposite the mathematics topics which appear at least once in the texts for that subject). There are 70 mathematics topics listed and it can be seen that 33 are used in biology (47%), 36 in chemistry (51%) and 70 in physics (67%), demonstrating that between a half and two thirds of the mathematics typical taught to 16+ finds a direct application simultaneously in the sciences. Many of the remaining topics will be used in more advanced work too. It was this sort of evidence which prompted the mathematician W.W. Sawyer to observe:

> Much of science still depends on the ability to use simple algebra as a language, intelligently and with understanding. This need is to be met ... by ... mathematics extremely well taught.

It is tempting, therefore, to suggest that the responsibility for this teaching lies solely with the mathematics department, being largely a service department, and that mathematics teachers should schedule their programmes to the needs of the science subjects. Various authors, working parties and the like have made such suggestions. A further Royal Society/Institute of Physics report (1986) recognised that it is not that simple, and it stresses in its fourteen recommendations 'the closest possible co-operation liaison between ... the subjects'. Nonetheless, it still saw the main problem as making the mathematics teaching more attuned to the needs of the physics course. It has to be remembered, however, that mathematics teachers have their own concerns. These include teaching mathematics for its own sake and teaching it successfully. To illustrate this, 316 heads of school mathematics departments in England were surveyed to investigate what influences their choice of teaching order in A level mathematics. They were asked to score twenty possible influences on their teaching sequence as follows:

1. Of no influence: I do not take this into consideration at all.
2. Of little influence: of secondary consideration only.
3. Very influential: I have to take this into consideration.

To represent the results, Figure 4.5 shows the distribution of scores for the top (*Quality of mathematics understanding*) and bottom (*To fit in with degree courses in mathematics*) ranked items, the needs of the physics course (ranked 13th) and of other subjects (ranked 18th).

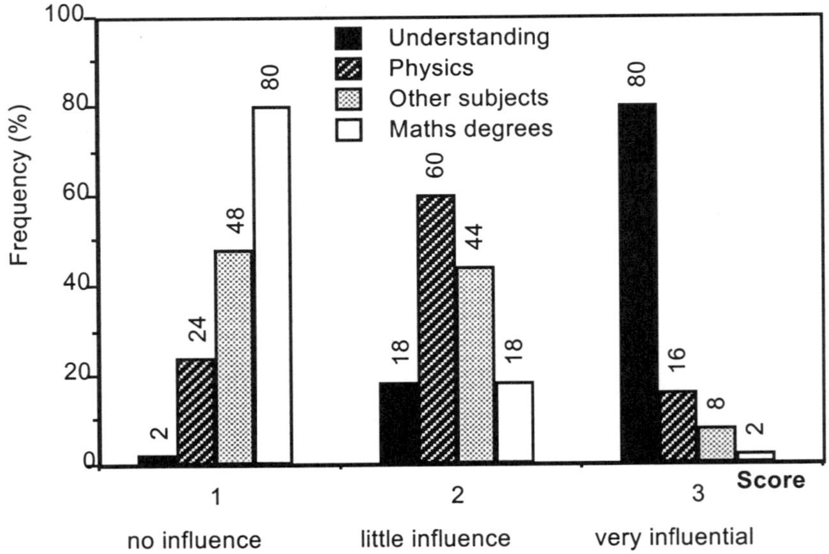

Figure 4.5. Percentage frequency responses from heads of mathematics departments scoring influences on the sequence of their A-level mathematics course

While they recognise only a minority will go on to take a degree in mathematics, the major concerns of the mathematics teachers, quite understandably, are the quality of understanding among their students, their abilities, mathematical background and so on. Next comes concerns about mathematics itself: the syllabus, to emphasise the unity of mathematics, mathematical rigour, and the like. The needs of the physics course does, however, come ahead of influences such as the order in the mathematics text book, but the needs of other subjects is only ahead of to make revision easy and to fit in with degree courses in mathematics.

However willing to co-operate they may be, then, mathematics teachers have their own higher priorities. Moreover, it can be assumed that mathematics teachers come from the pool of those who do well at A-level mathematics. Many of these do not also take physics, and of those who do about 40% get a low grade in it. There is no guarantee, therefore, that mathematics teachers are particularly well placed to apply their mathematics in scientific models. Close co-operation between the mathematics and the science departments is, of course, essential and it will help if there are common policies and practice in such things as shared schemes of work, agreed nomenclature, a negotiated teaching sequence, mutually acceptable teaching strategies and so on. The issue is not solely one of facilitating the transfer of mathematical skills from the mathematics to the science lesson, however. It has to be recognised that mathematical modelling is essential to scientific method; it is a necessary element in coming to understand science and what it is. To teach science as it is practised requires teaching mathematical modelling and it is up to the science teachers to do so.

Inevitably science teachers use mathematical models and teaching them might seem no more than making explicit what is already done. This will not do, however. For example, an analysis of ten textbooks on physics and chemistry yielded nineteen different ways of drawing an atom. Every text used more than one; half-included seven or more and one had ten. In one case, two different ways of picturing the atom appeared in the same drawing. None offered any explanation as to why it should change its iconic representation of the atom from page to page. It may well be obvious to the author that to model the science for the first process a specific iconic transformation is required and for the second another and this in turn demands differing iconic population models. It is far from obvious to the students, however, and unless they have been taught something about the nature of models and modelling it must seem most curious that the atom can change its appearance without comment. As with all teaching, this is not a one-off event, but will need to be a planned learning sequence deliberately sustained and including affirmative reoccurrence of notions in more and more abstracted form.

Little more than a sketch of how this might be done can be given here, but in their text for students on applying mathematics, Ormell et al. (1975) demonstrate how Bruner's sequence of enactive, iconic and symbolic can be harnessed to teach mathematical modelling. They illustrate the idea by recounting the action of the Oxford Council to determine where a monument might be placed. The particular monument is known as the Oxford Conduit and during a redevelopment was removed stone by stone from its original location ready to be reassembled elsewhere. A physical life-sized model was constructed as a painted canvas over a wooden frame and trundled to possible sites for relocation – an enactive transformation on a physical model. This action can then be replaced by iconic models, architects drawings, plans etc. and lastly converted to a symbolic model of co-ordinates on a map.

Similarly, introducing mathematical models needs to be a gradual and graduated sequence of small steps from the real world to the theoretical. Young children, for example, could act out being points on a graph which is then drawn on the floor beneath their feet. Processes such as interpolation and extrapolation can then first be physical movement of children to the appropriate points, and then be represented iconically by points, lines or curves drawn where the children have stood. Later the exercise can be repeated with model people on graph paper, then with drawings of people and finally by points and co-ordinates.

Older students also need exercises which juxtapose the real and theoretical worlds, but for them the real world is experimental science. An example of the staging they require is moving from actual electrical circuits to drawings of the apparatus to conventional circuit diagrams and then the equations. In order to make the modelling explicit at least two of these need to be co-present each time and the correspondence between them pointed out. Simple care in teaching is required; for example, ensuring that the direction and order of a circuit laid out on the bench is followed in the conventional diagram drawn by the teacher on the board or overhead projector transparency. Actions such as doubling the resistance should be carried out simultaneously with a real circuit and the diagram or equation. It is by having both and acting on both together which emphasises the correspondence between the model and what is being modelled. Too often that something can be modelled in a particular way is taken for granted and texts or teachers state 'a car travelling along a straight road …' but draw a point with an arrow or write an equation with the car represented by 'm' with no discussion as to the process or consequences of such abstraction. From time to time students should be asked to discuss 'which point on the car is this?', 'are there circumstances when this would not be an adequate representation ?', 'would there be a difference in exactly when in the real world two cars collide compared to the prediction in the model given by the intersection of two lines ?' and so on. The point is that mathematical modelling can be taught and so central is it to science that acquiring an understanding of what it is and how to do it ought not to be left to the student to pick up, *en passant* as it were, but be an essential element of what is taught integrated throughout all science courses.

NOTES

1. For example, the monumental work of B.I. Bleaney and B. Bleaney (1965) begins Chapter 10 with the subheading 'Maxwell's equations of the electromagnetic field' and the *Concise Dictionary of Physics and related subjects*, J. Thewlis (Ed) (1979) summarises the theory under the simple entry 'Maxwell Equations'.

2. This list has been the subject of a number of studies carried out for masters dissertations under the author's supervision on the M.Sc. course in Physics Education at the University of Reading. Some of what follows is based on a synthesis of a selection of them. The author wishes to acknowledge in particular Jonathan Cross (M.Sc. 1980) and Evelyn Mafeni (M.Sc. 1993) for their work analysing A level physics questions and John Babila-Njingum (M.Sc. 1986, Ph.D. 1996) for work on mathematics in science courses to 16+.

APPENDIX 1

The 'Mathematics Required' section of the University of London General Certificate of Education Examination Advanced Level Physics syllabus, repeated from a Royal Society/Institute of Physics Report.

Arithmetic

1.Make calculations involving addition, subtraction, multiplication and division of quantities expressed in decimal notation.

2.Make approximate evaluations of numerical expressions, using approximations such as $\pi \approx 3$.

3.Express small fractional changes as percentages, and vice versa.

4.Calculate an arithmetic mean.

5. Transform decimal notation to power of ten notation (standard form), and carry out calculations in standard form.

Tables

6. Use tables of squares, square roots, reciprocals, sines, cosines and tangents.

7. Multiply and divide using either logarithm tables or a slide rule, preferably both.

Algebra

8. Change the subject of an equation. Most such equations involve only the simpler operations, but do include positive and negative indices and square roots.

9. Substitute physical quantities into physical equations using consistent units so as to calculate one quantity. Check the dimensional consistency of such equations. (This formulation may present a difficulty in that many mathematicians regard a symbol in an equation as a number rather than a dimensional quantity.)

10. Solve simple algebraic equations. Most are linear, but they include equations involving inverse square relationships, and simultaneous equations.

11. Formulate simple algebraic equations as mathematical models of physical situations and identify failures of such models (applications include dynamics, electric circuits and kinetic theory).

Geometry and trigonometry

16. Calculate areas of right angled and isosceles triangles, circumferences, areas and volumes of rectangular blocks, cylinders and spheres.

17. Identify simple shapes whose areas approximate those of more complex shapes (mainly narrow triangles and areas of strips in integration).

18.Recognise applications of simple theorems: Pythagoras' theorem with application to the chord theorem for a diameter and perpendicular chord, congruency and similarity of triangles, angle sum of triangle.

19.Use sines, cosines and tangents in problems; recall or quickly calculate values at 0^0, 30^0, 45^0, 60^0, 90^0, 180^0.

20. Translate from degree to radian measure, and vice versa.

21. Use radian measure particularly in connection with trigonometric functions.

22. Recall $\sin\theta \approx 0$. $\cos\theta \approx 1$ and $\tan\theta \approx 0$ for small θ

23. Recognise and sketch graphs of $\sin\theta$, $\cos\theta$.

24. Sketch graphs of harmonically varying quantities,

e.g. $A \sin(\omega t) + B \sin(2\omega t)$

or $A \sin(\omega t) + B \sin(\omega t + \pi / 2)$.

Vectors

25. Find the resultant of two vectors, recognising situations where vector addition is appropriate.

26. Obtain expressions for components of vectors in perpendicular directions recognising situations where vector resolution is appropriate.

Graphs

27. Translate information between graphical, numerical, algebraic and verbal forms.

28. Select appropriate variables and scales for graph plotting.

29. Determine the slope and intercept of a linear graph in physical units.

12. Recognise the equivalent forms of the logarithms of each *ab, a/b,* x^n *and* e^{kn}.
13. Recall and use in the context of error estimation and other simple applications the expansions to one term in x of numerical instances of the form $(1\pm x)^n$, where n may be negative or fractional, but $0 < x \ll 1$.

<u>Symbols</u>

14.Comprehend and use the following symbols: $<, >, \ll, \gg, \approx, /, \propto$

(Graphs)

15.Test tabulated pairs of values for direct proportionality by a graphical method, or by constancy of ratio.

30. Choose by inspection a straight line which will serve as the 'least bad' linear model for a set of data presented graphically.
31. Recall the form $y = mx + c$.
32. Use logarithmic plots to test experimental and power law variations.
33. Sketch and recognise the forms of curves such that $y\propto 1/x$, $y\propto x^2$, $y\propto 1/x^2$.
34. Understand and use the slope of a tangent to a curve as a measure of rate of change.
35. Understand and use the area below a curve where the area has physical significance.

APPENDIX 2

Mathematics topics used in the sciences at 16+

MATHEMATICS TOPICS	*Biology*	*Chemistry*	*Physics*
1.			
(a) The ordinary processes of number manipulation	X	X	X
(b) Prime numbers, factors, multiples, indices	X	X	X
(c) Natural numbers, integers, rational and irrational numbers	X	X	X
(d) Weights, measures and moneys	X	X	X
(e) Simple and compound interest			
(f) Fractions, decimals, ratio, proportion and percentages	X	X	X
(g) Expressing numbers to a given degree of accuracy		X	X
(h) Numbers in standard form	X	X	X
(i) Logarithms to base ten	X	X	X
2.			
(a) Length, area, volume	X	X	X
(b) Mensuration of the rectangle, parallelogram, triangle, circle, cylinder, cone and sphere	X	X	X
(c) Length of an arc, area of a sector of a circle.			X
3.			
(a) The idea of a set. Set language and notation	X	X	X
(b) Union and intersection of sets	X	X	X
(c) Number of elements in a set			
(d) Complementary sets			
(e) Subsets	X	X	
(f) Universal set, null set	X	X	
(g) Venn diagrams, and their use in simple logical problems	X	X	X
(h) Use of symbols to represent sets	X	X	X
(i) Binary operations and tables; identity and inverse elements			
4.			
(a) The basic processes of algebra	X	X	X
(b) The symbolic expression of general results	X	X	X
(c) The construction, interpretation and use for formulae: their manipulation in simple cases	X	X	X
(d) The factorisation of simple algebraic expressions	X	X	X
(e) The manipulation of simple algebraic fractions, the denominators being numerical or linear		X	X
(f) Solutions of equations of 1st and 2nd degrees containing one unknown quantity	X	X	X
(g) Solution of linear simultaneous equations in two unknowns			X
(h) Solution of linear inequalities, and the representations of solutions on the number line and in two dimensional space	X	X	X
(i) The idea of a sequence			

5.			
(a) The idea of a function of a variable			
(b) Function as a mapping or as a correspondence between the elements of two sets			
(c) Domain and range of a function			
(d) Composite functions, inverse functions. Use of symbols to represent operators; transformations; functions; mappings			
(e) Rectangular cartesian co-ordinates	X	X	X
(f) Variation	X	X	X
(g) Graphs and graphical treatment of linear and quadratic functions	X	X	X
(h) The gradients of these functions by drawing	X	X	X
(i) Determination of gradients, rates of change, maxima, minima	X	X	X
(j) Applications to linear kinematics and to other simple practical problems			X
6.			
(a) Representation of data by a matrix			
(b) Addition and multiplication of matrices			
(c) Multiplication of matrix by a scalar			
(d) Unit, (identity) matrix and zero (null) matrix			
(e) Determinants, singular matrices			
(f) Inverse of non-singular 2 x 2 matrices.			
(g) Transformation of the plane associated with 2 x 2 matrices, combination of transformation			X
7.			
(a) Scalar and vector quantities. Representation of a vector by a directed line segment			X
(b) Sum and difference of two vectors			X
(c) Multiplication of a vector by a scalar			X
(d) Multiplication of a vector by a matrix			
8. Geometrical properties of Euclidean space, as listed below.			
(a) Angle properties of parallel lines triangles and Polygons	X	X	X
(b) Properties of the parallelogram, rectangle, square, rhombus, trapezium and kite		X	X
(c) Symmetry about a point, line or plane	X	X	X
(d) Use of Pythagoras' theorem		X	X
(e) Similarity, areas and volumes of similar figures	X		X
(f) Chord, angle and tangent properties of circle			X
(g) Loci in 2 dimensions			X
(h) Construction of bisector of an angle and of perpendicular bisector (mediator) of a straight line			X

9.			
(a) Use of sine, cosine and tangent of angles up to 180°		X	X
(b) Solution of problems in 2 and 3 dimensions by calculations and drawing			X
(c) Angles of elevation and depression			
(d) Bearings			
10.			
(a) Graphical representation of numerical data	X	X	X
(b) Determination of the mean and median of a small number of quantities	X	X	X
(c) Inter-quartile range and percentile			
(c) Determination of the mean of a larger number of quantities given in grouped frequencies			
(e) Standard deviation			
(f) Simple probability	X	X	X
(g) Sum and product rules of probability and their application to a simple problem	X		X

Section Two: The Development of Mental Models

Preface

The first Section of this book established the distinction between mental models in an individual's mind and those models which arise from expression in the public domain. These expressed models can be understood by other people (i.e. have mental models formed of them) through the various modes of representation in which they are put forward. This Section concentrates on mental rather than expressed models.

Chapter 5 begins with a review of the various ways in which the notion of 'mental model' can be appreciated. Abstraction techniques and thought experimentation, developed by Nersessian to model the processes by which a mental model is produced in science, is illustrated with a case study of Einstein's production of his mental model of 'general relativity'. The main features of mental models are then identified and used to provide a framework with which to understand children's mental models of 'the Earth in space'.

Chapter 6 takes up the theme, developed in the first Section, of the relationship of a phenomenon to mental models of it. The Chapter suggests that a target model of a phenomenon to be studied in the classroom should be constructed by researchers. This would allow the embedded interacting levels in systems, such as the circulatory system, to be defined and their parts analysed. This Chapter takes a model-based teaching and learning perspective. It suggests that mental model building requires teaching structures that allow students to engage effectively both with the phenomenon itself and with its representation in various modes. An analysis framework involving the structure, behaviour and mechanism aspects of any phenomenon is proposed. An analysis of the features of all representations, based on Goldsmith's scheme, allows potential semiotic challenges to be identified.

Chapter 7 expands the boundaries of mental modelling by reviewing the literature on creativity in Design and Technology education. It proposes that mental modelling is a critical means of achieving communication with self and with others in the production, development, and acceptance, of designs.

In analysing the nature of thought experiments in Chapter 8, the gestural bodily aspect of mental models is highlighted and the features of embodied cognition as non-propositional, not easily verbalisable, and requiring no explicit thinking, are discussed. The Chapter ends by suggesting that embodied cognition is a fundamental knowledge system and raises questions about how this relates to symbolic knowledge.

The use of computers in Chapter 9 suggests one way of integrating and using bodily learning. This Chapter provides a review of the use of computers in the development of mental models. It looks at the fast developing opportunities for accessing various expressed models via CD-ROM and Internet, at how learners can construct models on computers, and at the collaborative possibilities of model building.

The types of knowing involved in mental model building - propositional, imagistic and bodily - thread through these Chapters as does the crucial role of the generative nature of mental models identified at the start of the Section.

Chapter 5

Grasping Mental Models

Creso Franco[1], Dominique Colinvaux[2]
[1]*Catholic University of Rio de Janeiro, Brazil;* [2]*Fluminense Federal University, Brazil*

INTRODUCTION

Mental models have been approached from a number of different perspectives, from cognitive psychology to philosophy of science and science education. As a result, several definitions have been proposed that emphasise distinct aspects. It has also been suggested that researchers have access to people's mental models by means of the examination of individuals' expressed models. We argue that, in order to grasp mental models, a fruitful strategy involves developing two complementary approaches: one that focuses on how mental models are developed and the other one on their key features when people make use of them to think. In particular, we will deal with three basic issues: to what extent can we talk about mental models by examining expressed models? What are the tools individuals make use of in order to build mental models? What are the main features of mental models? Each issue is addressed below, starting from a critical analysis of existing literature so as to suggest and discuss a framework that could help us grasp mental models.

MENTAL MODELS AND EXPRESSED MODELS

According to Johnson-Laird (1983), mental models are structural analogues of the world as perceived or conceptualized. Gentner and Stevens (1983, p. 1) argue that mental models are related to human knowledge of the world and of how it works, i.e. to 'the way people understand some domain of knowledge'. Tiberghien (1994) specifies that modelling refers to a specific type of knowledge processing. Norman (1983) distinguishes between

J.K. Gilbert and C.J. Boulter (eds.), Developing Models in Science Education, 93–118.

individual's mental models of a target system, specialists' conceptual models of the same target and scientists' conceptualisation of people's mental models. For Gilbert and Boulter (1998), a model is a representation of a target which might be an object, event, process or system. They also introduce the concept of expressed models, that are placed in the public domain, in contrast with mental models, which are personal, private representations of the target.

Current studies on mental models derive mainly from the fields of cognitive psychology (as is the case of Johnson-Laird's study), artificial intelligence and human machine interfaces (e.g. Norman, 1983; Rouse and Morris, 1986; Gentner and Gentner, 1983), as well as from science education (Gilbert and Boulter, 1998; Tiberghien, 1994). The definitions above express this disciplinary variety: while Johnson-Laird (1983) focuses on the cognitive issue of how people make inferences without making use of propositional reasoning, Norman develops his views on mental models on the basis of his observations of how people handle devices such as calculators. Gentner and Stevens (1983) include AI-related issues on the use of mechanical devices as well as some studies in science education, the same field that underpins Tiberghien's (1994) and Gilbert and Boulter's (1998) views on the subject. Other research studies address the issue of scientific reasoning and refer to the history and philosophy of science to investigate the development of new theories and ideas (e.g. Hesse, 1963; Kuhn, 1970a; Nersessian, 1992a).

The very notion of mental models, increasingly pervasive in these different areas, has become multiple and thus difficult to grasp. Even so, however difficult the task may prove, researchers have attempted to investigate mental models and, in order to do so, they have relied on documents such as notebooks, published papers, scientific instruments and prototypes designed and/or used by scientists. Psychologists studying cognition have also conducted interviews, analysed the correspondent protocols and, maybe, checked their conclusions against observation of people acting freely.

In all these situations, mental models are studied via those models that are expressed by means of writing, drawing or other actions, such as creating or manipulating objects. The question which emerges from this scenario is the relationship between these expressed models and what was going on in the mind of the people who expressed these models. It is accepted that some sources are more reliable than others for the study of mental models. For instance, published papers are directly constrained by social conventions. Therefore, they shed more light on the process of rational reconstruction of scientists' work than on what was actually in scientists' minds prior to the decision of writing a paper for publication. This issue is at the root of the decision of some mental model researchers to look at scientists' notebooks (Gooding, 1992) or inventors' prototypes (Franco et al., in press) as the main source of empirical data on mental models. Gathering data through interviews also raises some questions of the complex relations between thought and speech. Considering that speech cannot be taken as a reliable mirror of thought processes, some procedures can and have been adopted that attempt to complement interview data. In these cases, speech might refer to prediction and problem-solving situations as well as being articulated with the elicitation of drawings.

Gilbert and Boulter (1998) stress the private nature of mental models. As a result, mental models appear to be inaccessible to the researcher, who will need to rely on some expressed version of the private mental models to infer what the latter may be. Norman (1983) expresses a concern that bears some similarity with Gilbert and Boulter's (1998) argument: he emphasises the distinction between individuals' mental models and the analysis that researchers can carry out regarding these models. As a result of their analyses, researchers will thus develop models of models. Although Norman and Gilbert and Boulter do not coincide entirely in their views, the point to be made here is that some common issues arise with respect to the investigation of mental models: the researcher's findings constitute a model of a supposed model, to which access is gained only through some expressed version of it.

Several methodological approaches have been devised to grasp mental models. Although there is no unique, consensually agreed, strategy, some criteria have been suggested that help to ensure sound results. The critical distinction between mental and expressed models requires in particular that the relationships between the two be investigated in their own right. Other issues also remain to be discussed before a set of methodological strategies can be agreed. In particular, we must look into the cognitive processes by which mental models are developed and into the defining features of mental models. To these we now turn.

MODELLING AND ABSTRACTION TECHNIQUES

To approach the issue of how mental models are developed, we start from Nersessian's (1992) construct of abstraction techniques and carry out a case study of a particular abstraction technique: coalescence, the process by which two existing concepts are merged into a new one. First Nersessian's views are presented and the construction of an abstraction technique is defined.

Nancy Nersessian proposed cognitive-historical analysis as an approach to understanding conceptual change in science. The underlying presupposition of this approach is that

> the problem-solving strategies scientists have invented and the representational practices they have developed over the course of the history of science are very sophisticated and refined outgrowths of ordinary reasoning and representational processes. (Nersessian, 1992a, p.5).

Therefore, cognitive-historical analysis combines traditional methods of history of science with analytical tools of cognitive sciences for making sense of how new conceptual representations can be generated from existing ones.

Nersessian developed her method by means of some case studies: the development of field theory from Faraday to Maxwell, Galileo's research on falling bodies and Einstein's work on the electrodynamics of moving bodies. The aim of her analysis was to overcome the limitation of approaches based solely on the role of induction or deduction in scientific practice since such approaches block the ability to make sense of actual constructive practices involved in science. Accordingly, Nersessian emphasied the role of abstraction techniques such as analogical reasoning, imagistic reasoning, thought experiments and limiting case analysis as tools used by scientists for building new knowledge.

Contrary to philosophers such as Duhem (1914), Heimann (1970) and Chalmers (1986) and in agreement with others such as Campbell (1920) and Hesse (1963), Nersessian (1992a, p.20) stressed that analogies

> are not 'merely' guides to thinking, with logical inferencing actually solving the problem, but analogies themselves do the inferential work and generate the problem solution.

Despite relying on Johnson-Laird's (1983) concept of *mental model* as a structural analogue to reality, Nersessian's focus on the dynamics of conceptual change in science lead her to the theme of *mental modelling*, which is developed by investigating scientists' use of abstraction techniques.

An important feature of abstraction techniques is that they bridge the gap between content-independent and content-dependent constructs for dealing with knowledge building. In other words, the set of abstraction techniques identified by Nersessian allows researchers to approach specific scientific practices in a way that, on the one hand, is more closely related to the core of the practice under investigation than would be possible on the basis of some general philosophical concepts such as induction, deduction, abstraction and generalisation. On the other hand, abstraction techniques such as analogy, thought experiments and limiting case analysis allow the development of approaches that are more general than those which are based solely on the subject matter under investigation. Therefore, each abstraction technique is both a *general construct*, in the sense that it goes beyond the specific case study that it helps the researcher to understand, and a *specific construct*, in the sense that different scientific practices may be approached by different abstraction techniques. This feature suggests that the examination of case studies other than those considered by Nersessian will allow us to identify other abstraction techniques. In the following section this task is carried out by examining a case study on the *coalescence* of the concepts of inertial mass and gravitational mass in the beginning of the 20th Century.

COALESCENCE AND THE DEVELOPMENT OF GENERAL RELATIVITY

Newtonian mechanics presents the concepts of *inertia* and *gravity*. The former concept is related to the property bodies have to resist attempts to change their state of movement: inertial mass is a measure of this resistance. The latter concept is related to the property of bodies to attract and be attracted by other bodies: gravitational mass is a concept that expresses a measure of this feature. From a conceptual point of view, there is no relationship between these pairs of concepts, inertia and gravity, or inertial mass and gravitational mass, within Newtonian mechanics. Despite this, the identity of inertial mass and gravitational mass is a well-established empirical law[1].

The identity of inertial mass and gravitational mass is closely related to Galileo's law for the fall of bodies and may be used to explain why the acceleration of falling bodies of different nature or different weights is the same. The argument is that a heavier body is, in fact, more attracted towards the Earth than a lighter one. However, a heavier body is more inert, therefore producing more resistance to be accelerated, than a lighter one. Given the identity of gravitational and inertial mass, the greater force attracting a heavier body towards the Earth exactly compensates the greater resistance offered by this same body to being moved, thus explaining why bodies fall with the same acceleration. In spite of being an explanation that is very comfortable for people, we must point out its circularity. In other words, the explanation does not provide a logical basis for the law of falling bodies because it makes use of an empirical law, the identity of inertial and gravitational mass, that has no rationale within Newtonian mechanics.

The complexity of the justification of Galileo's law of fall has a long history. Galileo (1954, p.166) himself wrote that 'the present does not seem to be the proper time to investigate the cause of the acceleration of natural movement'. This issue was later dealt with by Newton, who established the grounds for the idea of action at a distance and explained heaviness as a consequence of the interaction between bodies and the Earth. Despite this, a rationale for explaining why acceleration does not depend on the nature of the falling body remained an open question.

From 1905, Einstein tried to develop a gravitational theory coherent with Special Relativity. As did other scientists of that time, he

> tried to frame a field-law for gravitation, since it was no longer possible, at least in any natural way, to introduce direct action at a distance owing to the abolition of the notion of absolute simultaneity. (Einstein, 1979, p.306).

However, the attempts failed, for their results

> did not fit in with the old experimental fact that all bodies have the same acceleration in gravitational field. This law, which may also be formulated as the law of equality of inertial and gravitational mass, was now brought home to me in all its significance. I was in the highest degree amazed at its existence and guessed that in it must lie the key for a deeper understanding of inertia and gravitation (Einstein, 1979, p.307).

According to Einstein's autobiography (Einstein, 1970, pp.61-68), the lack of rationale for the identity between inertial and gravitational mass was at the root of his dissatisfaction with Special Relativity for this theory was incapable of dealing with gravity in a productive way. Einstein achieved the first step towards General Relativity in 1908, when he was able to conceptualise the relationship between inertia and gravity in a new fashion. This was carried out by means of a drive towards promoting the coalescence of the concepts of inertial and gravitational mass. In Einstein's (1970, p. 65) own words:

> Now it came to me: The fact of the equality of inert and heavy mass, i.e., the fact of the independence of the gravitational acceleration of the nature of the falling substance, may be expressed as follows: In a gravitational field (of small spatial extension) things behave as they do in a space free of gravitation, if one introduces in it, in place of an 'inertial system', a reference system which is accelerated relative to an inertial system.

Einstein's conceptions in this respect were explained in more detail elsewhere (Einstein, 1961, pp. 63-70)[2]. The argument makes use of a thought experiment in which an inertial system of reference consisting of a spacious chest resembling a room with an observer inside is located in a remote region of space. Einstein reminded us that, naturally, gravity does not act at this remote region of space. In the thought experiment, an 'immaterial being' starts pulling the chest 'upwards' with a constant force. The force is exerted by means of a rope linked to a hook located externally at the middle of the lid of the chest. Einstein invited us to consider what happens when the observer inside the chest releases a body that he had in his hand. Let us consider how Einstein summarized the observer's conclusion about this situation (Einstein, 1961, p. 67):

> If he releases a body which he previously had in his hand, the acceleration of the chest will no longer be transmitted to the body, and for this reason the body will approach the floor of the chest with an accelerated relative motion. The observer will further convince himself that the acceleration of the body towards the floor of the chest is always of the same magnitude, whatever kind of body he may happen to use for the experiment. Relying on his knowledge of the gravitational field [. . .], the man in the chest will thus come to the conclusion that he and the chest are in a gravitational field which is constant with regard to time. Of course, he will be puzzled for a moment as to why the chest does not fall in this gravitational field. Just then, however, he discovers the hook [. . .] and the rope which is attached to it, and he consequently comes to the conclusion that the chest is suspended at rest in the gravitational field.

The observer's conclusion outlined above provides a rationale for the conceptual linkage between inertial and gravitational mass. As stressed by Einstein himself, this functioned as a very important first step for developing General Relativity (Einstein, 1979), the theory in which the ideas of both inertia and gravity are subsumed within a new conception for the relationship between matter and space. It is well established that at the end of the process we have the coalescence of the concepts of inertia and gravity.

The analysis of Einstein's writings shows a case of mental modelling in science as an activity based on producing novelty. Coalescence, jointly with thought experiment, was a driving idea from the beginning of the process of building General Relativity. It is thus profitable to look at coalescence as a reasoning tool, belonging to a set of abstraction techniques that might be used for constructing mental models.

FEATURES OF MENTAL MODELS

Grasping mental models involves not only understanding the processes by which mental models are developed, as was just seen, but also requires capturing key features associated to the use of mental models. As previously noted, the issue of mental models is approached from several disciplinary fields and, as a result, many definitions are currently held that emphasise different aspects of mental models. We argue that to propose a unique definition of mental models would prove not only an arduous task, but also one that would probably limit the possibility of developing empirical studies that could help us further understand the issue of mental models. In our view, a more fruitful approach should focus on a critical set of key features of mental models. In this respect, our approach is distinct but complementary to Buckley and Boulter's analysis of the component features of mental models (see Chapter 6).

Research findings on mental models register a variety of features, some of which are repeated from one study to another. In our view, a preliminary set of key features includes those which help the researcher to identify and characterise the mental models used by individuals. According to this set:

- *mental models are generative*: in model-based reasoning, people produce predictions and new ideas;
- *mental models involve tacit knowledge*: the holders of a mental model are not aware of every constitutive aspect of their mental model;

- *mental models are synthetic*: in order to be efficient, mental models consist of a simplified representation of the target system or event which motivated the construction of the mental model in the first place;
- *mental models are constrained by world-views* that limit the range of models which could be built, depending on general belief-systems held by people.

Each one of these features is analysed hereafter in the following manner. They are initially illustrated by using one particular research study on children's mental models of the Earth (Vosniadou and Brewer, 1992; Vosniadou, 1994) and then discussed so as to put our views forward. This study was chosen because it offers a comprehensive analysis of mental models as well as a sound basis of empirical data; furthermore, the features we deem important to approach the subject of mental models are all dealt with in Vosniadou's study, even if they are not always conceptualised in the same way that we propose.

CHILDREN'S MENTAL MODELS OF THE EARTH: VOSNIADOU'S STUDY

Vosniadou's study investigates elementary school children's understanding of the Earth, its shape and the region on the Earth where people live (Vosniadou and Brewer, 1992; Vosniadou, 1994)[3]. Vosniadou's basic argument is that children's ideas about the natural world develop on the basis of a small number of *entrenched presuppositions* about physical objects, that are embedded in a *naive theory of physics*[4] (Vosniadou, 1994). These presuppositions, which appear to originate from everyday experience, are later articulated with culturally accepted views as they are expressed, among others, by adults. As a result, children develop new ideas that become gradually more consistent with such culturally dominant views.

The study aims at investigating children's knowledge about the concept of Earth and focuses on children's mental representations of the Earth, which are interpreted as mental models. Mental models are specifically defined as 'a dynamic structure created on the spot for the purpose of answering questions, solving problems, or dealing with other situations' (Vosniadou, 1994, p.414) and loosely referred to by Johnson-Laird (1983)[5]. Vosniadou thus distinguishes between conceptual knowledge, as incorporated in a naive theory of physics, and the mental models that children create, or make use of, we might add, when they are faced with particular situations and problems.

To characterise children's mental models of the Earth, a series of questions were asked about the shape of the Earth and the regions where people live; drawings were also elicited (Vosniadou and Brewer, 1992, pp.542-546). The questions asked included both *factual questions*, such as 'what is the shape of the Earth?' which can be answered by repeating acquired information without necessarily understanding it; and *generative questions*, such as 'if you were to walk for many days in a straight line, where would you end up?'. Generative questions do not refer directly to observable situations nor can they be answered promptly by repetition: they require thinking! To answer these kinds of questions, children need to refer to and use whatever relevant knowledge/experience they have, so as to create a mental representation that can help them form an answer.

A set of Earth models was identified in the US study. They are:

- *The rectangular Earth model*: the Earth appears as a flat, solid, supported object shaped like a rectangle.
- *The disc Earth model*: the Earth presents the same features as in the rectangular Earth model, only that it is shaped like a disc.
- *The dual Earth model*: this model includes two Earths, a round one up in the sky and a flat, solid and supported Earth - the ground on which people live.
- *The hollow sphere model*: the Earth is a hollow sphere with people living on flat ground inside it or it is made of two hemispheres, the lower one on which people live and the upper one with the sky like a dome.
- *The flattened sphere model*: the Earth is a sphere but flattened at the poles, or a thick pancake.
- *The spherical Earth model*: the Earth is a sphere with people living all around it on the outside.

These models are presented in Figure 5.1 below (reproduced from Vosniadou, 1994, p. 417).

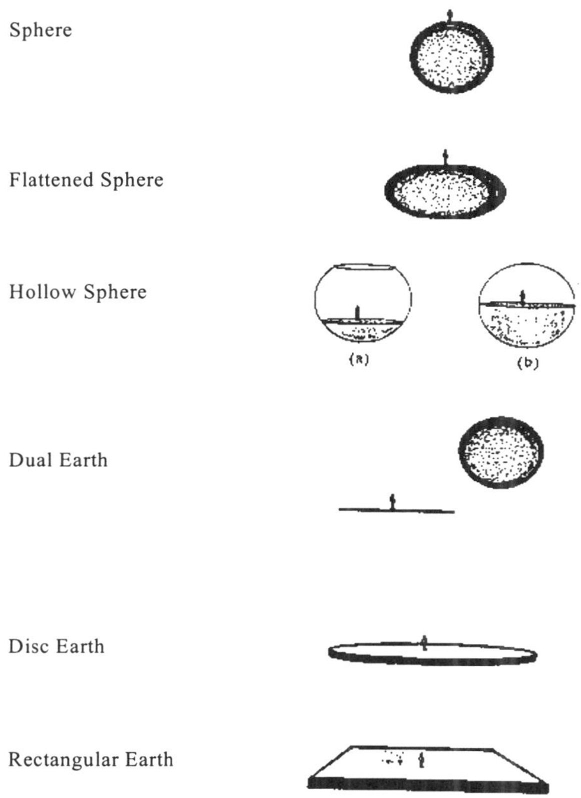

Figure 5.1. Earth Models (Vosniadou, 1994)

Having outlined Vosniadou's main research findings, which are supported by the literature on children's astronomical ideas (e.g. Franco, 1993; Nussbaum, 1979, 1985), we will now turn to describing how her study illustrates the four features that we propose.

MENTAL MODELS ARE GENERATIVE

That mental models are generative means that making use of a mental model produces new information. In Vosniadou's study, this feature is embedded in the definition of mental models itself, conceived of as dynamic structures that are created on the spot in order to meet the demands of the situation. Moreover, the methodological procedures she devises are particularly fruitful to obtain evidence about the generative feature of mental models. By asking generative questions such as 'if you walked for many days in a straight line, where would you end up?', 'would you ever reach the end or the edge of the Earth?', 'is there an end or an edge to the Earth?'; and also, when a child offers a positive answer to the latter questions, 'can you fall off that end or edge?' and 'where would you fall?', the interviewer puts the child in the situation of having to answer a question about unobserved phenomena for which there is no ready-made answer.

The different mental models of the Earth thus constitute the basis for generating specific sets of answers to the interviewer's questions. For instance, the disc, rectangular, and dual Earth, models imply that the Earth has an end or edge, from which one could fall off. The hollow sphere model suggests an end but, as people live inside the Earth, it is not possible for them to fall off. This is what Matthew, a first grade pupil, argues:

> *E:* *If you walked and walked for many days where would you end up?*
> *C:* *If we walked for a very long time we might end up at the end of the Earth.*
> *E:* *Would you ever reach the edge of the Earth?*
> *C:* *I don't think so.*
> *E:* *Say we just kept walking and walking and we had plenty of food with us.*
> *C:* *Probably.*
> *E:* *Could you fall off the edge of the Earth?*
> C: *No. Because if we were outside of the Earth we could probably fall off, but if we were inside the Earth we couldn't fall off.*
> (Vosniadou and Brewer, 1992, p.548).

That children are able to answer these questions in a fairly consistent manner is evidence that they produce novel information, that is, they are generating knowledge when answering the questions put to them. That they also answer the whole set of questions in a consistent manner is evidence that they are reasoning on the basis of a common and apparently unified mental representation, i.e. a mental model. In addition, it is relevant to note that Vosniadou also tested whether the apparent consistency of children's answer could be attributed to a methodological artifact and discarded this possibility (Vosniadou and Brewer 1992, pp.572-573).

Beyond Vosniadou's study, the notion that mental models are generative is recurrent and widespread, starting from Craik's (1983/1943) often referred-to ideas. For him, 'thought models, or parallels, reality' and thus it is possible for

> this reasoning process [to] produce a final result similar to that which might have been reached by causing the actual physical processes to occur ... (in Johnson-Laird, 1985, p.82).

The generative feature implies that using mental models means going beyond the level of description to predict and explain. In this sense, the generative feature stresses that mental models not only describe a state of affairs but are also used to infer information which is not explicitly or directly contained in the description itself. In this respect Johnson-Laird, based on the analysis of syllogisms, argues that people make inferences, not by using formal propositional reasoning, but by creating and testing mental models. The research studies that focus on human control of mechanical devices also stress the generative feature of mental models, which Rouse and Morris in their review article (1986, p.351) summarize as follows:

> Mental models are the mechanisms whereby humans are able to generate descriptions of system purpose and form, explanations of system functioning and observed system states, and prediction of future system states.

Some of the science education studies present a similar view when they suggest that mental models are used to describe a system and its component parts as well as its states, to explain its behaviour when changing from a state to another and to predict future states of the system. The generative feature of mental models as expressed by the science education literature is illustrated by Gentner and Gentner's classical study (1983) on analogical models of electricity. The study demonstrates that *analogies are conceptual tools* (op. cit., p. 125) that generate new information. Two experiments are carried out with high school and college students whose distinct inferences are derived from either one of two spontaneously used or learnt analogical models of electricity: the *water-flow model*, suggesting that electricity flows though wires in the same way that water flows through pipes; and the *moving crowd model*, in which electric current is seen as objects moving through passageways. Students' inferences are elicited in a series of simple problems involving serial and parallel combinations of batteries and resistors. The study predicts that the water flow model will make it easier to distinguish between batteries combinations whereas the moving crowd model will provide a better understanding of the serial/parallel combinations of resistors. The findings generally show that both predictions tend to be confirmed[6].

The generative feature is clearly illustrated by a student's protocol (op. cit., p. 115). When first using the water flow model to solve the parallel resistor problem, the student incorrectly predicted that there would be less current. When the experimenter suggested using the moving crowd analogy, the student derived the correct response of more current, as shown in the following excerpt:

> Again I have all these people coming along here. I have this big area here where people are milling around. ... I can model the two-gate system by just putting the two gates right into the arena just like that. ... There are two gates instead of one which seems to imply that twice as many people can get through. So that seems to imply that the resistance would be half as great as if there were only one gate for all these people. (op. cit., p.115)

Analogies, as discussed previously, imply that some aspects of the source domain are carried over to the target domain and it is precisely this mapping from one to the other domain which supports the emergence of new ideas. In Gentner and Gentner's study, both the water flow and moving crowd analogies support an inference process concerning electrical circuits, their properties and especially the interrelation between current, voltage and resistance. The generative feature of models appears therefore as a distinctive mark of this kind of knowledge processing.

MENTAL MODELS INVOLVE TACIT KNOWLEDGE

That mental models involve tacit knowledge means that the holder of a mental model is not (entirely) aware of the several component aspects of his/her mental models nor of how s/he makes use of it. This feature is mentioned by Vosniadou, although very briefly, when she claims that mental models are formed on the basis of entrenched presuppositions that form a naive theory of physics which, she argues, 'is not available to conscious awareness and hypothesis testing' (Vosniadou, 1994, p.413).

An analysis of the matter will help illustrate what is meant. All the mental models identified, except the spherical Earth model, appear to be formed and further develop according to two fundamental presuppositions. The first one is that the ground on which people live is flat (Vosniadou and

Brewer, 1992, pp.576-567; Vosniadou, 1994, p.418); and the second one is that unsupported things will fall down (Vosniadou and Brewer, 1992, p.577), which points to an up/down gravity (Vosniadou, 1994, p.418-419) where 'up' and 'down' are absolute notions that are defined taking the Earth as frame of reference. The influence of these presuppositions on children's mental models is fairly clear and will be later discussed with reference to the constrained feature of mental models. Here the example of the disc and rectangular models of the Earth as being directly derived from the presupposition that the ground is flat will be sufficient to illustrate the role of the first presupposition on the development of these models. The point to be stressed is that both presuppositions are indeed tacit: they are not explicitly referred to and, in this sense, they are not thematised by the children. In other words, children do not think *about* them but rather they reason *with* them.

Other evidence of the tacit nature of some aspects of mental models derive from the idea of an absolute verticality, previously referred to as the up/down gravity, which can be found in some children's drawings. One such example can be found in Nussbaum's (1985) study of children's understanding of the Earth in space. Here, the idea of an absolute up-down vertical direction permeates all of the Earth notions except the scientific one, from notion 1 of a flat Earth covered by a parallel horizontal sky, through notion 2 of a hollow sphere Earth, to the subsequent notions that bring out the idea of a spherical Earth surrounded by space. The initial flat-Earth referenced up-down direction appears to develop into an absolute vertical that is Earth-independent in the sense that it is not related to the Earth but determines, for instance, that stars will be placed on the upper part of the drawing (see Figure 5.4 below). It is interesting to note that Nussbaum, similarly to Vosniadou, stresses that 'the children themselves were not fully aware of their own belief sets' (op. cit., p.178).

These assumptions are not explicitly referred to, nor thematised by the children, but can be made explicit through their drawings, as will now be seen. Children were asked to predict the direction of falling objects presented on a picture of a round Earth. Leonidas, a 9-year-old boy, argues that the objects fall down and explains 'down' by drawing a line on the picture that is below the given round Earth. When the interviewer says 'Suppose you and I were in this picture. Draw two persons in the place where we would be', Leonidas completes his drawing by placing these two people on the 'ground', as shown in Figure 5.2 below (reproduced from Nussbaum, 1985, p.180-181).

Figure 5.2. Leonidas' drawing (Nussbaum 1985)

Franco's (1993) research findings on children's understanding of the shape of the Earth also include some drawings which express the tacit strength of the up-down assumption. For instance, Luis, a 6 year-old boy's drawing of people on the Earth assumes this same up-down direction in the way it places people (Figure 5.3). Roberta, a 7-year-old girl, recognises that people live all around the Earth, thus breaking up with the notion of an absolute vertical at the surface of the Earth, but still relies on an up-down direction when it comes to the stars, which all remain in the upper part of her drawing.

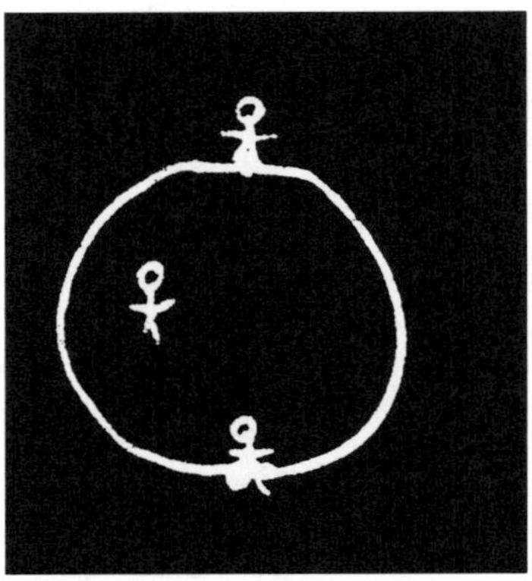

Figure 5.3. Luis, 6 years (Franco, 1993)

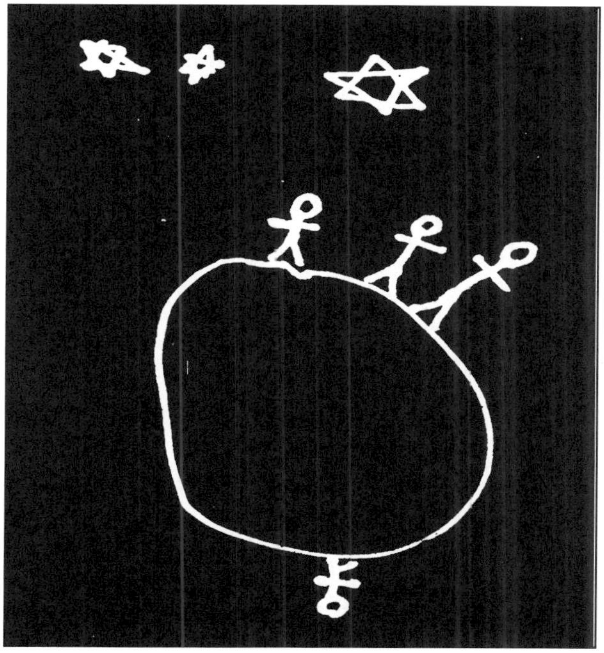

Figure 5.4. Roberta, 7 years (Franco, 1993)

That mental models involve a tacit dimension is a feature that is increasingly discussed in the literature. Although it is not specific to mental models alone, it appears a distinctive feature of this particular mode of representation and reasoning. Nersessian (1995), when discussing the implications for science education of what is known about scientific practices, focuses on constructive modelling, which she defines as 'a tacit dimension of the thinking practices of expert physicists'. Nersessian's argument that learning science requires being initiated to scientific practices seems to approach some of Kuhn's ideas (1970a), especially as presented in his Postscript to *The structure of scientific revolutions*. It is interesting that Kuhn specifies the meaning of paradigms as exemplars by referring also to science learning. As is well known, exemplars are taken as

> 'the concrete puzzle-solutions which, employed as models or examples, can replace explicit rules as the basis for the solutions of the remaining puzzles of the normal science' (Kuhn, 1970a, p. 175).

He then describes how a student would learn to *see* the *analogies* between problems that refer to different phenomena (free fall, simple pendulum, pair of interacting harmonic oscillators) under the same law $f = ma$, an example which he takes from a parallel development in the history of science (op. cit., pp.188-190). The important point made by Kuhn is that

> what results from this process is 'tacit knowledge' which is learned by doing science rather than acquiring rules for doing it (op. cit., p.191).

Kuhn then goes on to argue that the (individually or historically) acquired capacity to see phenomena in certain terms and to group problems according to similarities does not depend on the application of criteria and rules and it is in this very sense that this kind of knowledge is tacit. In other words:

> the practice of normal science depends on the ability, acquired from exemplars, to group objects and situations into similarity sets which are primitive in the sense that the grouping is done without an answer to the question, 'Similar with respect to what? (op. cit., p.200).

MENTAL MODELS ARE SYNTHETIC

That mental models are synthetic means that they represent only some aspects of the target rather than the complete target in all its possible interpretations. Our definition, however, does not coincide entirely with Vosniadou's views, as we will now see. In her study, Vosniadou establishes a difference between *initial*, *synthetic* and *scientific* models, which is related to the issue of how models change and develop (Vosniadou and Brewer, 1992, pp.578-579; Vosniadou, 1994, pp.419-420)[7]. Initially children form models on the basis mainly of a naive theory of physics that takes into account the two presuppositions mentioned above: these initial mental models include the rectangular and disc models of the Earth. But very soon children hear about the culturally dominant view that the Earth is round and so are faced with the task of reconciling their original model with this new piece of information. The dual Earth, hollow sphere and flattened Earth models are taken by Vosniadou to express these attempts and are characterised as synthetic models, which constitute intermediate steps between the initial models and the scientific, sphere model of the Earth.

Therefore, in Vosniadou's study, synthetic models are the result of children's attempts to reconcile their own ideas with the views presented to them by society:

> by forming these synthetic models children try to assimilate the information that the Earth is a sphere with their preexisting knowledge structures in a way that allows them to retain as many of their presuppositions as possible (Vosniadou and Brewer 1992, p.579).

In other words, these models propose a synthesis between the two perspectives of the child and of the adult world and, in this sense, the important feature of this process is its tendency to inclusiveness.

However, we argue that the notion that mental models are synthetic may present other meanings than the one proposed by Vosniadou. Two meanings in particular are outlined here: the first one points to the abstraction of certain aspects or parts of what is being represented, which results in an economic representation, while the second one stresses the systemic nature of the resulting representation.

Firstly, the notion of representation itself implies some kind of synthesis. A representation is never a complete reproduction of what is being represented but requires conscious or unconscious selection of what aspects will be represented and what other aspects will be left out of the representation. The idea here is that, in order to develop a representation of a target, some aspects or parts are isolated and abstracted because considered relevant in the context of the individual and the situation, and this implies some kind of simplification (which appears to be closely articulated, and interdependent, with seeing, visualisation or imagistic reasoning), a characteristic often referred to (see Kuhn above; De Kleer and Brown, 1983; Nersessian, 1992a). As a particular kind of representation, mental models would thus share the synthetic feature which characterises all forms of representation. Moreover, the synthetic feature, in the sense of simplification of what is being represented, suggests economy of cognitive resources. The issue that remains is whether, and how, this economy takes on specific aspects in the case of mental models. In other words, the question concerns what kind of economy mental models might provide to the holders of models.

Secondly, the synthetic feature points to a holistic kind of representation. The already mentioned case of how to organise bibliographical references illustrates this meaning and brings out the longstanding debate on the imagistic versus propositional nature of representation. A propositional approach to organising bibliographical references would imply remembering a long list of rules for each particular case (articles published in journals, books and book chapters, newspaper references, etc). This rule-based approach, adopted by many journals in their Norms for Publication to potential authors, explains that references start with the name of the author, written in capital letters, followed by a comma, then by the year of publication in brackets, and so on so forth. Differently, an imagistic-based representation would suggest a holistic model of the specific features of each kind of reference. In this case, the model would perform as an exemplar that captures the set of rules in an organised and imagistic manner. It appears to follow that, for the holder of a mental model, the synthetic feature implies a kind of unity in the representation.

That models offer a holistic, inclusive-kind of representation, as compared with other forms of representation such as concepts, is a point we argued in an earlier study on the invention of gliders and engine-powered flying machines (Franco et al., 1999). The case study shows that the attempts to fly at the beginning of this century involved building prototypes which, we stress, depended on inventors' mental models of what a heavier-than-air flying machine would look like and how it would be capable of flying. The several prototypes appear to present a common and persistent feature, namely, the introduction of an elevator at the front part of the aeroplanes to produce an upward force that could raise the aeroplane's nose and thus re-direct the propelling force. This feature recalls the well documented alternative conception in mechanics that associates speed and force (Viennot, 1979). The point is that, whereas the alternative conception focuses a specific, well-defined notion in classical mechanics, the mental models that support the development of flying prototypes embrace a set of several elements and relationships concerning, among others, the general shape of the prototype, the angle of incidence of the wings and the power of the engine. In this sense, inventors' mental models build up to a global, all-inclusive representation of the component parts of the flying machines and how they are put together, as well as how they take-off and keep in the air.

MENTAL MODELS ARE CONSTRAINED BY WORLDVIEWS

That mental models are constrained by worldviews means that people will develop and make use of mental models according to the general beliefs systems they hold. In other words, a set of constraints operates that limits the range of possible mental models a person will use. In Vosniadou's study, the idea of constraint plays an important explanatory role: the entrenched presuppositions of a naive theory of physics, which can be suspended and/or revised during development, will constrain the models actually formed by children and thus explain their specific features. As previously seen in the discussion of the tacit feature, two basic presuppositions bear a direct influence on children's models of the Earth. It is worthwhile returning to this issue.

All the mental models identified, except the spherical Earth model, appear to be formed and further developed according to the above mentioned presuppositions. For instance, the disc and rectangular models of the Earth are directly derived from the presupposition that the ground is flat. In a similar manner, the dual Earth, hollow sphere, and flattened Earth models are all three an attempt to maintain the presupposition that the ground on which people live is flat while at the same time reconciling it with the acquired information that the Earth is round. As put by Vosniadou herself, 'there would be no reason for children to form these systematic misconceptions if they did not believe that the Earth is flat in the first place' (Vosniadou and Brewer, 1992, p.576). The presupposition that unsupported things fall according to an Earth-centred view of up/down may be applied either to the Earth itself or to things and people on Earth. The first case appears to produce the rectangular, disc and dual Earth models which are usually associated to the idea that the Earth rests on water, ground or dirt (op. cit., p. 577). The hollow sphere model accounts for the second case, where the presupposition applies to people and objects on Earth, but not to the Earth itself: in this model, the presupposition appears to have been partially suspended inasmuch as it does not apply to the Earth itself but only to people and things on the Earth. The flattened Earth model offers yet another example: accepting that people live all around the Earth means breaking away with the absolute up/down presupposition, even though the flat/round paradox has not yet been entirely solved.

The Earth mental models identified by Vosniadou and her colleagues are therefore constrained by presuppositions that build into a naive theory of physics. Moreover, Vosniadou (1994, p.421) argues that the two presuppositions identified are both universal and domain-specific: since they apply to the physical world in general, they are directly related to the specific domain of physical knowledge; they are also shared by all human subjects whose experience of the physical world is necessarily the same.

The cross-cultural data, which brings out the specifically cultural aspects of children's mental models of the Earth, enlarges on the nature of the constraints that bear on mental models. Comparing the mental models of children from Greece, India and Samoa, as well as the United States, shows some variations (op. cit., p. 423 and following). Among these (cf. note 8), the ring Earth model from Samoan children is particularly interesting since it appears to reflect the physical and social space organisation in that culture. By making evident that children's understandings develop within, and according to, specific cultural settings, this example confirms our view that mental models are constrained by worldviews.

The idea of constraint could be interpreted within the currently dominant constructivist view according to which people bring to problem-solving and learning situations a mass of previous information, knowledge, experiences which will partly determine their understanding of the situation as well as how they engage in developing solutions. Norman (1983, pp.7-8), when discussing users' strategies for manipulating calculators, apparently makes this same point:

> mental models, ... will be constrained by such things as the user's technical background, previous experiences with similar systems, and the structure of human information processing system.

However, it is worthwhile stressing that the constructivist perspective, as it developed in the science education literature, took this particular point to mean that all observations are theory-laden. In our view, the proposition that mental models are constrained by worldviews, although it is not incompatible (but does indeed converge) with the assertion just mentioned, suggests something slightly different. We would argue that the idea of constraint refers to belief systems that underlie the process whereby decisions are made usually at a tacit level concerning which types of models, among classes of possible models, do make sense in a particular situation. In other words, constraints operate, often tacitly, to specify acceptable/unacceptable models, according to assumptions which are culturally shared.

Other examples also point to cultural constraints in the development and use of mental models. In a previous study, Franco et al. (1999) showed how the historical context in which Galileo Galilei developed his observations of the Moon determined his model of it. For instance, in his *Message of the stars*, he discusses and interprets his first telescope observations of the moon, fixed stars and Jupiter. He concludes that:

> After repeated inspection, we hold the conviction that the surface of the moon is not smooth, uniform, and of very exact sphericity, as it has been held for both the moon and other celestial bodies by numerous philosophers. On the contrary, it is uneven, rough and full of cavities and proeminencies, not diverse from the surface itself of the Earth. (Galilei 1968).

Our aim here is to stress that Galileo's model of the Moon is directly related to a particular purpose, namely, to show that there are no differences between the sublunary world and the heavens, which sets him in opposition to Aristotelian and Ptolemaic astronomical views. That is why he defends the existence of Earthly features such as atmosphere on the Moon. In other words, we argue that Galileo's investigations of heavenly bodies and the resulting models are constrained by a particular worldview which, according to historians of science, is based on the premise that the Earth and the Moon are similar, that is, on the assumption of a non-hierarchical view of the cosmos.

ABOUT THE SET OF FEATURES

Having specified a set of key features of mental models, it is necessary to discuss patterns of relationships between them. In this respect, we argue that the pattern for these general features is such that any particular feature – with the exception of the generative feature[8] – may be absent while the occurring features are sufficient to characterise the representation considered as a mental model. The idea that patterns arise, where not all component features need to be simultaneously present, is taken from the Optical Character Recognition studies, where the identification of a printed character requires the presence of some but not all features of the character. The conditions under which these features might or not appear in association with the development and use of mental models remains a question for empirical investigation.

The size of the proposed set of features also requires to be assessed. Although the four features discussed in this chapter are the more frequently found in the literature, researchers mention other features also. For Norman (1983), mental models are inaccurate, incomplete, messy and sloppy, features which point to the common idea that mental models are not precise, well defined mental entities. This feature, however, does not help to differentiate between mental models and other forms of knowledge processing. It is not specific to mental models only but could describe several other thought processes and, for this reason, it should not be included in the proposed set of key features. There is, however, a possible candidate that could integrate the original set of features: *visualisation*. To justify our view, this feature, mentioned by De Kleer and Brown (1983) among others, is related to the imagistic dimension of reasoning which, in its turn, is taken to characterise scientific thinking in several domains of knowledge (see e.g. Nersessian, Kuhn).

One more issue arises, related to the use of mental models. If, as we argue, mental models are generative, then the predictions and inferences produced by running mental models will allow us to test the accuracy and appropriateness of the models used. In other words, using mental models will lead to their confirmation or revision, therefore pointing to an issue that requires further investigation, namely, the issue of model change.

CONCLUDING REMARKS

Mental models are an elusive issue, even if it is one that is increasingly addressed in such diverse fields of knowledge as philosophy of science, psychology and science education, to name but a few. Our view is that, in the current state of affairs, to aim at proposing and debating unique definitions might not prove an adequate strategy to further develop our understanding of the features and role of mental models. Rather we have argued for a multiple approach in order to grasp mental models, that focuses on methodological considerations as well as the two issues of developing and using mental models.

The methodological strategies reported to investigate mental models present a composite picture although two issues arise as common concerns. The first issue refers to the fact that mental models are inaccessible as such. This means that mental models must be inferred from some expression of them, which might include material productions such as notebooks, diagrams and published papers, scientific instruments and prototypes as well as speech data obtained in interviews. In this respect, the context within which models are created and/or used, as well as the chosen mode of expression, bear implications on the thought processes and kinds of mental models analysed. The second issue, open to further inquiry, thus concerns the interrelationships and mutual determinations between mental and expressed models.

We have argued for a dual approach in order to grasp mental models. On one hand, it is necessary to look into the processes that underlie the development of mental models and, on the other hand, it is convenient to focus on the use of mental models so as to identify a preliminary set of defining features. In respect of the first issue, we build on and enlarge Nersessian's notion of *abstraction techniques* which we take as a theoretical construct that articulates two aspects: on one hand, as a general mechanism, abstraction techniques provide a powerful knowledge building mechanism but a mechanism, on the other hand, that takes on specific features according to the particular domain of knowledge in which it is used. As a result of this view of abstraction techniques, we developed a case study of *coalescence*, that is, the subsuming of the two concepts of inertial and gravitational mass into a new concept by Einstein. It is important to stress that, in this case, coalescence was not a mere ending-point of Einstein's General Relativity as presented in 1915. On the contrary, it was possible to establish that the perspective of promoting the coalescence of inertial and gravitational mass was a feature that shaped Einstein's scientific work as early as around 1908. Therefore, coalescence for Einstein was a reasoning tool used for exploring gravitation in such a way that it allowed him to create new representations and make inferences about the world. In other words, coalescence constitutes a further example of abstraction technique.

In respect of the second issue concerning a preliminary set of features, we suggest that mental models are generative and synthetic, that they involve tacit knowledge and are constrained by worldviews. However, these features do not need to be simultaneously present. The only necessary feature to identify a mental model is that it is generative: the remaining features, similarly to the Optical Character Recognition, might or might not be present according to particular situations. Finally, it is important to stress that this is an open-ended set of features, the completion of which depends on further empirical investigation. In this respect, we suggest that one additional feature should be considered as a potential candidate, subjected to further investigation: it is that mental models involve visualisation, a feature taken to support scientific thinking as well as the development and use of mental models.

We have argued that these two approaches, focusing on the abstracting techniques that underlie the development of mental models and on the key features of existing mental models, are complementary and provide a general framework that helps grasping mental models. In other words, they suggest a research programme that requires engaging in further empirical research. In this respect, a central issue arises here, concerning the conditions under which, and the processes by which, mental models are revised and changed. Finally, it is our view that the historical-cognitive approach provides an especially fruitful strategy that articulates historically-based case studies of specific scientific notions with cognitive analysis of the modelling processes by which new knowledge is developed.

NOTES

1. In fact, the ratio between inertial mass and gravitational mass is constant. From this, an adequate adjustment in Newton's gravitational constant makes the ratio equal to unity

2. It should be mentioned that the publication mentioned here is a popularisation book. However, as already discussed by Kuhn (1977a), the content of the argument is coherent with Einstein's scientific papers, with the advantage of spelling out the argument in a way that was as clear as possible.

3. The first study (Vosniadou and Brewer, 1992) presents research findings obtained with children from the USA whereas the second study (Vosniadou, 1994) includes intercultural data. Here we will refer mainly to the data obtained in the USA even though the general arguments apply to all the data.

4. In her 1994 study, Vosniadou refers to the pressuppositions of continuity, solidity, no action at a distance, gravity and inertia, which she takes from research studies carried out with infants.

5. More specifically, Vosniadou (1994, p.414) explains that 'We have adopted the construct of the mental model to characterise children's representations in observational astronomy. The construct of the mental model (...) is used here to refer to a particular kind of mental representation that differs from other kinds of representations in that it is an analog to the state of affairs (perceived or conceived) that it represents (see Johnson-Laird, 1983).'

6. There are some differences in the results obtained in the two experiments. The findings show that the first experiment, where students use the analogical models spontaneously, supports both predictions. However, in the second experiment where the models are learnt, only the second prediction related to the moving crowd model is supported by the data; the non-confirmation of the first, water flow-based prediction is explained by students' difficulties in understanding the behaviour of water on one hand and in accepting to use learnt analogical models on the other hand.

7. As Vosniadou is concerned with demonstrating the developmental path of children's mental models of the Earth, she discusses the role of the presuppositions and how they are suspended and/or revised, an issue which is not directly relevant to our present discussion.

8. In other words, we maintain that the generative feature is a necessary feature of a mental model and we do so because it is a feature that allows differentiating between mental models and other forms of knowledge such as e.g. concepts and schemas. The remaining features may or may not occur, depending on the particular situation that originated the mental model. In this respect, further empirical research is needed in order to analyse the conditions for the appearance or not of the several features included in the original set.

Chapter 6

Investigating the Role of Representations and Expressed Models in Building Mental Models

Barbara C. Buckley, Carolyn J. Boulter
The University of Reading, UK

INTRODUCTION

Chapter 5 has concentrated upon cognitive-historical studies in science and how these methods of analysis can provide tools for looking at the nature of mental models. Model-based teaching and learning within science education seeks ways of analysing dynamic systems. It sees representations and expressed models forming essential and accessible links between the many levels and contexts of learning. This chapter presents a method for analysing representations, illustrates it using particular models of the human heart and the lunar eclipse, and describes how they function in model-based learning. We focus on what aspects of the phenomenon are represented and how the particular features of the representations facilitate or hinder the learner's mental model-building.

In classrooms, museums, zoos and activity centres, teachers and learners make sense of the phenomena of the world through building and using models. These models are expressed in various modes of representation as we discussed in Chapter 3. They may be part of writing as graphs, pictures and formulae, or of discourse as metaphors and actions, or as concrete material objects. Whenever learners engage with phenomena, there are likely to be different modes of representations in use throughout the learning process as the teacher encourages learners to talk, write, draw, and interact with expressed models (Gilbert, 1993). Particular representations facilitate access only to selected aspects of a phenomenon and, therefore, contribute incrementally to the formation and elaboration of mental models of it. Thus, model-based teaching and learning, which recognises this intimate

J.K. Gilbert and C.J. Boulter (eds.), Developing Models in Science Education, 119–135.

connection between representations and phenomena in the building of mental models, is the cognitive core of science learning for individuals whether alone or in groups.

Since representations or expressed models are accessible in talk, text, gesture, and object and link the learning processes of the individual with the learning activities of the group, we needed a systematic method that will enable us to characterise, categorise and compare representations. Because we study how models function in science teaching and learning, the method needs to relate logically to the theoretical framework of model-based learning and to the phenomena that learners study. Furthermore, it needs to be sufficiently general so that it could be used in diverse contexts and science disciplines. The analytical framework we have developed draws most directly on prior research conducted in science classrooms by the authors (Boulter, 1997, 1992; Buckley, 1992) and on research into illustration conducted by Goldsmith (1984). We and our colleagues have used this framework to analyse not only paper-based illustrations, but also object-based, screen-based and discourse-based representations.

BACKGROUND

This section clarifies the conceptual context in which our analytical framework functions.

Phenomena, Expressed Models, Representations and Mental Models

The CMISTRE group uses the working definition of a *model* as 'a representation of an idea, object, event, process or system' (Gilbert, 1997, p. 2). *Mental models* are internal, cognitive representations used to reason about phenomena, and to describe, explain, predict, and, sometimes, control them (Gentner and Stevens, 1983; Johnson-Laird, 1983; Rouse and Morris, 1986). *Expressed models* are external representations used not only in communication but also in reasoning (Kindfield, 1993-1994; Larkin, 1989; Larkin and Simon, 1987). In our work we use the term *representation* to refer to external representations and expressed models. The interactive nature of the relationships among them is indicated in Figure 6.1. Mental models are used both to understand and to create expressed models. They influence our perceptions of phenomena, which in turn influence our mental models. Expressed models represent selected aspects of phenomena and of our mental models.

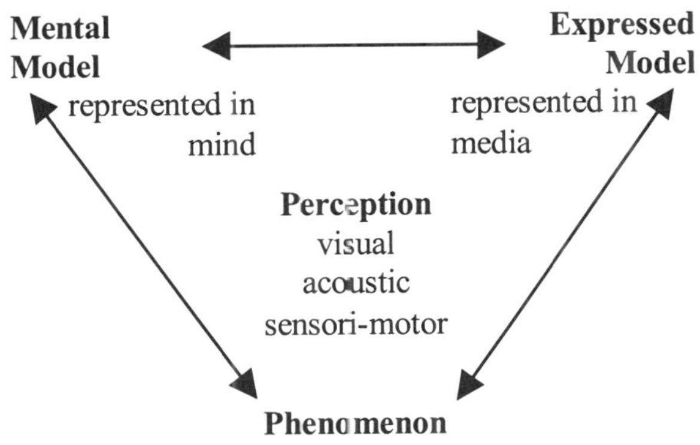

Figure 6.1. Interactive relationships among models and phenomena.

Model-based Teaching and Learning

Model-based teaching makes use of expressed models and diverse educational experiences to facilitate model-based learning. As witnessed by Section Three of this volume, it can encompass many different instructional strategies and environments but it has not yet been examined systematically across contexts. Our analytical framework is intended to support such systematic investigation.

Model-based learning is the construction of mental models through a recursive process of formation, use, revision and elaboration (Buckley, 1995; Clement, 1989; Stewart and Hafner, 1991). It can be considered a special case of generative learning in that learners use what they know to integrate new information and extend their knowledge (Osborne and Wittrock, 1985). This process is summarised in Figure 6.2. Mental models arise by a variety of processes such as induction from experience (Johnson-Laird, 1983; Norman, 1983) intentional model-building from pieces of information (Buckley, 1992, 1995), envisioning from components and causal principles (deKleer and Brown, 1981, 1983), and/or mapping from analogous models or phenomena (Clement, 1998; Gentner, 1983).

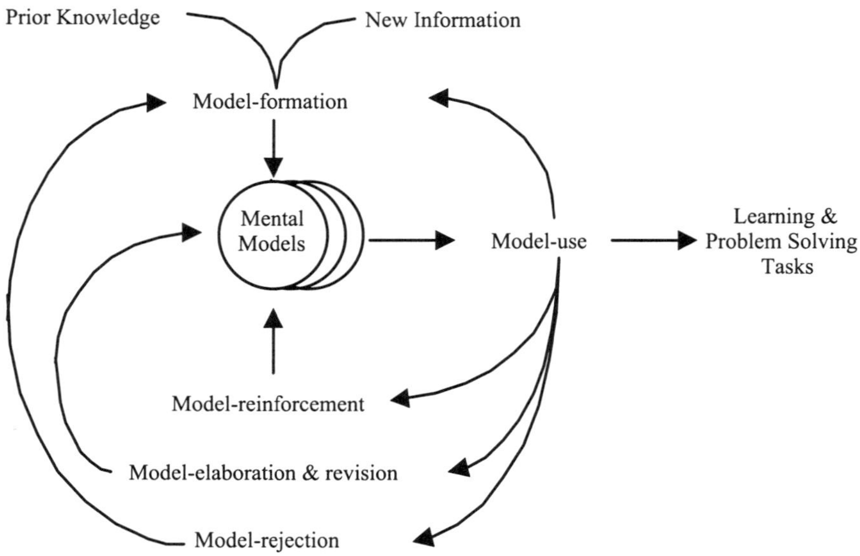

Figure 6.2. Expressed Model of Model-Based Learning

Mental models incorporate diverse knowledge from various sources. Direct experience with phenomena, vicarious or surrogate experience via video or simulations, or interacting with the many representations and expressed models we encounter in teaching, learning, and informal educational experiences, can all contribute to model building. With a mental model we have expectations about how the object, event or system looks and behaves. We can use it to generate expressed models in diverse formats ranging from transient verbal and gestural models during discussion (Crowder, 1996) to computer simulations of cardiac function or diagrams for use in problem solving (Kindfield, 1993/1994). We also use our mental models to understand and evaluate the expressed models produced by others. As we do so, we are testing that expressed model as well as our own mental model. Does either one enable us to understand, describe, explain, predict some specific instance? If not, what is wrong with the model and how must it be changed? Do we need to revise or elaborate the model or must we reject it and start again? When models, mental or expressed, are used successfully for the needs at hand, they tend to be reinforced and may become part of our repertoire of stable, precompiled, models that are readily available for use. We can then modify them to create transient, situation-specific models for reasoning about instances (Vosniadou and Brewer, 1992). Experts switch among representations depending on the task, representing only the aspects of the phenomenon relevant to the task at hand (Kindfield, 1994). The analytical framework is intended to help elucidate the development of this and other modelling skills in learning contexts.

ANALYTICAL FRAMEWORK

Taking into account our working definition of models, we initially selected expressed models of phenomena that can be considered an object, an event, and a system (see Table 6.1). We then used a range of representations of these phenomena to test and refine the criteria and our understanding of them. We will present only the analysis of the heart and eclipse examples.

Table 6.1. Phenomena and expressed models analysed

Phenomenon	*Heart*	*Eclipse*	*Greenhouse Effect*
kind of phenomenon	object	event	system
expressed model	animation from *Science for Living* (1990)	orrery	colour illustration (Miller, 1992)
representation type	animation	3D mechanical model	diagram with words

The phenomena are embedded in complex, dynamic systems. The global environment has long been viewed as a complex, changing system of interacting biotic and abiotic elements. *Objects* are parts of systems. The heart, for instance, is an object embedded within the circulatory system. However, it is also a complex, dynamic object and can itself be viewed as a system. This is an example of a parts-of hierarchy. *Events* such as the lunar eclipse are time-limited segments of (solar) system behaviour. They take place within the normal or abnormal functioning of the system, but span a limited time frame.

Since our analytical framework must relate to the phenomenon as well as the particular representation, two questions guide our analysis.

- What aspects of the phenomenon are represented?
- How does the representation facilitate or hinder comprehension?

Buckley's (1992) description of a learner's models of the circulatory system underpins the analysis with respect to the first question. The analytical matrix of Goldsmith (1984) forms the basis for answering the second question.

Aspects of the Phenomenon

At the basic level, an image of the phenomenon can be formed and its parts can be identified. Tversky suggests that parts are 'natural units for perception and natural units for function' (Tversky, 1989, p. 983). Since this chapter focuses on the functions of models, we broaden the examination of parts to include other aspects of phenomena (behaviour and mechanism), identified as important in model-based learning (Buckley, 1992). Buckley analysed learners' models of the circulatory system in a classroom of 28 high school biology students in terms of the parts of the phenomenon represented in their models. One student developed an integrated, useful, and extensible understanding that was both qualitatively and quantitatively superior to the others. Although one cannot generalise from a single individual, however systematically studied, one can create a working model for further investigation. Thus, our analytical framework is a necessary step for future research.

The top level of the learner's model of the circulatory system is shown in Figure 6.3. The model consisted of integrated pieces of knowledge relating to aspects of the phenomenon. These pieces included the structure, function, behaviour and mechanism of an anatomical entity (e.g. heart, blood vessels, and blood). They were integrated and embedded in a model of the circulatory system. *Structure* [S] refers to the anatomy of the system; *function* [F] to its role in the larger system in which it is embedded, and *behaviour* [B] to the dynamic changes in the entity. The behaviour of the circulatory system is explained as emerging from the interactive behaviours of its parts; this is termed *mechanism*. Thus, the function of the circulatory system (to transport cells and chemicals) is enabled by the pumping (B) of the heart (S), the blood vessels (S) carrying (B) blood, and the blood (S) carrying (B) cells and chemicals. Each of the embedded parts (S) can then be analysed in the same manner, as a dynamic system of interacting parts resulting in a hierarchy of models.

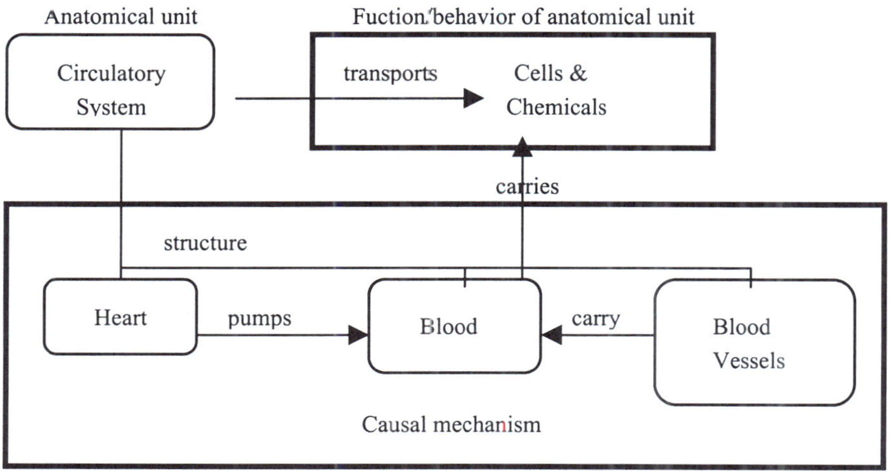

Figure 3. A model of the circulatory system.

Note that our focus here is on how the system works, a mechanistic, emergent-behaviour view of *proximate* cause (Mayr, 1982). Mayr contrasts this with *ultimate* cause, which refers to evolutionary pressure and the survival advantage conferred by a particular structure, function or behaviour; in short, why it is advantageous for the system to work that way. Although function is useful in understanding living systems and technology, it is problematic in other disciplines. We therefore omit function from our working definitions shown below.

> *Structure*: structural parts and spatial relationships.
> *Behaviour*: time-based processes and changes.
> *Mechanism*: Interacting behaviours of sub-components that produce the behaviour of the whole.

Features of the Representation

We have examined several conceptual frameworks developed by typographers and others who focus on creating and making sense of graphic images (e.g. Tufte, 1997; Twyman, 1979) in order to benefit from their analyses of representations. We found the analytical framework devised by Goldsmith (1984) to be most useful for our purposes because it includes both semiotic levels and visual factors.

The three **semiotic levels** are:

syntactic: perception of graphic signals as discernible images or sets of images;
semantic: recognition of the meaning of the image;
pragmatic: understanding the context of the image based on prior knowledge or experience.

These levels are not mutually exclusive. Prior knowledge (pragmatic level) may enable perception (syntactic level) of images or parts of images, but comprehension without prior knowledge requires both the syntactic and semantic levels.

Goldsmith (1984) crosses the semiotic levels with the visual factors of

unity: any area of an image that might be recognised as having a separate identity;
location: spatial relationships among images;
emphasis: hierarchical relationships among images;
text parallels: relationships among images and words.

Table 2 is our summary of the nature of the viewer's interaction with the illustration in each of the cells of a matrix devised by Goldsmith.

Table 2 Goldsmith (1984) analytical matrix

	Syntactic	*Semantic*	*Pragmatic*
Unity	perceive group of marks as an entity	recognise entity with the aid of relevant details	familiarity with situation enables recognition
Location	depth cues without meaning	recognition of the relation of parts via physical cues	familiarity with situation enables recognition of structural relationships
Emphasis	attention directed by sensory factors	attention directed by human experience	attention directed by cultural conventions (e.g. reading direction, colour coding)
Text parallels	physical relationship of text and images	naming consistency, mapping between image and text	familiarity with situation allows text parallel to work

Goldsmith uses this matrix to examine the communication value of illustrations and to analyse problems in interpretation that arise during field tests of illustrations. We find it works equally well with other representations and is particularly helpful when considering what the learner makes of the particular aspects of the phenomenon represented.

Analysis of Examples

When using the framework to analyse representations, we complete the following steps for each representation:

1. Create a target model of the phenomenon (Norman, 1983).
2. Identify which aspects of the phenomenon are represented and which are not.
3. Complete a Goldsmith matrix for the representation.
4. Identify aspects that may present difficulties to learners (semiotic challenges).
5. Summarise the aspects represented and missing and the semiotic challenges posed.

We illustrate these steps in the following two examples.

Example 1. The Human Heart as represented by an animation

Science for Living (1990)

Figure 6.4. Frame from heart cycle animation

The first representation to be analysed is an animation of the heart cycle from *Science for Living* (1990), an interactive multimedia resource prototype developed at Stanford University with funding from Apple Computer and the Carnegie Corporation. Imagine the sketch shown in Figure 6.4 as a two-dimensional colour animation with red and blue representing the oxygenated and unoxygenated blood. The lines representing the walls of the heart and the heart valves move in a co-ordinated fashion to represent the movement of the heart and the flow of blood through the heart. The simultaneous audio narration describes the heart cycle in terms of systole/diastole or the squirt phase and the fill phase and the associated heart sounds.

Target Model

A target model of the human heart (the phenomenon) (Figure 6.5) identifies the aspects of the phenomenon to be represented: the heart (S), its behaviour (pumping blood) and the causal mechanism that produces that behaviour. The pacemaker (S) stimulates (B) cardiac muscle (S) to contract (B) squeezing (B) blood (S) through heart valves (S) that open in one direction only (B).

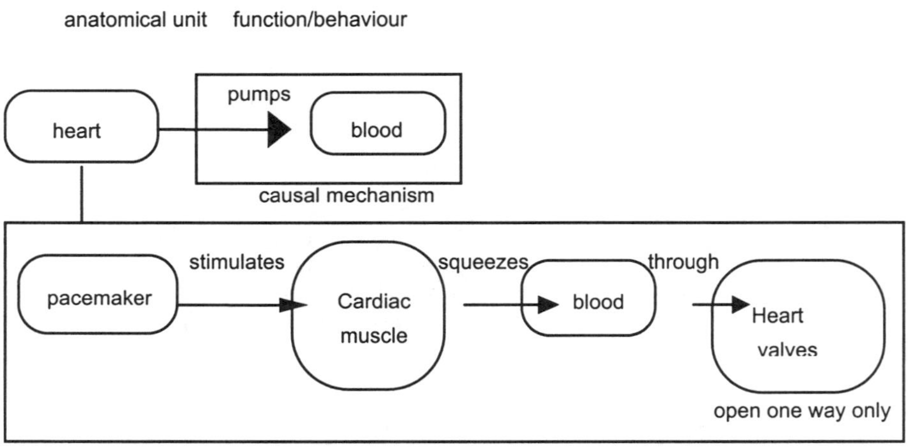

Figure 6.5. Target model of Heart

Aspects Represented

Observing the animation and listening to the audio narration enabled us to create the following chart of which aspects of the phenomenon were represented in the segment.

Structures	*Behaviour*	*Mechanism*
heart walls (cardiac muscle) heart valves blood	walls move in and out valves open and close blood flows	walls moving in squeeze blood through one way valves into the next chamber; valves keep blood flowing one direction

Analysis of Representation

Examination of the graphic elements employed in the representation enabled us to complete the Goldsmith matrix shown below.

	Syntactic	*Semantic*	*Pragmatic*
Unity	thick and thin lines shadings patterns of movement		recognition of heart walls, valves, blood enabled by school and human biology context
Location	no depth clues	contiguous, no scale clues	familiar interactions among physical objects and fluids
Emphasis	movement colour differences	movement of walls, blood, and valves	audio and motion draw attention to heart valve action
Text parallels	simultaneous audio and visual text	audio and video text names parts and processes	audio and video text establish context for interpretation

Combined Analysis of Heart Animation

When we compared the aspects represented with the target model shown in Figure 6.5, we observed:

- the structure is incomplete because the pacemaker system is missing;
- mechanical behaviour of the heart is shown schematically, but there is no explanation of the causes of the behaviour.

Summarising the Goldsmith matrix,

- Lines adequately represent the structural relationships of the heart anatomy.
- Co-ordination between the motion of the walls and valves is good.
- Shadings do not adequately represent flow of blood. It almost looks as if blood flows out of the chamber and back in.

- There is insufficient detail to suggest that cardiac muscle contraction and relaxation stimulated by the pacemaker system is causing the movement.
- Names of parts are not connected to their visual representations.

In overview, this short animation can contribute to an understanding of some structures and some behaviours to a learner's model of the heart. Not all structures are shown, nor is the mechanism that causes the behaviour. Features of the representation that may present difficulties to learners include the loose coupling between the names of the parts and their images and the potentially confusing patterns representing blood flow.

Example 2: Lunar Eclipse as represented by an orrery

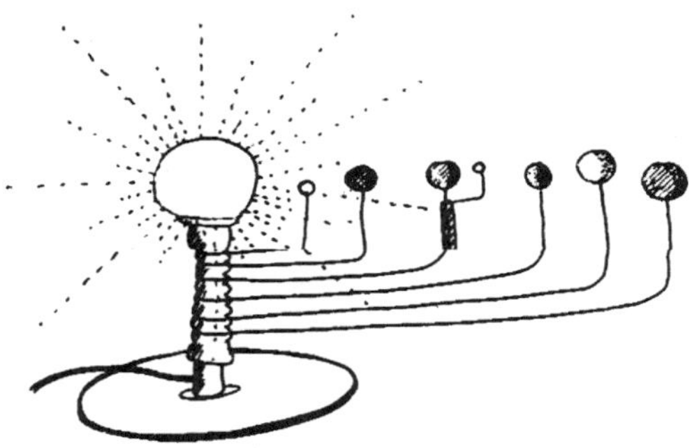

Figure 6.6. Sketch of orrery

The mechanical orrery shown in Figure 6.6 is our next example. It is a 3D concrete model composed of coloured balls of varying diameters fixed by wires to a central axis topped by a light bulb. The balls can be individually moved by hand around the central axis as can the ball attached to the axis of the third ball from the light.

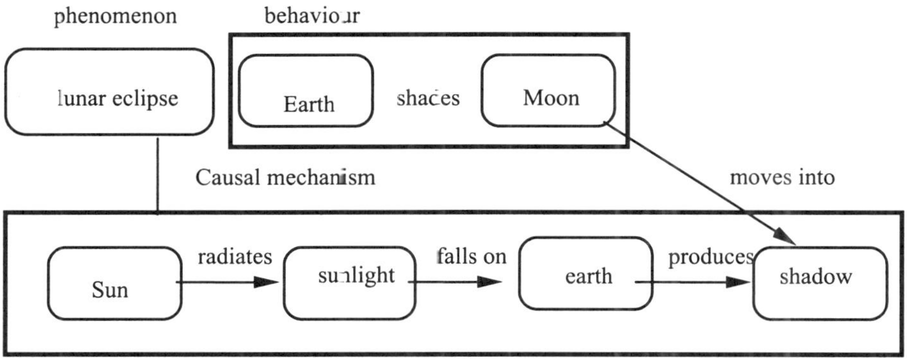

Figure 6.7. Target model of Lunar Eclipse

Aspects Represented

Analysis of the structure and use of the orrery enabled us to create the following chart of which aspects of the phenomenon were represented.

structures	sun = central light planets and moon = spheres of different colour and radius
behaviour	Planets can be moved around the sun in a planar circle. Moon can be moved around the blue planet in a planar circle.
mechanism	The spheres representing the earth and moon can be arranged so that the earth creates a shadow on the moon by blocking light from the central sun.

Analysis of Representation

Examination of the elements making up the orrery enabled us to complete the Goldsmith matrix below.

	Syntactic	*Semantic*	*Pragmatic*
Unity	discrete spheres of different colour central light spheres supported on wires, fixed centrally	central light = sun. Spheres identified by colour and distance from centre	familiarity with representations of solar system and memorisation of the order of the planets from the sun enable identification
Location	varying alignments possible. Planar circular rotation	order of planets is shown, size of planets (spheres) and distance from sun are not to scale	familiarity with objects spinning about a central hub, such as a wheel or toys.
Emphasis	relative sizes colour light	sun/central light is a large bright object and noticed first followed by earth and moon, then other planets.	focus on sun, earth and moon as most familiar.
No text			

Combined Analysis of Orrery

When we compare the aspects represented with the target model shown in Figure 6.7, we concluded that:

- Concrete elements of the solar system are represented in the correct order but spatial relationships among the planets are not represented.
- Planets can be moved around the sun. Since the wires are of fixed length, the paths are all circular and planar. This behaviour differs from the behaviour of the solar system.
- The behaviour of light and the effects of varying alignments in producing a shadow on the moon are adequately represented.
- Other phenomena of an eclipse such as the reddish appearance of the moon at the beginning and end of the event are not represented.
- The mechanisms that account for the shadow on the moon are represented. However, the mechanisms that account for the behaviour of the solar system are not.

Examining the completed Goldsmith matrix reveals that

- Colours of the spheres relate loosely to familiar conventions, but the spheres are unnamed.
- The wires represent just one fixed dimension of gravitational attraction, that between the sun and each planet. They cannot represent the noncircular orbits resulting from the gravitational attractions among the planets as they pass near each other.

Taking all of these factors together we concluded that without text labels, a legend, or an explainer, to support mapping of the spheres to planets, it would be very difficult to make sense of the orrery unless the user had prior knowledge. With such support the mechanism of the lunar eclipse could be understood from the model, but not the mechanisms of the solar system that produce the particular alignment of sun, earth and moon present in a lunar eclipse.

DISCUSSION

Representations and expressed models contribute to model-based learning by providing pieces of information about the structure, behaviour, and mechanisms of phenomena. However, our analysis has illustrated some ways in which the particulars of a representation may present semiotic challenges for learners' sense-making. Phenomena may be hidden within or may be too small, too large, too fast, or too slow for humans to see. Even when phenomena are within the range of the human perceptual system, it can be difficult for learners to detect the parts of a system or model. This is especially the case when directly observing phenomena or images thereof. Nature doesn't come with labels and boundaries between parts are often indistinct. Therefore, the syntactic level of 'seeing that something is there' can be problematic. Visual factors such as unity (seeing the entity) and location (structural relationship among entities) are essential components of discerning the parts and structure of a phenomenon, relating not only to the structure *per se* but also to the causal mechanisms that account for its behaviour. Behaviour is notoriously difficult to portray in static illustrations. Narrative text and small multiples can be used to create a temporal sequence (Tufte, 1997), but dynamic representations such as video, animations, simulations, and mechanical models do the job more clearly, but not, as we have seen, without potential problems. The represented behaviour may be oversimplified and unrealistic, relevant detail may be missing or difficult to see, or the causes of the behaviour may be unclear or missing. Explaining the cause of the behaviour (mechanism) often requires additional representations that focus on the interactive behaviours of the embedded structures. Since integration of structure, behaviour and mechanism are essential for model-building, all of the above present semiotic challenges to the learner.

A variety of representational techniques can help learners overcome the semiotic challenges of a given representation. An outline, overlay, colour coding, highlighting, or some other form of emphasis can help learners 'see' the parts of phenomena by explicitly pointing out and defining the entities in an image (Buckley, 1998; Buckley and Boulter, 1998). Tightly coupled text parallels can add semantic information such as names for the parts. Zooming in and spotlighting can serve to emphasise relevant details that are important in recognising phenomena (Salomon, 1980). Parallel text in audio tracks can also enhance understanding images and phenomena by filling in details that are difficult to portray in visuals. Parallel text can also explain causal mechanisms. Interactive media that support the embedding and linking of different representations. These facilitate not only text parallels, but also the representation of causal mechanisms of phenomena (Horowitz et al., 1998).

From our work in classrooms, however, we have seen that while it may be necessary to provide necessary pieces of information, this is not sufficient to ensure model building. Despite access to a wide range of representations of the circulatory system, only one student in Buckley's (1992) study engaged in model building. She expressed a desire to have all the pieces of information, posed questions about the structure, function, and interacting behaviours of the circulatory system, and reasoned with her mental model about the expressed models and representations she encountered in her studies (Buckley, 1992, 1998). This case study highlighted the importance of learning strategies and constructive engagement (Chan, et al., 1992) with the representations. Similarly, when students studied the lunar eclipse in a primary classroom, they reasoned about the different representations, critiquing how well the orreries and other representations fit their mental models of the event (Prain et al., 1998). This study highlighted patterns of participation, persuasion, and model-building in collaborative classroom learning. Both of these studies situate the focus on representations within the larger context of classroom learning. In addition, subsequent research efforts have explored the impact of the interface on reducing semiotic challenges with representations carried in interactive media (Buckley and Boulter, 1999).

When considering model-based teaching in classrooms, conflicting concerns generate constraints on the use of representations. On the one hand, we want the learner's mental model to become as extensive as possible. On the other hand, being selective about what one presents to the learner at a particular time helps to focus attention on particular aspects of the phenomenon (Dwyer, 1978; Joseph and Dwyer, 1982). Work by Gobert and Clement (1999) demonstrates that asking learners to create spatial (structure), dynamic (behaviour) and causal representations of plate tectonics fosters their construction of mental models. This directly parallels the questions posed by the model-building learner (Buckley, 1992): What are the parts? What happens? How do they work together? The most appropriate use of representations is likely to depend upon the state of the learner's knowledge and the nature of the learning task. However, we could also teach students learning strategies conducive to model-based learning, such as posing the questions above and let them seek out and evaluate and integrate the information they need to construct mental models.

Chapter 7

Modelling and Creativity in Design and Technology Education

Roger Elmer[1], Trevor Davies[2]
[1]*King Alfred's College of Higher Education, Winchester, UK;* [2]*The University of Reading, UK*

INTRODUCTION

The overarching aim of this chapter is to examine critically modelling and creativity in design and technology education. To locate this inquiry, and to assist readers less familiar with this curriculum area, the first section gives a brief history of modelling in design and technology education whilst the second section is dedicated to a general review of creativity followed by a focus on creativity and children. The very limited usage of the term in design and technology education is discussed and the section finishes by comparing creativity with a frequently used term in design and technology, innovation. The purpose of modelling, both in industrial and educational settings, and creativity are subsequently explored, and the chapter finishes with a discussion and issues for future research investigation.

The chapter draws mainly on the United Kingdom (UK); both its historical perspectives and much of the literature sources, particularly the modelling literature. The justification for this is twofold: the UK has been at the forefront of developments in design and technology in general education; modelling has been a central, albeit at times confused, theme within this development. Many countries, including USA, Canada and Australia, have embraced the centrality of modelling but inherited some of the confusions surrounding its usage. Increasingly the research literature reflects an expansion from a sole UK source; recent examples are Liu (1996) and Welch (1998).

J.K. Gilbert and C.J. Boulter (eds.), Developing Models in Science Education, 137–156.

Over the last 30 years design and technology curricula have been developing in many countries but frequently with significantly different conceptual frameworks. For example a technology curriculum has been developing in the USA for the last 20 years but it does not equate with that in the UK even though there is a tradition of technology innovation influencing some local practice in the USA (Sharpe, 1996). The complexity of policy and decision-making processes existing in the multi-tiered management of education result in great diversification of approaches between states and school districts. In recent times, however, the International Technology Education Association has been constructing a set of National Standards for Education under the auspices of a project called 'Technology for all Americans' which is loosely related to the UK National Curriculum.

A BRIEF HISTORY OF MODELLING IN DESIGN AND TECHNOLOGY EDUCATION

Diversity and change have been the defining features of the importance given to models and modelling during the relatively short curriculum life of design and technology. Despite this short history three distinctive periods are usefully detected in their importance. During the first period, coincident with the genesis of the subject, there is, with one very notable exception, extremely limited use of the terms. Eggleston (1976) draws heavily on the work of the two Schools Council research and development groups yet the sole reference to 'model' is the most familiar, the construction of a model steam locomotive by a group of sixth form boys and their enthusiastic teacher. In 1985 three General Certificate of Secondary Education (GCSE) examinations were implemented at 16+. Modelling is referred to in the GCSE National Criteria (DES, 1985) under Making Skills but with no definition or explanation. A support book for GCSE (Secondary Examinations Council/Kimbell, 1986) makes no mention of the term whatsoever. The notable exception is the work of Archer (1979). He looked at the features of designing that distinguish it from, as he saw, the other two cultures of the humanities and the sciences and proposed parallels between mathematical notation as the language of science, natural language as the language of the humanities, and modelling as the language of design.

The second period is characterised as modelling in ascendant. No definitive evidence can be presented for this change but significant contributory factors were a wider recognition of Archer's work coincident with increased questioning of the educational aim of design and technology in general education. This change in importance is readily illustrated by contrasting its absence in (Secondary Examinations Council/Kimbell, 1986) compared to the second Assessment of Performance Unit (APU) report one year later (Kelly, et al., 1987). The author of the former book is, although not writing in this capacity, the co-director of the latter project where models and modelling take on a central importance (op. Cit., p.15):

> The third broad intention involves the generation and development of potential solutions and this is the creative heartland of design and technological procedure. The activities here are informed through investigation and achieved through imaging and modelling.

The National Curriculum Design And Technology Working Group Report (DES, 1988) were strongly influenced by the work of the APU. Their final report (DES, 1989) identifies sixteen Programme of Study (POS) headings (which received much criticism), four of which make explicit reference to models and modelling: Exploring and Investigating, Imaging and Generating, Modelling and Communicating, and Making.

The third period is characterised as bifurcation, between a reduced emphasis given by those conceiving and implementing policy in schools and an increased emphasis identified by research and development groups, the APU in particular. At the core of this bifurcation is the subject's educational aim. As a blunt and very broad generalisation, for the policy makers it is a 'quality product', for others it is a 'quality process'. The revised orders (DFE, 1995) for the curriculum give a reduced importance to modelling, whilst in sharp contrast, the ideas expressed in the 1987 APU report are expanded and given even greater emphasis in the 1991 report (Kimbell et al., 1991, p.21):

> It is our contention that this inter-relationship between modelling ideas in the mind, and modelling ideas in reality is the cornerstone of capability in design and technology. It is best described as 'thought in action'.

Summarising, in a relatively short period of time the role of models and modelling in design and technology education has gone from obscurity to the potential of allowing, through the relative accessibility of 'thought in action', access to meta-cognitive activity in learners. For many this would be the Holy Grail for education in the last years of the twentieth century: the nurture of transferable or core skills.

A BRIEF HISTORY OF CREATIVITY IN DESIGN AND TECHNOLOGY EDUCATION

Little systematic work was done on creativity prior to J.P. Guilford who in 1950 (Guilford, 1950) passionately stated before the American Psychological Association that he believed that creative thinking was the most vital resource available to the country. His own work, based on factor analytical approaches and psychometric testing, set the vision for the next thirty years. Developments, led and funded by the American military, included extensive testing programmes linked to the forces unleashed by the Cold War. The central concern was the identification of 'giftedness' particularly in science and technology.

Torrance Tests (Torrance, 1962) subsequently became a vehicle for entry to attempted accelerated educational programmes. Research by Getzels and Jackson (1962) examined the links between intelligence and creativity with a particular emphasis on gifted children which was a major focus at the time. In the 1960's, Feldman et al., (1994), amongst others such as Sternberg (1988), began the struggle to identify ways of identifying, harnessing and promoting creativity. In latter years the concerns have become associated with the nature of creativity in specific domains and have become psychologically driven by development issues, cognitive processes and social context matters.

There is an agreement amongst seminal writers (Koestler, 1964; Feldman et al., 1994, ch.4), about definitions of terms associated with creativity. For them, creativity exists when an individual moves the boundaries of a domain of knowledge. Additionally, those who have recognised authority in the field (who know the rules of their domains and act as gatekeepers to them) endorse the movement concurrently, or eventually. A form of leadership is exercised by those individuals who make special contributions to their field. The individual becomes respected by the members of the field, who are themselves stakeholders in it. Such authority is gained through perceived excellence and / or novelty in the creative acts of the individual who is also able to communicate and promote their value and worth. The circumstances under which creativity is recognised varies according to the domain and the field, e.g. in fine art, personal statements and interpretations are valued much more than in industrial design where creativity is perceived through the relationship between form and function of a product. In attempting to locate the origins of creativity, psychological and sociological factors are perceived to influence outputs and approaches.

Research by Getzels and Jackson (1962) and more recently by Gardner (1995; ch.3) proposes that personality is a more common identifier of creative individuals rather than cognitive factors. Particularly creative individuals are often very demanding of themselves and committed to their task. They are often 'difficult' individuals, sometimes surrounded by tragedy and often marginalised from 'ordinary' communities. Selfishness, intolerance and stubbornness are often present and there is the enjoyment of complexity and asynchrony which, if not present, is sought. Often, those who contribute creatively to a domain hold values about aspects of their domain in higher regard than social and economic values, e.g. their work is considered to be more important than materialistic concerns, or concerns for their own well-being. This might well favour them in the eyes of the field. In some cases, members of the field might be encouraged to be more sensitive to them and reflective about their work, as a result of observing their lives, values and behaviour traits.

Csikzentmihalyi in Feldman et al. (1994; p.147) believes that focusing on the individual alone when studying creativity is like: 'studying how an apple tree produces its fruit by only looking at the tree and ignoring the sun and the soil'. The social and economic cultures in which individuals work play an important role in the recognition of creative contributions to any field in addition to the essentially psychological factors that identify particular individuals. Judgements about creativity cannot be separated from the more general norms and value judgements in a culture. Csikzentmihalyi, again in Feldman et al., (1994, ch.6, p.145), concludes that: 'creativity is not an attribute of individuals but of social systems making judgements of individuals'. It is thought that creative individuals show the desire to create new order from breaking down existing order and this takes place through: constructing and testing new knowledge; holding notions of changeable reality and working with detail and complexity within a domain. New individual and social realities are constructed and reconstructed in the remorseless change construed as culture. It is through culture that we judge the qualities associated with creativity.

From this general discussion attention is now directed to children and creativity. Children begin their lives with little knowledge or understanding of the world around them but with a strong disposition to explore and develop. Links are forged between the components of their understanding as it evolves and the desire to experiment and gain wide experiences in divergent ways. Hudson (1968), Torrance (1962) and Guilford (1957) explore the nature of convergent and divergent thinking and their relationships to creativity. Guilford (1957) posited that young children are largely locked into divergent thinking, which is essentially creative. As children's understanding within domains and of fields increase through their greater experience of the world and their perception of boundaries, so does their propensity for analytical thinking. Beetlestone (1998) builds on this and refers to 'little' creativity or 'creativity for all' which is particularly pertinent in classroom situations:

> In considering creativity it is important to establish that all children have equal rights to be creative and to have full access to opportunities within the creative areas of the curriculum (Beetlestone, 1998, p.34).

She recognises that creativity has cultural dimensions, that children do not all respond to creativity in the same way, that we do not perceive that all children have equal gifts and indeed we may ascribe differences to:

> perceived notions of ability, class, race, gender and able-bodiedness (Beetlestone, 1998, p.34)

Piaget (1962) struggles to explain creativity but recognised that he had failed, largely as a result of his inability to cope with the humanly constructed world. The explanations he proposes were based on a structured, stable world, not one with unpredictable change and lack of universalisability. Vygotsky (1978), however, recognises the developmental nature of knowledge and that individual's capacities for developing knowledge are also developmental. It is recognised by Feldman, et al., (1994) that to be creative, individuals have to come to believe that they can change the world and add to knowledge. Intrinsic motivation in addition to supportive frameworks at least creates the opportunity for individuals to realise their creative potential. Great creative acts often occur, when, at crucial times, the appropriate support for an individual's ideas through effective mentorship is upheld and developed.

The opportunity for children to grow creatively in classrooms would appear to depend critically upon how 'scaffolding' is maximised through teachers, peers and parents. This is an area little researched and understood but referred to generally by Vygotsky (1978), Gardner (1995), Rogoff (1990) and in specific classroom terms by Fryer (1996) and Beetlestone (1998). Characteristics associated with creativity, particularly in the affective domain, such as doggedness, single-mindedness, can result in disruptive classrooms and teachers can find it difficult to sort out more random disruptive behaviour from that associated with children taking up creative challenges. If teachers are limited in their own creativity, what impact will this have on children? Given the domain specific nature of creativity, what relationships and values develop as a result of children's experience of the different subjects of the curriculum which is asynchronous to, e.g. Gardener's forms of intelligence for which he posits separate 'types' of creativity? Moyles (1989) believes that teachers need to be involved in observing, initiating, participating, encouraging, maintaining and extending children in order to stimulate both psychological and sociological factors in classrooms to promote creative responses. Educational writers such as Alexander et al (1992), Beetlestone (1998) and Shallcross (1981) recognise that the ultimate endeavour of teachers is to promote creative acts and release creative potential. They however recognise the levels of difficulty in achieving the conditions to do this with respect to the characteristics of individual children, the domain within educational settings and the values associated with related fields.

Use of the term creativity in the design and technology education field has been very limited. One rare example is Design Education at Secondary Level (Design Council, 1980). This is a radical and far-sighted document, calling as it does for, amongst other things, that creativity should be seen as a central aim of Design Education.

> We consider that design should be an essential part of the education of all children at all stages of secondary education up to the age of sixteen. It should encourage creativity and develop the skills of problem-solving, decision-taking and evaluating, all of which are valuable in adult life, while generating an awareness of the qualities of the made world (Design Council, 1980, p.5).

There is little evidence that this vision is in the process of being fulfilled. In school settings the emphasis has remained firmly on developing and refining craft skills and the knowledge base associated with a narrow range of materials that it is practical and affordable to use within schools. Innovation can be both expensive in time and resources and difficult to manage. As Kimbell reflects on worldwide comparisons between technology in the UK, Germany, USA and Taiwan, he notes that:

> ...in all countries the technology curricula is struggling to establish itself by transforming existing traditions, typically craft traditions (Kimbell, 1997, p.229).

A major problem concerning continuity from primary to secondary design and technology education is that imaginative activity in design and technology is often not celebrated or valued in favour of organisational, skill-based activity in schools (Kimbell et al., 1996). Indeed, teachers often perceive imaginative pupils as the most ineffectual (Fryer, 1996, p.57). Fryer (1996) also points out however that there are many famous examples in the arts and the sciences where great breakthroughs have been made through imagination, e.g. Einstein imaging himself travelling on a beam of light when devising his theory of relativity. All school subjects including design and technology can be taught mechanistically but it takes a special sensitivity by teachers to recognise and encourage imaginative activity at all stages of the design process. Fryer (1996, p.44) says:

> Given a long history of neglect in formal education, we may not yet have discovered all the uses to which our imaginative capacity can be put. For instance we do not know a great deal about the use of 'visualisation' techniques.

Currently, in UK schools, in line with Fryer's (1996) findings limited opportunities in design and technology exist for learners to be creative. There are some encouraging signs that this deficit is recognised at a political level within the UK. In 1997, Ken Robinson, Professor of Arts and Education, was appointed by the Secretary of State for Education and Employment as Chairman of the Advisory Committee on Creative and Cultural Education. This was largely in response to the Government White Paper of 1996 'Excellence in Schools' (DFE, 1996). The Committee brought together leading specialists from the worlds of science, business, the arts and education; people who have developed the potential for creativity within their own careers. Their mission was to consult widely and advise the Government on what could be done to provide for the creative and cultural development of young people.

Invention and Innovation

The two terms, invention and innovation, are more frequently utilised than creativity in design and technology. Although both are particularly associated with commercial fields and production activities, innovation is distinct from invention. Invention takes place when individuals or teams work on developing new ideas and proving that exploitation is possible through the creation of prototypes or models. Innovation is the translation of new or novel ideas and information, such as market research, into products or systems that are commercially viable. Many inventions are not commercially viable and remain undeveloped until innovation, in e.g. materials technology, allows an invention to become commercially viable. Innovation in business and/or manufacturing systems is then often required to allow new products to be developed and marketed. Both invention and innovation have clear links with creativity. They both make demands of a cerebral nature but are there intrinsic differences. Creativity is concerned with fundamental human potential whilst invention and innovation relate to the specific context in which that potential is realised.

In the modern world there is a strong desire for invention and innovation driven through market-place demands and responses of a consumer society, but inertia and caution also lead to invention and innovation being a fragile, high-risk activity for individuals and companies. Consumers are continually looking for new styles and novelty in the full range of products and artefacts used in homes, in industry and society at large. These desires are created by concerns for efficiency, style and fashion, sustainability and, more fundamentally, fitness for purpose. Changing the values of consumers about the products they use is nevertheless a challenging task. James Dyson has gained a reputation as someone able to both invent and innovate through his introduction of the dual cyclotron vacuum cleaner into the international market-place. This has the potential to replace the vacuum cleaner with a dirt bag collector which until recently has been the mainstay for cleaning household carpets. The invention of James Dyson was reported by *The Times* to be:

> the most inspiring business story of the late 20th century. Knocked back at every turn by multi-national giants who ridiculed his invention, plagiarised by international business villains, plagued by debt as he sought to pursue his vision in a country reluctant to fund and research development, he worked alone for 14 years, from the concept of the machine to its appearance in the shops, clinging relentlessly to his dream (Coren, 1996).

Once his products were accepted as a real alternative to the traditional cleaner, consumers recognised the considerable functional and aesthetic improvements associated with the products and sales mushroomed. Public recognition followed, fame and fortune resulting in a growing domination of the market.

THE PURPOSE OF MODELLING IN DESIGN AND TECHNOLOGY EDUCATION

This section builds on and extends the discussion introduced in Chapter 1 and in so doing draws heavily on the idea of audience. At first sight the purpose of modelling may appear unproblematic: to assist the process of moving from intention to achievement. Various authors have proposed lists of more specific purposes (see Chapter 1) and it would be easy to debate redundancy and omission in such lists. The Assessment of Performance Unit (APU) (Kelly et al., 1987) adopts a different approach. Rather than attempting to catalogue the myriad tasks that modelling might serve, the APU focuses on the essential need for communication and proposes that modelling has two main purposes: to communicate the emergent idea to oneself and to communicate it with others. The idea of purpose linked to audience will now be explored, initially in industrial and then educational settings.

Modelling in Industrial Settings.

The audiences of industrial designers' modelling would appear to be easily defined, either the designers themselves, utilising a mix of mental and expressed models (see Chapter 1), or the client at reviews and presentations, when expressed models alone would be utilised. The form the modelling took would change. from the apparent informality of 'back of the envelope' when talking to themselves or others 'in the know', to the measured perspective and numerical analysis of a client presentation. Clarity of audience and purpose are needed to assist the designer to meet the needs of the client. In the 1960s many designers, witnessing the failure of the various design professions (architecture in particular) to identify and meet users' needs, began to attempt to bring users themselves into the process. This resulted in a number of techniques being proposed to provide newer frameworks for designing. One seminal book was *Design Methods* (Jones, 1970), with its sub-title of 'seeds of human futures'. An intention of Jones and others in the design methods movement was that any proposed design should be communicated to those affected by it and should allow them the opportunity to influence the choices that were being made. Such a change would mean that the public effects of designing could become the subject of public debate.

If so, a designer would need to consider not only the form of modelling used in order to meet the needs of the client's project but the form of modelling most appropriate to communicate design proposals to this wider audience. Some forms of modelling, the symbolic calculations of the structural engineer, the orthographic drawing of the draftsperson, are only accessible to small and specialised audiences. Baynes (1992) uses the term 'ethics of representation' and acknowledges that questions such as these begs other ones including: Does the client or sponsor making the proposals actually want them widely understood? Is there a tendency to adopt the apparently more 'scientific' at the expense of the more understandable model? Which forms of modelling are more understandable by a specific audience? At about the same time as the emergence of the design methods movement, the multi-disciplinary design team came to be recognised as a necessary adjunct, if not alternative, to the single, sometimes autocratic, designer. Baynes (1992) considers the forms and purposes of modelling in multi-disciplinary teams, in particular their crucial importance in team building and communication between teams. He echoes a number of themes of Jones (1970): the desire to make the design process more transparent in order to increase participation, that models can mystify rather than demystify, they can close down debate rather than augment it, but he believes that the emphasis of the design methods movement was on techniques rather than an exploration of attitudes and values. Summarising thus far, if the purpose of modelling is inextricably linked to its audience, designers working in an industrial context can be faced with ethical and political dimensions to their modelling if they allow the possibility of multiple audiences.

Modelling in Educational Settings

Chapter 1 introduced the fundamentally different purposes of the setting of design and technology tasks in industry compared to educational settings. For the industrial context the purpose is to achieve a workable product or system; in a learning context it is to achieve learning. In the terminology of Downey and Kelly (1986), professional education has instrumental, or extrinsic aims, whereas general education has to pursue intrinsic aims; ones that are somehow inherently good for the individual. This confusion can be exacerbated as, although the outcomes of design and technology activity in industrial and educational settings are similar (most frequently two- or three-dimensional objects), the underlying aims of the activities are so fundamentally different:

> In quite a unique way it [design and technology] promotes the development of a combination of personal, intellectual, social and physical capabilities. This is its educational raison d'être, and in the schools context we must see the outcome of the activity not as three-dimensional artefacts but as enriched and rounded young people (Kimbell et al., 1991, p.18).

What constitutes the nature of Kimbell's 'enrichment' will be contested by those with different motives for design and technology's inclusion in a curriculum. Extending the 'process' argument, Kimbell et al. (1996) draw on the research work of the APU and the Understanding Technological Approaches project (both of which Kimbell and Stables were principal authors), and see the central curriculum justification for the study of design and technology as the uniqueness of the language it employs (op. cit., p.23):

> the language of technology is indisputably a concrete one of images, symbols and models. Without this language it is just not possible to conceive of technological solutions.

They use 'concrete' in the sense of modelling 'outside of the head' and they distinguish it from modelling 'in the mind's eye' or cognitive modelling (op. cit., p.24). (In CMISTRE's terminology it is the distinction between *expressed* and *mental* models.) This 'concreteness' of language prompts the claim that design and technology has the potential to give access to students' thought processes, in particular how they have gone about the learning task (op. Cit., p.31):

> design and technology not only enhances the thinking and decision-making powers of young people, it also enhances their *conscious awareness* of those thought processes. They not only learn to think and make decisions, they also know (and can see) that that is what they are doing.

Thus models have the potential to be provide externalisations of cognition; through them the concepts that were modelled are able to be captured and communicated (op. Cit., 1996, p.97): 'it provides pupils with "a concrete lever" that can expose and get a purchase on their thought processes'.

If design and technology education seeks to give access to learners' thought processes then the trace of that thinking can only be captured in learners' modelling. Learners will utilise mental and expressed modelling to bring their ideas into the world and test them; teachers have only the expressed modelling to nurture and assess the learner's design and technology capabilities. A priority for learners is that the expressed modelling represents as fully as is possible to their facilitators and assessors the nature and quality of their mental modelling. Three issues arise from this, the discussion below drawing on Gilbert and Boulter (1998).

First, this is the reverse priority of importance in the different *social setting* of professional practice. Professional designers utilise certain expressed models to communicate with their client *audience*. If the aims of the two settings are seen as identical, then this highly stylised, narrow, range of expressed models can be seen as the proper outcome for the different social setting of education. Second, learners (and their assessors) are surrounded by the very tangible manufactured products (*consensus models*) of an industrialised society and their intentions are influenced by them (Elmer, 1996). *Consensus modelling* can be similarly pervasive and persuasive. The marker-rendered perspective, the computer-aided orthographic, can carry with them a status far higher than other forms of modelling, e.g. the freehand, annotated, sketch. This leads to the third and most important point. Modelling for private use is very individual; how one person makes sense of and finds patterns in diverse and shifting 'ill-defined problem spaces' can be very different to another person. For the manager and assessor of design and technology education there is a need to access this private world of sense-making in a relatively short period of time. This conflict of purposes can be manifest in a variety of ways but frequently modelling for private or collaborative use is subverted to modelling for judgmental use.

Chapter 1 introduced the use of modes of representation to place models in the public domain. Utilising this and the discussion above allows an exploration of measures to support the development of learners' skills in modelling. There is the obvious development of technical skills (e.g. rendering with markers etc.), which are the target of the majority of text-books, but a far higher order skill is concerned with the exercise of judgement. Which mode of representation best allows the learner to test an emergent idea? Which mode of representation best allows the learner to communicate an emergent idea to others? The exercise of such judgements is not straightforward. An idea will have embedded within it a number of interlocking issues. Each issue may require a different mode of representation to allow it to be more effectively tested and/or communicated, to self and to others. Thus a crucial skill that needs to be developed by learners the ability to utilise multiple representations when modelling.

Summarising, the purpose of modelling has been approached through linking purpose with audience. An examination of modelling professional settings has raised ethical and political issues. Although educational settings have very different aims, there would appear to be a similar issue of whose interests are being served by the modelling a 'politics of representation'. If modelling is seen as the trace, the frozen thinking, that allows the tracking of meta-cognitive activity in learners, then certain forms of expressed modelling will give extremely limited access to that meta-cognition, but other, potentially richer, forms of expressed modelling can have a lower status among learners and they are far more difficult to access by others. This discussion is neatly encapsulated in the question: 'Whose problem is the modeller trying to solve: their own or their assessor's?'

THE PURPOSE OF CREATIVITY IN DESIGN AND TECHNOLOGY EDUCATION

There is a range of purposes concerning the place of creativity in schools. On the one hand it is recognised that when learners undertake creative work it can be a powerful motivating force for teachers and learners, a vehicle for high levels of individualised achievement and offer clues to learner development patterns. On the other hand, creativity is frequently of low priority as there are so many other things to be learnt in classrooms, e.g. basic knowledge and skills, codes of response and behaviour, conduct of relationships. Fryer (1996, p.57) quotes Erikson (1970) in making the case for the classroom difficulties posed by highly creative pupils:

> Not all teachers like highly creative pupils. In a rare comprehensive study of teachers' views about creativity, Swedish educators expressed ambivalent and negative attitudes towards the pupils they thought were creative. They described them as a worrying element, wanting to do everything differently, unwilling to co-operate, adjusting badly to conventional tuition, troublesome in class, egocentric and egotistical, listless at the prospect of some subjects, cheeky, careless, coming up with strange ideas and disobedient.

To deal with creative pupils and more importantly to stimulate creativity in others requires 'high risk' teaching strategies, with a concern for a 'long term view' of learner's potential, a willingness to wait for results and the confidence to act intuitively at times. A great deal is demanded of the personal and professional qualities that teachers hold in order to develop the appropriate climate. Such a climate is difficult to attain in UK schools at the present because of the emphasis on individual pupil and school performance in tests and examinations, whilst it is recognised that the features and consequences of creativity are notoriously difficult to evaluate and assess. Fryer (1996, pp.57) quotes a female primary teacher in emphasising, however, the importance of the role teachers play in promoting creativity:

> Teachers can be either the most significant positive factor or the main hindrance.

Teaching is about maximising success with learners in accord with its accountabilities, including parental and learner expectations, national and local legal frameworks, examination board criteria, employer expectations, further education and higher education requirements and others. Creativity is one virtue amongst many which has to be balanced. Depending upon the context and circumstances there is the potential for any student to act creatively in the way in which they behave and interact with their world in the process of gaining experience and understanding. The quality of the creative output from relevant acts will depend upon the orientation and nature of personality factors with creative learners, such matters as 'being comfortable with uncertainty'. Much work has been done over the last 50 years identifying personality characteristics associated with creativity, e.g. Koestler (1964), Stein (1984), Torrance (1962) and Gardner (1995). In design and technology, it is important that the learner has command of knowledge and skill in the area of work and interest, in addition to having the opportunity to take responsibility and make decisions about the nature and directions of their work. Without such resources the impact of any creative act will be limited and recognised only for its worth to the creator.

Learners, in the main, dislike mistakes and failure and are motivated by success. To be creative requires them to take risks, in terms of their relationships with peers and teachers, with time and resources and also with their feelings. The climate for learning requires the presence of a supportive culture (school, home etc.), exposure to the field and access to feedback of a supportive but critical kind.

In design and technology imagination is frequently employed when discussing creativity. Imagination feeds mental modelling as it is both a growing store of perceptions based on real and distorted perceptions of the world and the mental ability to form images of external objects and events not actually present. Such images are predominantly visual but can also include all our other senses, particularly auditory. Everyone has the ability to form, hold, manipulate and record strong visual images, through the ability to create new ideas and speculate on their potential. Most people have simply not practised and developed their ability since childhood. Design and technology commentators would say that practising is not that difficult, it is a matter of creating and concentrating on situations in one's mind's eye (Archer, 1980), picturing participation in the scene and paying close attention to the sights, sounds, tastes and smells. Mental role-playing can be a good starting point. The purpose of imagination in design and technology is that a wide range of possible courses of action is identified: to establish new connections, to transform existing ideas into new approaches. Imagination in school settings is generally undervalued and often thought of in a derogatory fashion even though 'imaging' is a term frequently used and linked to the development of images in the 'mind's eye'. With the focus in school design and technology on realising 'products', there are difficulties for learners within cost and time frameworks to draw on their imagination and realise a functional product which results in what McCormick and Davidson (1996, p. 232) refer to as 'process as ritual'.

To summarise this section; in design and technology education the creative act contributes to the realisation of a product behind which lies an accumulative, evolving decision-making process. There are rational and affective aspects to this process; the way learners construe the problem affects the personal resources they are willing to bring and the risks they are willing to take. The rationale is the explicit, frequently technical, requirements of the evolving specification. The affective will be located in the nature and context of the problem as perceived by the learner and the social context of the designing. The learner needs to possess the confidence, interest, or mandate to attend to the problem; hopefully supported through some form of mentorship based on a relationship with a significant other, frequently but not necessarily the teacher.

DISCUSSION AND ISSUES FOR FUTURE INVESTIGATION

As can be seen from the preceding, there has been some enquiry into modelling within design and technology education but creativity has received far less attention. In particular there appear very few studies which start to explore a more direct juxtaposition between the two terms (Davies, 1996, 1997; Liu, 1996). This concluding discussion will commence from a highly abbreviated starting point: that modelling is the testing and communication of ideas and that creativity is the degree to which an idea 'fits' existing cultural norms. The physical setting of formal learning, whether classroom/studio/workshop is relatively unproblematic but the cultural norms, the network of moral and mental affiliations, within which such settings are framed, are far more complex.

In design and technology education learning is prompted by the identification of a need in the made-world and then enabled by learners attempting to meet that need. Thus two significant cultural determinants for a learner are: the purpose and nature of the task; who defines these and the subsequent activity. This duality of outcome, and a concomitant duality of who controls the process, is articulated in various publications but succinctly in Kimbell et al. (1996, p37). Figure 4.1 (op. cit.) is reproduced below as Figure 7.1.

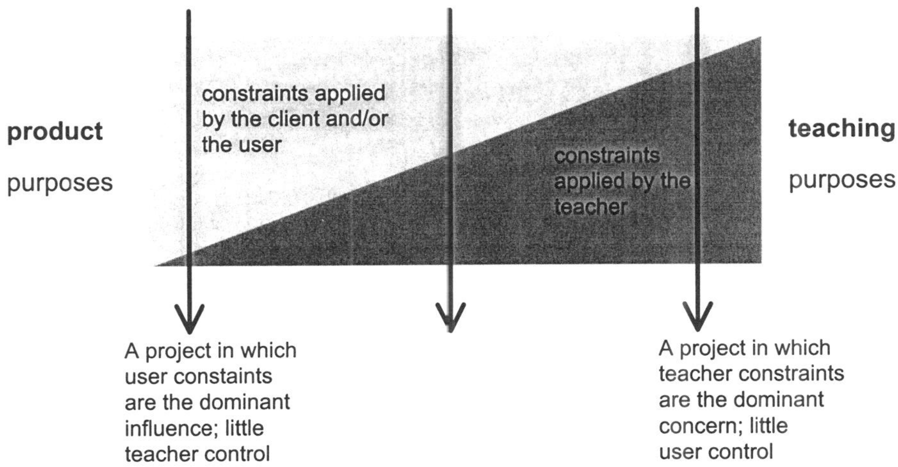

Figure 7.1. The Dual Purpose of Tasks. From Kimbell et al., (1996, p.37).

Utilising this diagram in principle but shifting the emphasis from teaching to learning, from constraints to needs, the diagram has been reconfigured to form Figure 7.2. This is seen as a proper pedagogic shift, not solely a superficial, terminological one.

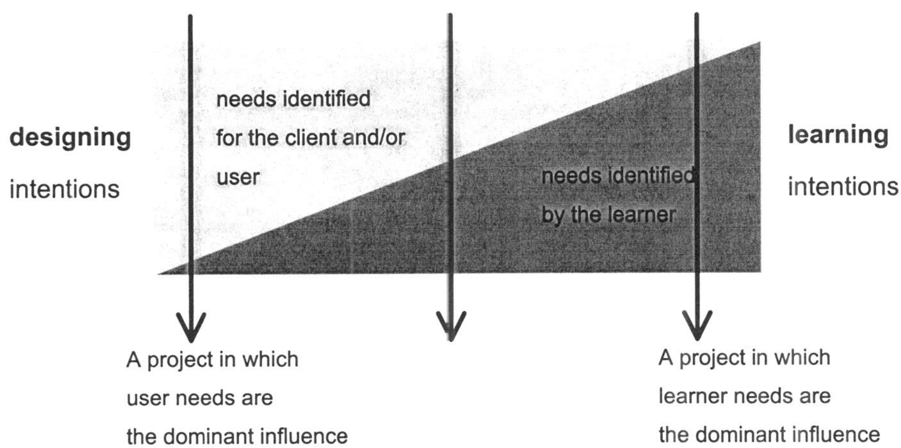

Figure 7.2. A reconfiguration of Figure 7.1.

However, this diagram still omits the degree of freedom available to the learner at any point on this continuum of designing/learning intentions. Designing intentions may be dominant but the degree of freedom may be minimal: the Design Brief calls for a teenager's CD rack but the exercise of judgement is limited to the shape of the back or the colour of the wood stain. The teacher will play a central role here but others (e.g. peer, parent) may influence the desire to extend or restrict the problem space. In design and technology activities the Design Brief, and its subsequent interpretations, plays a very significant role in the setting.

How does this assist with an appreciation of the role of creativity in design and technology? We propose that if intentions are tightly constrained in a project then the opportunity for learners to challenge and move outside cultural norms, for the learner to exercise some degree of creativity, is severely hampered. In order for learners to experience a range of design and technology activities, there should be a commensurate range of degrees of constraint. The more constrained design brief poses less risk of failure for learner and teacher but it has less potential to encourage and engender creativity. This tension is mirrored in the emergence of design and technology in the UK. A detailed historical review is outside the scope of this paper (detailed reviews can be found in Penfold (1988), Layton (1993) and Banks (1994)) but *Design Education* began to emerge during the late 1960's in the UK and its protagonists identified the educational potential and value of a broad, cross-curricular approach to design and designing. *Attitudes In Design Education* (Baynes, 1969), *Design Education: Problem Solving And Visual Experience* (Green, 1974), and *Design Education* (Baynes, 1976) are examples of this vision of a new multi-disciplinary approach. The less constrained brief is strongly promoted including the role of playfulness when designing. A challenge to this approach is seen in the present National Curriculum orders (DFE, 1995).

How do these issues, in particular the prompting of creativity in a learning setting, affect learners' modelling? They may be mutually supportive, the development of creativity might be encouraged by a wide range of modes of representation when modelling; some forms of modelling may inhibit creativity. What is clear is that very little is known about modelling in general and its interaction with creativity in particular; prompting agenda for future investigation.

There is general agreement that there is a conspicuous lack of research in design and technology education. Priorities for research have been proposed, in general (Archer, 1991; McCormick, 1991; Anning et al.; 1992) and, far fewer, for modelling (Roberts, 1992) and for creativity. Much has been written about what learners should do in design and technology education but the over-arching research imperative identified by the authors of this chapter is that there is a need for studies which provide description and analysis of what does actually happen. This chapter suggests that there are some very diverse intentions underlying learners' activities, influenced in particular by social context.

Specific research objectives for modelling are:

1. To establish the purposes that pupils have when they utilise a particular form of modelling during a design and technology project. To identify and analyse the criteria used in making that choice.
2. To establish teacher's perceptions of the purposes of modelling in general, and compare these with the specific examples from pupil work utilised in Objective 1.
3. To analyse dissonance between pupils and teachers' perceptions of the purpose of modelling.

Whilst for creativity, research objectives are:

1. To analyse the limitations of creative activity in design and technology settings at different ages and stages of development.
2. To establish any relationship between the roles of knowledge and skill and creativity in educational settings.
3. To relate the personal creativity of teachers to their ability to promote it in classrooms
4. To examine the impact of social and cultural settings on the creativity of learners.
5. To explore the role of the imagination in imaging and designing.

To summarise, this chapter has sought to give a broad overview, concentrating in particular on the historical emergence and purpose of modelling and creativity in design and technology education. Within its literature, both that of policy making and research and curriculum development, modelling has received far more attention than creativity. This imbalance of emphasis may now change with the influence of, amongst other things, the increasing economic importance given to creativity in post-industrial societies. Despite the relative attention given to modelling there is certainly no consensus as to its role. For the policy makers it is a means to achieve the end of a 'quality product', for others modelling is the end itself through its ability to provide externalisation of the learner's cognitive processes. The chapter finishes with a set of issues for future investigation. This would be expected for any scholarly endeavour, but these issues need to be addressed with urgency in order that design and technology in general education can make a contribution to curricula commensurate with its potential.

Chapter 8

Thought Experiments and Embodied Cognition

Miriam Reiner
Department of Education in Science and Technology, Technion, Haifa, Israel

INTRODUCTION

The question raised here concerns the validity of a mental model generated through a reasoning heuristic, known as a thought experiment (TE).

At least two types of knowledge are involved in thinking through a TE: logical-conceptual and embodied. The latter is based on visual schemas (e.g. pictures) and muscular knowledge, e.g. knowing how to handle the body, so that it efficiently catches a ball. This study mainly looks at how embodied knowledge is involved in TE's applied to problem solving in physics. I analysed students' discourse while collaboratively designing an optimal strategy to win a bike race and found that bodily knowledge is extensively used in problem solving situations in physics. Two interwoven schemas of bodily knowing are identified: 'balance schema' (three types) and a 'symmetry schema' (four types). These are non-propositional, not easily verbalised, and require no explicit thinking. I further show that in order to make prediction about the impact of forces acting on an object, students use an imagery strategy – they imagine as if they are in the position of the object.

These results show that there is a basic, insistent, and efficient, mode of learning that is not based on symbolic communication and logical-conceptual reasoning. It shows that this kind of knowledge is elementary, postulate-like in the sense that it needs no justification, therefore easily accepted, extensively used in making sense of situations, and is powerful as a predictive mechanism.

J.K. Gilbert and C.J. Boulter (eds.), Developing Models in Science Education, 157–176.

The Chapter concludes by discussing theoretical implications for cognitive processes and applied implications for designing learning environments.

The problem is – how do I know the results of an experiment carried out in thought?

In thinking about mental models one could look at their epistemological nature, structure, expressive affiliation (Chapter 5), processes of refinement, relations to learning environments, and implications for cognitive processes. This Chapter deals with a possible heuristic for generating and refining mental models. I suggest that a thought experiment (TE), a reasoning heuristic well known in the history of physics, is a tool for generating, testing and refinement of mental models. I further suggest that this is a two-way process. While experimenting in thought, mental models may be triggered. A mental model may be based on mental schemas. I assume that mental models are generated through the process of experimenting in thought. I show that a TE is based on particular schemas. Through TEs I relate mental models to schemas.

Watching a baseball game reveals a surprising process: the response time of the player is shorter than the 200 ms needed to process visual information of the ball in the air (Allard, 1993). How does the player know how to respond 'intelligently' without any complex calculations of the rate of change of the position of the ball? Somehow, the body-acts are performed in an accurate manner, without explicit thinking. One explanation is that the player runs a mental animation in his mind, based on bodily cues of the player who throws the ball, predicting the future orbit of the ball (Allard, 1993). Predictions are based on the features of the given situation. Hagarty (1993) found a similar effect: her subjects were able to predict future positions of inter-connected rotating wheels without explicit calculations of the rate of change of the position of the wheels. In this sense, there is some 'knowledge' that is accessed, becomes available, in imagined situations, such as mental animations, but hidden when verbal representations are used.

A popular and effective way for training in sports is based on imagery techniques. Basketball players are asked to imagine dynamic situations including the position of the players of both sides; then they are asked to 'imagine' the course of events under varying conditions; finally they are asked to imagine their physical response. This process includes both a bodily-based memory and a conceptual analysis of the situation.

Two points are of importance in understanding the above example. First, in predicting future events based on a mental animation, the thinker carries out a reasoning heuristic, well known in the history of physics and philosophy: a thought experiment (TE). The above example does not follow a particular TE in physics, in the sense that it is not a logical-derivation structure, but rather an implicit process. Yet, as it will be shown later on, it includes all the ingredients of such a heuristic. Second, predictions are made possible by using accumulated bodily experience. The player recruits tacit knowledge about motion of flying objects, time factors, and impact of the object on his hand. He implicitly 'knows' how to respond without any explicit physics calculations. This kind of body knowledge is reflected in the dynamic manipulation of objects, in swimming, skiing, walking and any other motor acts. The person integrates bodily knowledge of a visual and haptic nature in order to respond through bodily acts. I suggest that a similar kind of knowledge is accessed for making implicit decisions about 'results' of a TE in physics learning.

Most studies look mainly at the logical structure of a TE (Brown, 1986). I suggest that an additional type of knowing is implicitly involved in thought experimentation: tacit knowledge, accumulated through bodily experience, used mainly to decide about the course of events during an experiment run in thought. It may be of a visual, or of a muscular-haptic nature.

The applied importance of this result has implications for designing learning environments, especially consider interface design. Most technological learning environments are traditionally based on visual and symbolic representations. None involve any haptic sensations that may trigger the learner's bodily knowing. The technology for haptic interfaces is currently in development. Simple devices, such as joysticks that exert forces on the user's hand, are already available. The results of studies, like those presented here, may provide some leads to the cognitive principles of design of bodily interfaces that support construction of mental models that arise from bodily knowledge.

The Chapter evolves in four parts. The first explores the nature of a thought experiment. I look at mental models expressed in the imaginative processes of a TE. The second looks at the nature of embodied, tacit, non-propositional knowledge. The third is a case study analysis that shows how embodied knowledge is reflected in TEs while modelling a situation in a problem. I conclude by discussing the implications of this study both for designing learning environments and for learning in physics.

WHAT IS A THOUGHT EXPERIMENT?

Thought experimentation, as a reasoning heuristic, is used not only for major insights in physics and philosophy (Sorensen, 1992; Brown, 1991; Kuhn, 1977d) but also for making sense of everyday situations (Wilkes, 1989). It is a way of thinking that integrates knowing in physics with everyday intuitive knowing, therefore is central to this study.

TEs are part of the reasoning culture in physics and are the basis for many innovations. (Sorensen, 1992): Schrodinger's-cat (Gribbin, 1988); non-simultaneity (Einstein 1960), relations between weight and average free fall (Galileo, trans. 1974) are only a few examples. These are mainly consensus TEs, and correspond to consensus mental models. TEs are also used in collaborative problem solving in classroom situations. Similar to mental models, these are the collaboratively expressed TEs, and include children's mental models, but also act as triggers for testing and changing these models. Results show that conversations gradually developed, constructing a logical structure of a TE. The TE is then an evolving pattern, triggered by each of the students' narratives, identified as such only when completed. It is not a pre-designed, well thought-ahead-of-time, logical structure, but rather a series of mental models, that are changed according to the conclusion from a TE.

It's hard (or impossible) to define what a TE is (Brown, 1991, 1986). I can show that it involves constructing imaginary situations and 'running' experiments in thought (Sorensen, 1992). The results of such experiments were often a basis for drawing conclusions. What is then the validity of conclusions made on the basis of an experiment carried out in thought only? What can I learn about the physical world from an experiment carried out in thought?

Structure of a TE

A thought experiment is a reasoning design (composition) that aims to convince or raise a problem. Sometimes, it is seen as an experiment that purports to achieve its aim without the benefit of execution (Sorensen, 1992; Wilkes, 1989). It mainly focuses on the generation of new 'facts' by mental imagery or propositional processes, or both. A stereotypical thought experiment is autonomous in the sense that it is not subject to limitations of physical set-up, is set in an ideal world, sometimes of bizarre features (Wertheimer, 1945; Holton, 1988). Two examples can be used to analyse the role of bodily cognition in TES: Galileo's TE (Galileo, trans. Stillman, 1974), and Einstein's TE on simultaneity.

Galileo's TE:

Two objects, of different weights, W1 and W2, are dropped simultaneously to fall freely from the same point. Assume that the heavier object, W2, falls with an average velocity bigger than lighter objects.

According to the hypothesis,
if W2>W1 then V2(av)>V1(av).
Suppose the two objects are tied together as described in the following figure:

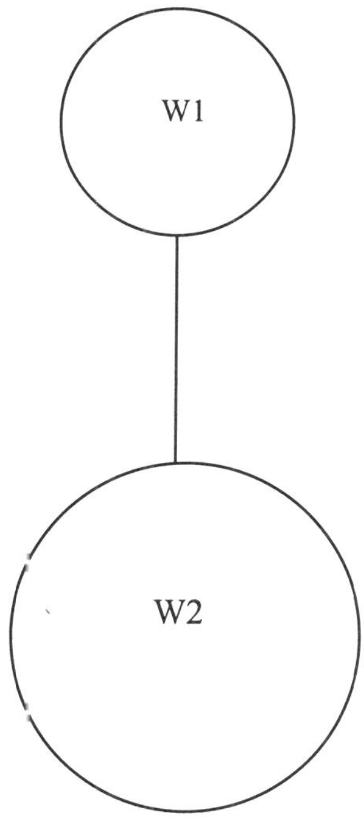

Figure 8.1 Galileo's system

A new object of weight W1+W2 is created.
According to the hypothesis:
If W1+W2>W2, then: (1) V(W1+W2)>V2.
But:
If W1 is slower then W2, then, it slows down W2 (similar to the effect of a parachute).
Thus: (2) V(w1+w2)<V2

(1) and (2) are contradictory statements, thus, the hypothesis is rejected.

Analysis of this thought experiment reveals the following components:

1. *An imaginary world*, in which mental constructs represented in graphical or other representations, such as *objects* are generated.

2. *A problem*: Test the hypothesis that:
   Heavier objects fall with an average velocity bigger than lighter objects.

3. *An experiment*:
   Two objects, of different weights, W1 and W2, are dropped to fall freely from the same point, at the same time.

4. *Experimental 'Results'*:
   Based on everyday embodied knowledge, students create a naive mental model of the situation (see Chapter 5): *objects* have *properties* such as weight and shape, can be tied to each other, objects fall 'down' with a particular velocity, the path of the objects is visible in the mind's eye even though there are no specifications, objects tied together fall differently than free objects, a slower object slows down the faster object, two tied objects are heavier than each of them, the velocity changes according to their weight. The results are then based on a set of tacit, elementary assumptions drawn from embodied experience. Modelling of the situation is used to construct the imaginary parts of the TE. The results of the TE are a basis for refining and re-generating a more 'fit' mental model of falling bodies of varying weight.

5. *Conclusion* based on logical derivations from the 'results'.

The above features are typical to a TE. The following is of a different nature, goal, content, context and conceptual background. Yet it shares a similar structure to the previous and is as deeply based on embodied cognition as the previous.

Einstein's TE

The railway thought experiment was part of Einstein's argument to establish the idea of simultaneity in different frames of reference (Einstein 1960):

> Lightning struck the rails at two separated places along the track in a railway station.

Two observers watch the lightning.

Observer T is on a train travelling with constant velocity from A to B.

Observer P is on the platform.

Problem: Given that the observer in the platform frame perceives the lightning instances as simultaneous, will the lightning instances appear simultaneous to the observer in the train too?

To make the question clear I need to pause and define simultaneity. Simultaneity is defined by:

Flashes are simultaneous if a light signal travelling out from a midpoint M, of AB, to arrive at A and B when the lightning strikes them.

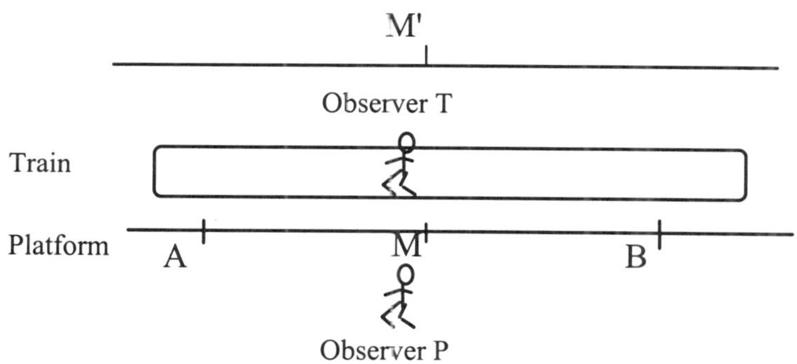

Figure 8.2. The Train TE

In the frame of observer P:
signal emitted from A when lightning occurs
and
signal emitted from B when lightning occurs
arrive to M at the same time.
This satisfies the definition of simultaneity.

Therefore: a lightning happening at A and lightning at B are simultaneous in the frame of observer P.

If M' is the midpoint of the distance AB on the train (see Figure 8.2) then, as judged from the platform,

M' and M are symmetrical relative to points A and B.

In the frame of observer T:
The observer T located at point M',
is rushing towards B away from A.
Therefore: signal emitted when the lightning strikes A
and signal emitted when the lightning strikes B
cannot be received by T at the same time.
This is because the velocity of light is finite.

Therefore, simultaneous events in the frame of observer P are not simultaneous in the frame of observer T.

This TE includes the same five components:

1. An imaginary world: a railway, a train, observers, lightning. These have *properties* and their behaviour is predictable as if observed in nature.
2. An hypothesis and a problem: Hypothesis: The observer in the platform frame perceives lightning as simultaneous. Problem: Are the lightning strikes simultaneous in the frame of the observer in the train?
3. An experiment: lightning strikes the rails …
4. 'Results': based on experience of speed, of 'arrival time', of distances covered by the moving object, envisioning spatial relations of the set-up, for events to happen at the 'same time', of what may be visible from T's perspective or P's perspective.
5. Conclusion based on the 'evidence' and 'results' ('simultaneous events in the frame of observer P are not simultaneous in the frame of observer T').

The two imaginary worlds discussed above differ in objects, rules of behaviour and conclusions (see Chapter 6). In both, graphical representations are part of the visualised world. In spite of the difference in details, both have the same structure, and both rely on experiential embodied knowledge.

THE RESEARCH QUESTION – VALIDITY OF STUDENTS' TEs

Out of the five components, three may involve bodily knowledge: constructing an imaginary environment, running an experiment, and 'seeing' results. The question of the validity of TE mentioned in the introduction is relevant to each these three components. I will then re-phrase the question so that it incorporates each of the above components: what is the validity of the results of an imaginary experiment run in an imaginary world?

EMBODIED COGNITION

Mach coined the term *Gedanken experimente* (Mach, 1943). He claimed that the justification power of TE's originated from the fact that minds may mimic the patterns of nature, thus

> grounded the similarity between our inner private world and the outer public one, in the biological necessity of conforming thought to environment (McCormack, 1943 as cited in Sorensen, 1992, p.51).

He further suggests that a somewhat constant environment allows the development of constancy of thought.

> By virtue of this constancy, our thoughts are impelled to complete all incompletely observed facts (Mach, 1943).

This constancy, originating in the resonance between cognitive processes and events in the environment, lays the ground for completion of events in a thought experiment. The completed 'facts' draw upon 'instinctive knowledge', a cognitive reflection of the environment

> ... a natural outgrowth of primitive experimentation, lead by the.... Darwinian necessity of adapting thoughts to facts (Mach, trans. Williams 1976).

Mach did not comment as to the roots of 'instinctive knowledge'. Piaget suggested that the roots of higher operational functions were the sensory-motor bodily experience (Piaget, 1971, 1976).

Bodily knowledge is the kind of knowledge reflected in motor and kinesthetic acts. When imagining an experiment, such as being in an accelerating car, riding a bike on a curve, being pushed by a crowd towards the exit, the learner recruits his knowledge of bodily movements.

A person playing tennis, whilst observing the flight of the ball in the air, will be manipulating the racket so that it hits the ball at exactly that angle and velocity to direct the ball towards a particular, predetermined, place on the court. How does that person know exactly how to manipulate the racket so that it imparts the correct direction and magnitude of velocity to the ball? Obviously it is not processing of physical formalism. It seems that the body somehow, through accumulated experience, knows how to manipulate objects in space in an efficient manner. Furthermore, when achieving at a particularly high level, an athlete seems to disconnect bodily performance completely from overt cognitive control and the body 'takes over' (Allard, 1993). It seems as if the body 'knows' something the player 'does not'. Rather than rational, propositional, knowledge being used, some sort of imagistic, embodied, form of knowledge, which is not 'registered' in the conventional manner, is being employed.

Clement's (1988, 1989) findings support this view. He showed that embodied intuitions about forces have a role in understanding physics situations. He suggests that knowledge embodied in perceptual motor intuitions are used for physics problem solving by experts and novices.

It seems that, at the moment, there is no sound theory to account for the relations among the forces felt and our cognitive interpretations of them. As a common type of knowledge I know how to attach meaning to forces felt such as the vibrations of an engine, the tremors of an earthquake, the gentle swing of a boat on quiet water. This is done unconsciously. I have no syntax or dictionary of meaning to attach to patterns of forces (Johnson, 1987; Lakoff, 1987). I suggest that images of forces are triggered and used without any external stimulus, but through the intention and needs of the TE alone. These images carry tacit knowledge used in a thought experiment. Learners know what the results of an experiment in thought may be by using this innate non-verbal knowledge, that 'comes to life' through the images of forces associated with the particular situation in the TE.

Johnson (1987), in an attempt to explain this type of knowledge, claims that 'human bodily movements, manipulation of objects, and perceptual interactions involve recurring patterns' (Johnson, 1987) which carry meaning. These patterns are termed an 'image schema':

> because they function primarily as abstract structures of images. ...They are gestalt structures consisting of parts standing in relations and organised into unified wholes, by means of which our experience manifests discernible order.

Image schemas, a structure of embodied imagination, is then a frame of interpretation of the environment. These structures are non-propositional, figurative, pictorial, dynamic, associated with particular situations. Through a cognitive cycle of linking past-experienced-situations with prediction of future events, mimicked imaginative worlds are created. Being a image of past experience these imaginative worlds carry a similar nature and 'rules' of behaviour, a product of the learners implicit, tacit knowledge (see Chapter 5).

The next section brings evidence collected and analysed in a learning experiment. Results show how students, in an attempt to solve a tricky problem, recruit their body knowledge to respond to questions of a 'what if' type. 'What if' questions are typical to a TE (Wilkes, 1989). It requires that students imagine a situation which is not physically accessible. In the following, I show how students construct images of riding a bike down a steep dust road, so that they can complete the race. Parameters of the situation are changes, continuously altering the students' inner reality. These inner worlds sometimes result in impossible states, never to occur or to be investigated in a physical 'reality'. I further show how students run imaginary experiments that reflect an inner consistency shared by other participating students, and by me as a researcher. They describe results, elementary in the sense that these need no justification – in both their world and the researcher's world. Both their colleagues and I, the external detached viewer, find their suggestions plausible, needing no questioning or explanation.

I suggest, then, that validity of TEs stems from the human ability to mimic whatever is captured as the external world – objects, events and processes. This Chapter further suggests that this ability, to construct valid reflection of our worlds, is not a result of symbolic processing system, an outcome of logical inferences, but rather a process based on triggering tacit, implicit, non-propositional knowledge, accumulated through experience, organised an image schema.

EMBODIED COGNITION AND STUDENTS' TEs – A LEARNING EXPERIMENT

This section describes a learning experiment aimed at identifying TEs and instances of reasoning based on embodied knowing. I analysed students' discussions that took place in a problem-solving context. Twelve students were separated into four groups, and presented with a task, designed to provide opportunities for imagining situations, related to physics-conceptual thinking:

> You are racing in a bike race. Obviously (...) you need to carry water. The organisers provide you with four uncovered buckets full with water up to about 3cm from the margin, each of about 30N. Two dust roads are available for your choice. The first is relatively short, narrow, rocky, very steep, straight. The second is windy, long, of moderate slope, smooth. The winner is the one who arrives first. You cannot afford to lose the water. What route would you choose so your chances to win are optimal?

Students participating in the study were 15-16 years old, had no background in physics other than a limited background on fluids (no background in mechanics), and were experienced with biking.

RESULTS

Overview of Results

I identified two main factors in students' discussions around a TE. The first relates to types of bodily knowing. The second relates to methods used by students in order to know. The first, bodily knowing, can be roughly categorised according to 'balance schema' (three types) and according to 'symmetry schema' (four types). The two appear together in the sense that bodily force considerations are related to tacit underlying symmetry considerations. In analysing symmetry and forces, students show various levels of a globality in treating the system. This changes according to the context, and sometimes students' reason about one particular component of the system. Segments exemplifying each are given in the next section. Each instance is characterised by the type of balance schema, symmetry schema, and globality.

The second factor is related to how students know. Since it shows how students refer to objects, I term it as a referential strategy. I identified three types of referential strategies: students either imagine themselves as being in the object's place in order to 'feel' the forces, or they comment directly about forces on the object, or on their own body. This suggests that students use an imaginative strategy their own body substitutes the object in order to predict the impact of the acting forces.

The first factor provides evidence that students use bodily knowledge. The second shows that this knowledge is of a non-symbolic nature, and that it is often based on imagining the feel of force.

I conclude the results section by showing some implicit propositions that seem to have the state of 'primitives'. I call these implicit rules, constructed by no logical process, not a result of a domain specific culture, but rather an outcome of students bodily knowing.

Balance Schema and Symmetry Schema

I identified several types of embodied justifications. The major scheme that is reflected in students' considerations is a feel for *'balance'*. I term this a '*balance schema*'. Decisions about what situations are balanced are based on the their own feel of forces. If they feel for instance that a heavy load of the water behind will disturb the balance, students suggest loading the front, or distributing the water buckets differently – an act they feel is a solution. Similar to axiomatic reasoning, students do not provide additional justifications other than their own 'feel'.

The balance schema involves the feel of forces and their impact on the balance of:

1. the system of biker-bike-water as a whole – system thinking;
2. each of the components: biker, bike, water, each as a separate entity – local thinking;
3. the environment: biker-bike-water-rocks-slope-curves – global thinking.

I also show that the balance schema is related to the *symmetry schema*. Each of the instances given in the following implies tacit considerations, not necessarily visual but related based on different types of symmetry. However, the symmetry considered is a symmetry of the feel of forces. Instances of each of the categories are provided in the following:

A. *Bodily knowledge used to identify forces and their impact on the system of biker-bike-water as a whole:*

Instance 1: balance based on left-right symmetry of forces.

E: I am very good with bikes. I am for the short one.
T: where would you put the buckets?
E: two on each side, tied behind the seat. Basically it makes no difference. I won't feel the weight. Will be pushed the same on both sides.
Y: this will keep the bike straight [vertical].

Instance 2: balance, based on spatial symmetry of forces.

E: We can cover the buckets
T: against the rules. It won't help anyway – I mean the water moves inside all the time so it will move you too…

Instance 3: balance considerations based on front-back symmetry.

Y: the water will pull you backwards. you'll fall. Too much weight in the back.
E: why would I fall? I am pulled equally on both my sides…
Y: No… I mean backwards. Imagine like a heavy weight on your back. It lifts the bike in the front….like the smallest rock will lift your front and you fall backwards. [positioning his pencil in a horizontal state, pointing at the back to show what happens if you load too much, and lifting the front of the pencil]

Instance 4: front-back symmetry of forces.

Y: Also the buckets will tilt backwards…like in a speeding car [probably mean accelerating]. You are thrown backwards.
T: no…
Y: yes … especially if you plan to be fast … [probably means 'accelerate'].

Instance 5: spatial symmetry of forces.

E: …two buckets in the front, two in the back…
T: the water will move, in the front and in the back, so you'll be very non steady.

B: Bodily knowledge used to identify forces and their impact on each of the components (biker, bike, water, each as a separate entity):

Instance 1: balancing the water, up-down symmetry.

T: you'll spill the water. Because of the rocks.
E: the water will just vibrate in the bucket.
T: the water will 'jump' out. Like when you step with all your weight on a rock you feel the force through to your neck.

Instance 2: balancing the body, back-front symmetry.

E: I can balance myself. I'll just tie a rock to the handle-bars…
T: or you can just lean forwards. .you need to lean until you feel steady
Y: you are not heavy enough.

C. Bodily knowledge used to identify forces and their impact on the environment: biker-bike-water-rocks-slope-curve:

Instance 1: Spatial symmetry.

T: you'll be too unstable. rocks and the buckets and speed, you won't be able to control the bike, need to be very fast and see each pebble, the water will 'dance' which will make you even less stable. Slim chances…

Instance 2: Leaning the bike on a curve , symmetry around a slope.

E: the other road will force me to lean. Otherwise it throws you out. Because of the curves.
T: why do you have to lean? Right, on a curve…
E: The water will spill out… because you lean.
Y: I don't think so, it is still balanced, the force ... when you speed… holds the water. just as it holds you. I mean you and the bike [on a curved road], you won't fall even if you lean….it's correct [that the biker does not fall] only when you have speed [on the curved road], otherwise if you lean, you fall.

Instances A-2, B-1 and C-2 are straightforward examples for the main claim in this paper both about embodied cognition and elementary. In A-2, T had no background in the conservation of angular momentum, which is the physics explanation of the fact that the moving water makes the whole system move in order to maintain the sum of angular momentum of the system as zero. Yet is was accepted without any questioning. Similarly for C-2 and B-1. It had an implicit appeal, considered correct on its own, similar to a postulate, which needs no justification. The principle of elementarity appears in all the instances, in the sense that bodily knowing is easily acceptable.

Students' bodily thinking, reflect four types of symmetry relative to the bike: left-right, front-back, around a point in space, and around a tilted line. The term symmetry was not mentioned by the students, symmetrical considerations were tacit, termed as such by the researcher, recognised as such only by the external observer. As to the students, symmetrical considerations were implicit, non-propositional, unrecognised as such.

Referential Strategies, Methods used by students to identify forces and their impact on objects

Bodily knowing varies also according to the method used by students in order to raise justifications. Basically, they used three types of referential strategies:

1. about their own body's state based on bodily experience;
2. about particular objects' behaviour by imagining themselves instead of the object;
3. about objects' state without comparing it to their own body, but by using bodily descriptions.

Categories (1) and (3) are not surprising, a sort of everyday wisdom, examples for which are innumerable. There is no need for research to show that these exist. The more interesting bodily prediction strategy is the second. The instances given above reveal an 'as if' strategy – students imagine being the object: for instance T's explanation to the water spilling out of the bucket once the bike hit a rock:

> ...the water will 'jump' out. Like when you step with all your weight on a rock you feel the force through to your neck.

Students use thus a muscular sense in order to predict possible events:

> Imagine like a heavy weight on your back. It lifts the bike in the front ... like the smallest rock will lift your front and you fall backwards.

This strategy, immersing oneself in an imaginary 'as if' situation, seems to be used by students in order to know what the forces and impact on the object are. As mentioned in the introduction, it is frequently answered by a TE (Wilkes, 1991). It is a strategy used to recruit knowledge which is non-propositional, triggered by patterns of forces, felt or imagined. I have no evidence, but the two other categories may be related to this one. Students predict situations based on the forces felt by their own body. In order to make sense of other objects' behaviour, they imagine themselves as the object, and finally comment about the behaviour of these objects.

A possible explanation is that schemas evolve based on past experience. These are employed in modelling situations, constructing mental models, which are further refined according to experience and reasoning heuristics such as TEs. The constructed mental models are used to predict effects on object in various situations. The way schemas are used for modelling and constructing mental models is by substituting the object with their own body. This allows students to trigger existing embodied schemas. There are two levels of bodily thinking: the first is built through embodied substitute of the object; the second is automated, an evolution of the first, refined and tested, thus needs no further the strategy of substituting the object. It seems that the three categories are stages in the evolution of one single category – in which the third is an automated stage of the second. The first is the base, given capability of translating force patterns into modelling situations.

Elementary Embodied 'Axioms'

Students' knowing through elementary bodily knowledge can be represented as a set of implicit axioms, elementary assumptions that need no proof, and need no other justification. These elementary assumptions are applied by students to model the situation. The following are a few examples of such assumptions:

> [on a curved road], you won't fall even if you lean
> [biking on a curved road]will force me to lean. Otherwise it throws you out. Because of the curves.
> [while accelerating the bike down the road] ... the buckets will tilt backwards … like in a speeding car.
> or you can just lean forwards…you need to lean until you feel steady.
> Like when you step with all your weight on a rock you feel the force through to your neck.

These rules are elementary, similar to axioms, in the sense that they need no explanation or justification. They are readily accepted and no objections are raised. The process of constructing rules is then through the sensory accumulated information reflected in image schemas. Some are not consistent with the science community conventions which were largely identified in the literature during the last thirty years. For a while these were termed 'misconceptions' to show their variation from the scientific knowledge, then these were termed 'alternative frameworks', 'alternative science', 'children's ideas' etc., in an attempt to reveal the fact that these are legitimate constructions through interaction with the environment. The work here suggests something about the cognitive roots of these ideas in the bodily tacit knowledge through implicit learning.

SUMMARY AND CONCLUSION AND IMPLICATIONS

This paper shows that the validity of thought experimentation is rooted in human capability to mimic the physical world in a way that is communicable. As such, thought experiments reflect students' mental models, but also are a tool for testing and re-constructing mental models that fit students' environment better than those previously held. I further show that students' construct imaginary worlds in a TE, run experiments and observe results of an imaginary experiment, by recruiting tacit knowledge, through muscular and other bodily-sensory descriptions.

By analysing students' narratives, I found the following main results:

- type of 'bodily schema' - balance and symmetry schemas applied in a global or local frame of thought;
- type of 'referential strategy' used by students to recruit a body schema - imaginary substitution;
- tacit rules, similar to axioms in their nature, reflected in students' thinking.

Each student's narrative can then be characterised by four factors: the situation examined; kind of schema applied; type of strategy used; rules applied. This suggests a new way to analyse students thinking based on bodily knowing: by analysing four types of resources.

It was somewhat surprising to find that the balance schema is intertwined with the symmetry schema. It seems that both force sensations and the sense of symmetry, probably based on the human body's symmetry, are somehow considered together in order to make sense of situations, and predict future behaviour. Another surprising result is that such bodily thinking often addresses the behaviour of complex systems such as bike-biker-water cans-road curves. In this sense thinking is often global. It does not exclude predicting the behaviour of components of the system. In this sense thinking is sometimes global, sometimes local, focussed on a single component, according to the nature of the context. Many researchers show how complicated and difficult global system thinking is for students. It is surprising then that bodily knowing is so 'naturally' global or local according to the situational needs.

What do these results basically say? They say something about the origins of students' elementary epistemology. They suggest that knowledge is not necessarily symbolic. It may emerge from bodily motor acts and sensory interaction with the environment. It further suggests that sensory accumulated experiences are shaped and organised in dynamic schemas, non-propositional in any sense of this term. These schemas are components in mental models. These schemas are tested in mental models of situations. They evolve, are refined and further tested in modelling situations until these models best fit the environment and the predictions made. Just like symbolic knowledge, it relies of basic axioms. Another surprising result is related to global thinking. Expressing global views of a system is one of the more complicated tasks in science and technology problem solving. Surprisingly, analysing situations through bodily knowledge on many occasions involves a global view. In this case it seems to be inherently embedded in the bodily knowledge. This may be related to the fact that the body itself acts as a system.

Playing the devils' advocate, I may ask whether the origins of the knowledge that students show are not in past symbolic learning. Furthermore is it not the case that the students indeed know, but that this is not due to bodily experience, but because they have read this information somewhere, or heard of it form a friend, or seen it on TV. I suggest that the nature of the students' justifications shows that these cannot be symbolic, in that the knowledge expressed is based on imagining bodily substitution in order to 'feel' the forces. The knowledge is elementary, intuitive and reflects body experience. Three factors show this:

1. The use of naive, non-scientific terminology.
2. The lack of need to justify propositions, that would otherwise require an extensive formal proof in scientific language.
3. The terms used reflect bodily experience, describing a motor-sensory (mainly force) feel.

I wonder as to the relations between symbolic and bodily tacit knowledge. Obviously, at the end, a verbalised mental model may be constructed. Yet the major question now is how to design experiences that allow students to construct symbolic knowledge.

Implications for Design of Learning Environment

These results suggest a new, additional channel for communication for learning, not only symbols but also through embodied tacit knowledge. The components of a learning environment need to include sensory inputs of bodily features such as force sensations and tactile input as well as symbolic representations. The existence of such knowledge has major implication on the current thought of what I term as learning activities and learning environment. Is the classroom environment designed to provide children with an optimal learning experience? I claim that classroom environments currently address only conceptual learning, often ignoring the underlying insistent bodily mode of learning. Recognition of the role of embodied cognition has also an impact on design of technological learning environments. Virtual reality may especially be influential in providing students with the opportunity to both 'feel' and 'see' events and manipulate representations. According to these results, interfaces have to be based on both conceptual and bodily communication channels. Thus interaction with the Internet needs to be based on visual representations and simultaneously attach the feel of force to a situation for instance, playing tennis over the Internet.

Chapter 9

Computers and the Development of Mental Models

Patrick Carmichael
The University of Reading, UK

INTRODUCTION

This Chapter reviews some of the ways in which the introduction of computers into classrooms may encourage the expression and elaboration of mental models in science. Programming languages such as Logo (Papert, 1980; Resnick, 1994) have provided a basis for a range of modelling environments; computer-generated imagery has allowed the visual representation of increasingly complex modelled environments; the development of graphical user interfaces has made it possible for even young children and others with limited programming skills to develop computer-based models; and the development of the World Wide Web has provided a medium for dissemination and discussion of models and model-building tools.

A MODELLING CURRICULUM

What Mellar (1994) calls a 'modelling curriculum' might now extend to include the use of both text-based and graphical 'expert systems' such as: Expert Builder (Webb, 1994); 'rule-based' modelling environments such as KidSim (Gilmore et al, 1987) in which users create models representing the behaviours of real or imaginary objects by describing their responses to different sets of conditions; 'alternative reality kits' such as Thinkertools (White and Frederickson, 1998) in which users can conduct 'virtual' experiments which would be impossible in any 'real' laboratory; 'microworlds' such as Tierra (Ray, 1997) in which virtual inhabitants, sometimes with 'real world' equivalents, but often without, can interact,

J.K. Gilbert and C.J. Boulter (eds.), Developing Models in Science Education, 177–189.

grow, compete and evolve; and collaborative programming environments such as Moose Crossing (Bruckman, 1998) within which many users may model and interact simultaneously. Within this range of virtual worlds, which can represent the very large, the very small, and the spatially or temporally distant, the distinction between 'thought experiments' and empirical enquiry is blurred, as model builders have opportunities to define unusual objects and materials, redefine or ignore physical constants and ascribe to the 'observer' powers ranging from microscopic size to the ability to alter genetic code.

THE VIRTUAL CELL, AS A COMPUTER-BASED MODEL TO EXPLORE

An example of a popular and well-constructed computer-based model, constructed using the POVRAY (Persistence of Vision Ray-tracing) 3-D modelling environment (http://www.povray.org) and the Perl programming language (http://www.perl.com), and which is available via the World Wide Web at the University of Illinois at Urbana-Champaign, is Matej Lexa's 'Virtual Cell' (http://ampere.scale.uiuc.edu/~m-lexa/cell/) (see Figure 9.1). This is described as a collection of still images, texts and movies covering the structure and functioning of a typical plant cell around which the learner can navigate using a graphical user interface, sectioning and inspecting cell structures and organelles. Lexa describes this World Wide Web-based resource as incorporating

> things (that) should look real, however they are models that are only an approximation of the real things. And since we do not want to deceive you, but teach you, we included real electron micrographs. Look at them and keep in touch with reality. (Lexa, 1997)

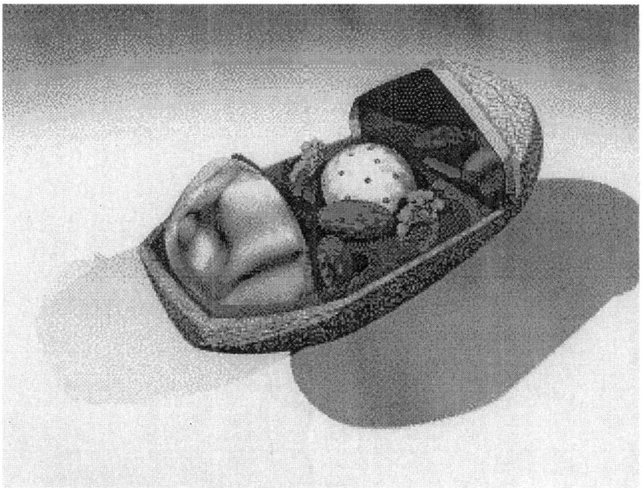

Figure 9.1. Illustrations of Virtual Cell

This explicit statement of the status of the model as an 'approximation of the real thing' echoes the point made by Bliss (1994): that 'models are always simpler than reality: in building a model, a lot has to be left out'. The 'Virtual Cell' presented onscreen - albeit in association with electron micrographs of cell structures - is necessarily a simplification. That said, Lexa is careful to make clear that some apparent simplifications are in fact accurate representations, the caption for the external view of the nucleus reading

> If you think the nuclear pores are spaced too regularly, check the real electron micrograph of a nucleus.

Ogborn and Millar (1994) identify this as a characteristic of many models which

> are based on ... assumptions, the only check on them being against empirical observation.

Lexa describes his virtual cell as a model and offer learners the opportunity of checking of his representation of objects against electron micrographs, thus making this a valuable aid to learners' visualisation of a complex three-dimensional structure. But the learner in this case is not the modeller: they have the opportunity to explore and manipulate elements of Lexa's model, but it was the modeller/author alone who obviously made two sets of decisions: about the selection and representation of cell structures, and about the extent to which the learner visiting the website would be able to interact with the model.

There are two standpoints from which we may assess Lexa's cell model. In terms of the aspects of modelling exemplified by the UK National Curriculum in IT (DFE, 1995c) the 'virtual cell' offers learners the opportunity to 'explore' and, to a limited extent, manipulate, a model. Despite the richness of the modelled environment, the 'modelling' experience of the user is limited.

THE VIRTUAL CELL AS AN EXPRESSED MODEL

An alternative interpretative framework is provided by Gilbert and Boulter (1998) in their differentiation of 'mental' and 'expressed' models:

> A model is a representation of an object, event, process or system; mental models are personal, private representations of the target, and expressed models are placed in the public domain.

Thus Lexa's 'virtual cell' is an 'expressed model' reflecting not only the mental model of the modeller, but preferred interpretations of cell structure and the perceived needs of the students for whom the model is intended. Its role is not to impose a particular interpretation of cell structure on learners, but to assist them in formulating their own mental model which may, in turn, become expressed differently (and possibly through a medium other than ICT) depending on the understanding and intentions of the learner and the perceived needs of their audience.

In most classroom settings, as a teacher expresses a model, there is the opportunity for interaction between teacher and learners which, ideally, should lead either to the development of the learners own models or to the establishment of a consensus model. Learning from a pre-constructed computer model differs, however, in that a key individual or group, the modeller or modellers, are absent. Their ideas (mental models, intended learning outcomes and assessments of audience needs) are represented in the expressed model they have produced: the situation when learners are confronted by 'interactive' museum exhibits without the benefit of their constructor is rather similar.

THE SCOPE AND ASSUMPTIONS OF ON-SCREEN MODELS

In some cases, the modeller provides clear information about the scope of their model, citing areas where assumptions were made and simplifications were necessary. In some cases, the entire code-base is provided for inspection and development. In many models, however, the user has to explore the model in order to discover its scope and, in some cases, identify the modeller's assumptions. This is to some extent encouraged by the belief that young children in particular should be encouraged to 'explore' the on-screen environment: many proprietary software products provide little documentation and cite 'learning through doing' as their reason for doing so.

Mindscape's popular 'virtual pets' simulations Catz and Dogz (sic) offer models of animal behaviour, but these are simplified and selective: the animals neither excrete nor die, the cats never kill the mouse which periodically runs across the screen and their 'learning' is guided by rewards (biscuits) and sanctions (water sprays). Current research with young children using these simulations suggests that they initially 'explore' the model, discovering features (many of which are undocumented), but once they have gained a degree of familiarity, may display interest in the 'accuracy' of the model provoking the pets in order to see if they will respond, depriving them of food to see if they starve and 'catching' the mouse and placing it in the cat's food bowl to see it is eaten. This may not be the outcome intended by the modellers, but reveals how learners are prepared to 'test' models and even to attempt to assess how the model has been programmed despite the fact that this can only be inferred from on-screen experiments. As one eight-year-old commented,

> There's nothing in the computer to say 'if you don't eat for ten or twenty or some days then you die', it just says 'if your bowl's full eat some food'.

LEARNER CONTROL AND DEVELOPMENT OF MODELS

Other models allow some control over the behaviour of model components, although only certain parameters may be changed. Interestingly, the Mindscape virtual pets do allow some changes to be made but these are trivial (cat colour, background pattern, mouse active or dormant) in that they obviously do not affect the significant behaviour-modelling components of the model. Maxis' SimAnt is a sophisticated model of the development of ant colony which may be used as a game or in 'experimental' mode offering more control and no criteria for 'winning'. In the latter mode it has considerable value as an aid for teaching about animal behaviour. Despite the fact that, as with Catz and Dogz, the coding is unavailable for inspection or development, it is possible for the user to adjust a range of parameters, the tendency of the ant queen to produce different kinds of ants, the amount of food in the environment and the presence of competitors and predators, for example. This degree of user interaction with the software allows sustained, non-trivial, experimentation to take place, in which users can apply 'real world' knowledge in the model and potentially gain insights into the behaviour of real insects.

Once learners are able to control and develop existing models in this way (rather than simply 'exploring' them) or to develop their own models, they become involved in an 'expressive task' (Mellar, 1994). The true value of this expression of a model and of subjecting to it public scrutiny is that the modeller may, in the process of expression, also become aware of hitherto unrealised potential of, or deficiencies in, their mental model which may then be refined and developed as a result of the expression.

This view of computer-based modelling echoes Hesse's (1989) discussion of the 'interaction' analogies, which may involve 'selection, emphasis, suppression and illumination' of elements of the analogy. As the computer model and the mental model it reflects interact, the modeller's understanding of the nature and scope of both are liable to change. Once in the public domain, an expressed model may be subjected to further scrutiny and may be further refined by either the original modeller or by others. Hesse's (1989) 'comparative' analogies, on the other hand, in which systems are simply compared on a point-by point basis and a literal statement of similarities and differences made, are, according to this interpretation, more likely to occur when users of existing computer models assess their 'accuracy' in representing 'real' situations. Thus a learner might assess the Catz program in the restricted (comparative) terms of 'being like a real cat in that ... but not like a real cat in that ...' while a learner programming 'cat-like' behaviour into their own model in Logo, for example, would be forced to make explicit their understandings both of cat behaviour and of the way in which computers represent real-world situations, with the potential outcome that both might be refined and developed, involving Hesse's (1989) 'interaction' view of metaphor.

EXPRESSING MODELS THROUGH NETWORKING

Networking, either locally or over the Internet, offers exciting opportunities not only for collaboration on modelling projects, but also for expression of models in a supportive public domain. A recent development which offers opportunities for collaborative and negotiated development of models are online multi-user domains ('MUDs') within which users can program objects, their surroundings and other virtual 'inhabitants'. Users can also interact with each other via online 'chat' or email, and may seek support and advice on programming projects.

The two best-developed projects of this kind are Amy Bruckman's 'Moose Crossing' (Bruckman, 1998) based at Georgia Institute of Technology and Austina de Bonte's 'Pet Park' at Massachusetts Institute of Technology (de Bonte, 1998a). Initially developed for members of after-school clubs and home users of the Internet, Moose Crossing has subsequently been implemented in classroom settings (Bruckman and de Bonte, 1998). At present the primary objective of both projects is to develop children's programming skills and to explore the potential of MUDs as a means of developing collaborative learning, and little modelling of scientific phenomena takes place. In 'Moose Crossing', children are encouraged to construct fantastic environments in which normal physical conditions are regularly violated. The resulting world, which may be explored and extended by users, owes much to the alternative visions of Greek Mythology, Lewis Carroll's 'Wonderland', Terry Pratchett's 'DiscWorld' and Disney and Warner Brothers' cartoons.

De Bonte (1998b) writes that the intention of the developers of these projects was not to set up environments in which real-world situations could be modelled. Despite this, however, the programming environment of 'Moose Crossing', in particular, provides ample opportunities for users to develop elaborate models. Many of the children construct 'pets' with which they and others can interact. While many of these have unusual characteristics (the power of speech, for example, is commonly attributed to dogs, cats and even plants), it is obvious from examination of the code developed by children that many of them spend time refining their on-line creations so that they match the 'real world' behaviours of their own pet animals. It is obvious that the technology used in MUDs such as Moose Crossing could potentially be developed so as to provide a more consistent virtual environment devoted to the definition, application and elaboration of scientific models.

THE INFLUENCE OF GENERIC SOFTWARE

Perhaps the most significant contribution to the development of Mellar's (1994) 'modelling curriculum', however, is not the development of MUDs or Java applets, but the increasingly widespread availability in schools of 'generic' or 'content-free' software with integrated programming capability. Recent years have been marked by the appearance in schools of industry-standard IBM compatible computers running 'Office' software suites produced by Claris, Lotus or Microsoft. Many schools have, at time of writing, at least one computer capable of running Microsoft Excel, which, together with Microsoft Visual Basic for Applications, offers significant modelling opportunities. Brosnan (1994) and Atkinson (1997), for example, describe how spreadsheet modelling may be used to address learner difficulties in understanding complex chemical processes, particularly those with a number of independent variables.

Despite the opportunities for modelling offered by products such as Excel, it must be recognised that the education sector forms a small part of the global ICT market, and products specifically developed for the 'educational' market offer only limited revenue-making potential. This is reflected by the number of promising modelling environments which have failed to be developed to their full potential despite obvious educational applications. Those developers using Apple's Macintosh Operating System have had the opportunity to use and develop applications using Apple's HyperCard authoring environment together with the HyperTalk and Applescript languages (what Coulouris and and Thimbleby (1992) term 'Hyperprogramming'). The range of educational resources and models that have been developed using HyperCard, development of the system, however, is currently 'stalled', despite Michael Swaine's (1999) assessment of it as 'a powerful power-user-level development system with no equal on any other platform, and the reason many people stay with the Mac'. At the same time, Apple's Cocoa modelling environment which was developed from the earlier KidSim system described by Smith et al. (1994) and Gilmore et al. (1997) is no longer supported by Apple although another software developer is continuing to develop this and other products.

Swaine (1994) concludes his discussion of HyperCard with a plea to 'let us have [the long overdue] version 3.0 or let us have the source [code]'. The notion of application code being released 'open-source' so that users are able not only to use a software application but also to examine its workings and use the code in their own projects has important implications not only for professional software developers but for those involved in computer-based modelling.

Open-source software is, at time of writing, receiving a good deal of attention for a number of reasons, notably the market share achieved by the Linux operating system, the decision of the Netscape Corporation to release the source code for its Communicator World Wide Web Browser and the enduring success of the Perl programming and web scripting language used in many World Wide Web applications. 'Open source' is, however, not merely a programme for effective software development (its protagonists, notably Raymond (1998), argue that it is the most effective approach); it also provides a framework for description of a range of different software with the potential to be used for modelling, and of computer-based modelling applications and projects.

PROGRESSION WITHIN A MODELLING CURRICULUM

The requirements of the curriculum imply a progression from 'code-free' 'exploration' of pre-constructed model environments through manipulation of specified variables and rules, to the development of code to either describe or define the operation of the model. Some modelling environments, such as the Worldmaker package described by Boohan (1994), allow learners to progress in this manner in the course of working with a single software package, but the number of such products is limited. It is instructive to examine a number of currently available modelling environments, as the parallel between their 'openness' and the potential they offer the modeller is clear. Stagecast software (http://www.stagecast.com), who licensed the Cocoa modelling environment from Apple and have developed it into their own product range, are quite explicit both about the extent to which code is made available to users and the modelling opportunities it provides:

> Stagecast Player ... The lightweight player will run any simulation created with the authoring tool, and will allow limited modifications to the simulation ...
> Stagecast Creator ... The (Java-based) authoring tool will feature a user-friendly graphical interface that allows a user to create simulations that can be played as Java applications or applets...
> Stagecast Extender (allows) Java programmers (to) enhance Stagecast Creator beyond the current feature set and increase the types of worlds that can be created (and) provides tools and sample code for programmers.

The first of the products allows the user to 'explore' existing models, but in order to manipulate and develop models, the 'Creator' and 'Extender' products which offer greater modelling potential and access to code must be used.

This pattern is repeated in other modelling environments. Superscape's (http://www.superscape.com) three-dimensional modelling tools, used in a number of commercial, games and educational products, range from a commercial developer's modelling environment most recently used for the popular Lego Creator product to a more limited (and very much less expensive) product, DO3D, which is aimed at educational users. This product, however, offers only limited modelling potential since it includes only set 'libraries' of objects and behaviours, and it is not possible to define new objects or behaviours.

An 'open-source' solution to the problem of having only limited access to a potentially rewarding modelling resource is David Baum's NQC (Not Quite C) programming language. This is designed to extend the applications of the Lego RCX system (http://www.legomindstorms.com), which is designed to use Lego Logo, a dialect of Logo developed at M.I.T. (Martin and Resnick, 1990; Resnick et al., 1996). While Logo itself is an 'open' language there are proprietary versions, but others are freely available, and 'Java-Logo' dialects for inclusion in web-pages are currently under development, the Lego product presents the user with a graphical user interface which offers only a subset of the full Logo command-set. In response to this, Baum and his associates produced an open-source version of the C programming language with which they have then used to model a number of target systems using the hardware components of the original RCX system, but which would be impossible to model using the limited version of Logo provided with the commercial product.

Even those modelling environments which do offer the user the opportunity to develop their own models often do so through graphical interfaces which, while they allow description of parts of the model, may not encourage or allow its definition. This is a key distinction made by Gilmore et al. (1997) in their assessment of KidSim, in which they discuss the problems children had in establishing transferable rules to handle events. They particularly draw attention to children's approach to programming a behaviour such as 'stair-climbing': rather than defining a simple conditional instruction which would be generally applicable, they tended to write many similar sets of instructions which the authors characterise as 'animations'. Brand et al. (1998), in their discussion of children's work with Cocoa, also differentiate between models which incorporate general causal factors and 'descriptive models' in which the software was used primarily as an animation tool. Brand et al. (1998) also describe how those who developed only descriptive models stated that they had learned little science from the experience and that they seemed to see the modelling software principally as a tool for presentation of their existing ideas.

My own experience in working with children using Bruckman's Moose Crossing online environment confirms this distinction: children's early attempts at 'building' an object initially focus on its appearance, subsequent developments often being descriptive sequences of instructions, with definitive rules incorporating 'if ... then ... else' structures only emerging after tuition by other users online or 'in real life'. De Bonte (1998a) describes similar patterns in the work of children using her Pet Park graphical programming environment.

In many cases, however, the inability of the user to express models in which behaviour of entities is defined rather than described depends not on their experience of programming or on the level of tuition which they receive, but on their level of access to programming code. This may be denied entirely (as in Mindscape's Catz and Dogz); is limited and mediated through an interface (as with Stagecast's Stagecast Creator), or comes at considerable cost.

The lack of an appropriate modelling environment which offers this, however, is, in some respects, a blessing in disguise. Despite the need for an 'integrated modelling framework' (Cox and Webb, 1994) to provide a basis for development and sharing of resources, this does not necessarily mean that a single integrated programming environment needs to be selected. The fact that modellers are working to express their mental models in the more 'ope' context of Microsoft Excel and Visual Basic, or in the platform-independent environment of the World Wide Web, may mean that they are encouraged to define explicitly the mechanism, scope, and role, of those models. The fact that the code used to define models is in the public domain along with the models themselves should be the basis of fruitful interaction between modellers, and between modellers and users of the models.

THE INFLUENCE OF THE WORLD WIDE WEB ON COLLABORATIVE MODELLING

The rapid expansion of the World Wide Web and the increasing numbers of teachers and learners with access to internet-connected computers provides new opportunities for the development of communications between model-builders and model users, and for the establishment of collaborative modelling projects. Dillon (1998), in reviewing educational developments in telematics, notes the emergence of projects in which the networking of computers is seen no longer solely as a means of dissemination of materials, but as a framework within which a community of learners share applications and work collaboratively.

One has only to visit one of the websites or Usenet newsgroups devoted to the Linux operating system or the Perl programming language to appreciate the extent of the collaboration, constructive criticism and mutual support within these online communities. Most critically, however, the interactions between community members amount to much more than invitations to view completed projects; rather, they include calls for assistance and collaboration, posting into the public domain of useful programs, modules and 'patches' to software, and broader discussions about the future scope and application of the operating system or language itself. Bruckman (1998) and De Bonte (1998a) both identify the same patterns of interaction between learners in their collaborative programming environments within which, 'open-source' is the norm: learners are encouraged to present their completed projects, but also to share programming experience and code with others. De Bonte (1998a) writes:

> Plenty of individuals are around, eager to share their knowledge and help work through problems ... Individuals who are skilled builders earn respect and admiration. Having a common goal gives members of the community something substantive to talk about, and an endless stream of projects they can work on together to construct.

It is not difficult to envisage, then, two tiers of collaboration between model builders and model users. Developers of modelling platforms have much to learn from the community aspects of the open-source movement. The World Wide Web needs to be recognised as much more than a means of publication of completed modelling projects: it can, with the appearance of 'groupware' products such as BSCW (Basic Support for Collaborative Work) (Bentley et al., 1997), also provide a platform for collaborative project development by geographically dispersed groups.

INVOLVING TEACHERS AS RESEARCHERS AND DEVELOPERS

The involvement of teachers as researchers and developers in such communities provides a vital link to a second tier of collaboration, with networking making the task of maintaining those links easier than has previously been the case. This involves teachers and learners in schools, colleges and elsewhere also working within 'open' environment at school or class level, or even informally in smaller groups. While not every school has the facilities to maintain an integrated and networked environment along the lines of 'Moose Crossing', it is possible for teachers to establish shared workspaces on school networks or even on single computers where computer models built in Logo or Excel, for example, or using a platform such as Stagecast Creator are freely accessible together with advice, documentation and evaluations rather like the Web-based open-source communities in microcosm. The most critical element in such an environment, however, is not hardware or software, but the establishment of a user community of what Kemp et al. (1998) call 'critical friends' to whose scrutiny learners are happy to submit queries, evaluations and models at different stages of development in addition to completed projects. Ultimately, successful collaborative development of models may depend on learners being prepared to share problems or even (echoing Swain's (1994) plea for the code for Apple's HyperCard) relinquish control of projects which they feel they can take no further. When learners are prepared to follow Raymond's advice that 'when you lose interest in a program, your last duty to it is to hand it off to a competent successor', then a community of computer modellers can truly be said to exist, and the prospects for the establishment of a 'modelling curriculum' will be much improved.

Section Three: Teaching and Learning Consensus Models

Preface

This Section examines the various ways in which the consensus models of science, or, perhaps more likely, curricular models derived from them, can be taught and learnt.

It begins, in Chapter 10, with a review of the nature of explanations and how different types of teacher explanations answer different types of questions and require different models and modes of representation.

As, in many classrooms, textbooks are used to present the consensus views of science, Chapter 11 examines representations of chemical kinetics and the atom using a Lakatosian perspective. It introduces the novel idea of 'hybrid' models which use parts of different scientific and historical models merged together as if they related to the same 'hard core'.

Chapters 13 and 14 deal with the consensus, curricular, and teaching models used in explanations of light and various biotechnological processes respectively. Chapter 13 considers the competing wave and particle models for light and colour. It shows how texts and teachers tend to delay teaching this duality, first dealing with the ray diagram model and using simple mathematical equations to explain light phenomena. Chapter 14 shows how teachers construct teaching models to bridge students' understandings towards consensus models and to mediate an understanding of the phenomena of small microbes, of microbial techniques, and of manufacturing processes.

Chapter 15 turns to classroom talk and uses an analysis of collaborative talk to examine how a dialogic from of classroom argumentation allows pupils to debate their expressed models for an Eclipse and to examine the scope and limitations of their models. It shows how there is a balance to be drawn between teaching the consensus models of science through didactic and questioning discourse. This balance allows pupils to debate their own models which the teacher may mediate towards the consensus view over a period of time through dialogic talk.

Learning consensus models of molecular structure through computer modelling is the topic of Chapter 16. It explores both the improvement in students' visualisation skills which result from their use of a computer package and the nature of teachers' perceptions of the effectiveness of that package. The context is that of students striving to understand consensus models of molecular structure in chemistry.

The penultimate Chapter in this Section is a discussion of the notion of 'models of pedagogy', which arose from an enquiry into how teachers explain their own classroom practice. A model of pedagogy is seen to have three component elements: a model of the nature of science; a model of the learning of science; a model of the teaching of science. Chapter 17 also discusses the significance of the coherence between these three elements for the model of pedagogy that a teacher holds and the consequences of mixed and incoherent combinations.

This last Section thus explores teaching and learning consensus models from a number of perspectives and with a number of ages and using a variety of projects. The significant themes which emerge are picked up and discussed in Chapter 18, where issues connected to the contexts within which teaching and learning takes place are examined.

Chapter 10

Explanations with Models in Science Education

John K. Gilbert, Carolyn J. Boulter, Margaret Rutherford
The University of Reading, UK

INTRODUCTION

Consensus models which are important outcomes of science play a major part in providing the explanations are at the heart of the scientific enterprise. The business of science may be said to be the production of explanations of the natural world. The problem with such a definition is that the nature of 'a scientific explanation' is left unaddressed. Martin (1972) has pointed out that the word 'explanation' can have any one of five meanings in science:

- the clarification of what a word means in a scientific context (e.g. 'friction is the force between two solid surface in relative motion whilst in contact');
- a statement of why some belief or action in science is reasonable (e.g. 'placing an oil film between two solid surfaces makes relative motion easier by reducing friction') ;
- a causal account of some state, event, or process, of interest to science (e.g. 'friction is caused by the interlocking at atomic level of two solid surfaces') ;
- the attribution of function to an object which is the object of scientific enquiry (e.g. 'the heart pumps blood around the arterial system');
- the citation of a theory from which a law may be deduced (e.g. 'the billiard ball theory of matter enables the Universal Gas Law to be mathematically deduced').

This last interpretation comes closest to the meaning most widely used in the philosophy of science, which is due to Hempel (1965). It is that an

J.K. Gilbert and C.J. Boulter (eds.), Developing Models in Science Education, 193–208.

explanation is the application of a law (a universal generalisation) under stated conditions which predicts, either deterministically or probabilistically, the behaviour of the phenomenon under enquiry. For example, the application of the Universal Gas Law to a enclosed volume of gas at a relatively low pressure, under the conditions that the temperature remains constant whilst the volume decreases by a specified amount, predicts (explains) the observed value of the increase in pressure. However, Bird (1998, pp.72-79) has pointed out the many weaknesses with this interpretation of explanation. Bird (1998) identifies two major problems. First, that cause and effect are often transposed in such a representation of an explanation, thus a decrease in volume is most commonly the result of an increase in pressure and not its cause. Second, that the approach gives no explanation for the actual occurrence of an unlikely event, thus 'a build up of electrostatic charge causes a lightning strike' but that explanation gives no account of when or where it actually happens.

Arriving at an agreed meaning of ' scientific explanation' seems to be difficult. If scientists and philosophers of science haven't produced a solution or set of solutions so far, why should science educators persist in trying to do so? The reasons have to do with obligation and purpose. Science education is now a compulsory part of the formal education of all young people in most countries. Science teachers have a obligation to make the best use of this opportunity. There are increasing calls for such an education to provide a secure foundation for the voluntary, non-formal, lifelong, science education of the global adult population (Millar and Osborne, 1998). This can only take place meaningfully if the school science education that is received is 'authentic' (Roth, 1995), i.e. it conforms as closely as possible to the actual practice of science. This implies that all aspects of the latter must, wherever possible, be made explicit, including that of explanation.

THE NOTION OF 'EXPLANATION' IN SCIENCE EDUCATION

If a definition of explanation is required, then it should be one which is 'authentic' and which can be readily used within the social arrangements (e.g. large classes, a fixed curriculum) which shape formal science education. It should also be capable of extension into non-formal provision. Building on our earlier work (Gilbert et al., 1998a) we suggest that *an explanation is the answer provided in response to a specific question*. The term *explanation* (meaning 'that which is provided') must be distinguished from the term *explaining* (meaning 'the physical act of providing an explanation') and the pedagogic devices used (Ogborn, et al., 1996). There are two 'parties' to any explanation: the 'questioner' and the 'explainer'.

Throughout the discussion below both of the terms 'questioner' and 'explainer' may apply to any participant in an educational discourse: a student, a teacher, or some object acting as a teacher, e.g. a textbook, a CDROM. However, the roles of question poser and explanation provider are all too often just a manifestation of the power distribution in a teaching/learning situation. As Scandamalia and Bereiter (1991) have pointed out, the initiative both in posing questions and in providing explanations all too often lies with the person or object in nominal charge (the teacher).

The *appropriateness* of an explanation is an overall evaluation of it, arrived at by conjointly considering three components. The *suitability* of an explanation is a statement about the relationship between the type of the explanation provided and the type of the question posed. The *relevance* of an explanation is a measure of the extent to which it meets the needs of the questioner. The *quality* of the explanation provided is a measure of its scientific standing. A judgement of appropriateness is initially made when the question is asked and the explanation is provided, but may be changed later in the light of further consideration of its value and significance. Such judgements are made by both the questioner and the explainer, although it is the view of the former that must prevail, given that the purpose of science education is the learning of science.

THE NOTION OF 'SUITABILITY' IN EXPLANATIONS

A Typology of Explanations

If the suitability of an explanation rests on the use of the correct type of explanation, then a classification of explanations is required. A five category typology of explanations can be constructed based on the nature of the questions which give rise to them (Gilbert, et al., 1998a):

- An *intentional* explanation is response to the question 'why is this phenomenon being explained?' The explanation will include a statement of the purpose being addressed and give some idea of the importance of the phenomenon. It necessarily also carries some definition of both the occurrence, the scope and boundaries, of the phenomenon. The work of researchers at the frontiers of science involves the frequent formulation, restatement, and reformulation, of intentions. An example might be 'to identify the nature and mode of operation of the AIDS virus, which now causes the premature death of a wide range of people in many countries, with the intention of enabling individuals and societies to take effective preventative or curative action'. Science education spends too much time, in our view, on phenomena where the intentional explanations have long ceased to be important. In any event, students are rarely introduced to intentional explanations, whatever their historical or contemporary scientific significance may be.

- A *descriptive* explanation is the response to the question 'what are the properties of this phenomenon?' Although philosophers of science (e.g. Hempel, 1965) do not usually classify descriptions as providing explanations, we have done so because they are often the first product of any initial investigation, whether in science or in science education. A descriptive explanation is a summary of the measurements that are made. A phenomenon which is apparently not changing is measured repeatedly to obtain a value, which is as certain as the measuring instrument will allow, together with an estimate of the error bias. Where the phenomenon is changing, several cycles or examples of the change will be monitored to obtain values of a parameter at convenient time intervals. Much of primary school science education, where phenomena of interest to science (and, we hope, to the students!) are first met, consists in producing or being introduced to simple descriptive explanations.
- An *interpretative* explanation answers the question 'of what is the phenomenon composed?' Such an explanation often postulates the actual existence of entities which are incapable of direct observation, what Harré (1986) has called 'policy realism'. Statements are made not only about their nature but also about how they are distributed in space and how that distribution changes with time. For example, an interpretative explanation of the allotropy of tin is that an allotrope consists of atoms of tin which exist in a particular geometrical arrangement, that arrangement making a specific change at a specific temperature (a 'transition point'). School chemistry must seem to students to consist almost entirely of interpretative explanations: abstraction starts early.
- A *causal* explanation is a response to 'why does the phenomenon behave as it does?' A mechanism is proposed by means of which the phenomenon produces the observed behaviour through the operation of cause and effect on the entities of which it is composed. A mechanism can operate either deterministically, in that one cause produces a single, definite, event: for example, the persistent application of sufficient force to an object will cause it to accelerate. The mechanism may operate probabilistically, in that one cause may produce many, possible, events. For example, a change in the physical environment may cause the reproductive status of

individuals within a species to alter. This is because there are a number of reproductive characteristics, randomly spread across the population, which are relevant to that environmental change. Science places the highest value on causal explanations (Deutsch, 1997) and hence so should science education.

- A *predictive* explanation answers the question 'how will the phenomenon behave under other, specified, conditions?' This might be thought of as a subset of descriptive explanation. Again, philosophers tend not to include it as a separate type of explanation because it is derived from and produced after interpretative and causal explanations (Deutsch, 1997, p.8). For example, the rate of a particular chemical reaction at some future moment in time can be successfully predicted because models of chemical kinetics have identified both the nature of the entities involved and have postulated the mechanisms by which change takes place. However, we have included it because predictions play an important part in the evaluation of the explanatory adequacy of theories and models and hence are of paramount importance in science education.

All explanations make extensive use of models. This is because the latter are simplified representations of specific aspects of phenomena produced in order to facilitate that production of visualisations which is at the heart of addressing any particular scientific purpose. Indeed, an explanation is often just the verbal presentation of a suitable model. As was stated in Chapter 1, there are five possible representational modes that a model can take. An example of the *concrete mode* is the plastic and wood small-scale representation which has been built of the facilities available at the Olympic Games 2000 at Sydney. Examples of the textual (or symbolic) modes are the mathematical equations used to represent the flow of visitors around the venues for the events at the Games (in the *mathematical mode*), the graphic schematic diagrammatic map which shows, in a very formal, non-geographical, way how these venues are linked together by the public transport system (in the *visual mode*) and the *verbal mode* used by the tour guide to describe the layout of the facilities. The tour guide may use her/his arms (in the *gestural mode*) to enhance this description.

Concrete models are particularly helpful in the provision of descriptive explanations. Thus a model of a jet engine, which is much smaller than the original, enables an explainer to give an account of how fuel and air are mixed and ignited in the combustion chamber, such that the resulting gases are propelled out at high temperature and pressure through the exhaust pipe. A model of the human body, which is the same size as the original, with the blood vessel system emphasised and the two 'types' of blood colour coded, enables the flow of blood around the system to be explained. A model of a human cell, which is much larger than the original, enables an explanation to be given of the way that a virus enters it though the cell wall.

At the opposite end of complexity of explanation, the symbolic modes (the mathematical, visual, and verbal modes) and the gestural mode are of most use in providing predictive explanations. In a mathematical model, this is because all the variables thought to be salient are included, such that the behaviour of the whole equation can be derived by changing the value given to one variable. In particular, predictions can be made of behaviour under extreme conditions. For example, the universal gas equation (PV=RT) was devised as being applicable to systems at low pressure and at near room temperature. After predicting the consequences for volume of a progressive increase in either pressure or temperature, the predictions can be empirically tested. The point where the predictions are not borne out in practice indicates the limit of the applicability of the Universal Gas Law.

There are many sorts of graphic schematic representations. In a diagram, the entities depicted can be any of objects (e.g. parts of a car), events (e.g. the stages in construction of a car), or systems (e.g. the transmission train of a car). The links between the depicted entities can be of any propositional form, e.g. ' is next to' (the arrangement of sub-assemblies in a car engine), 'changes to' (stages in the production of a car part), 'is joined with' (the linking of car parts to form operational sub-assemblies). This versatility of diagrams is particularly useful in facilitating interpretative and causal explanations. Entities can be imagined and the causal relations between them shown visually. To take a different example, a number of types of meteorological map are available. They are built around a range of entities, ranging from abstractions, e.g. 'fronts', 'isothermals', to pictorial evocations, e.g. ' clouds with rain coming out', 'bolts of lightning'. This range of representational conventions, which are often mixed in a forecast, enables the weather to be presented to and discussed with diverse audiences (Lowe, 1993).

The Suitability of Different Types of Explanation

The choice of a suitable explanation, where there is correct match between type of explanation and type of question, initially depends on the nature of the question asked. For example, a suitable explanation produced in response to the question 'Of what is this material made?' (a question about entities) is 'It is thought to consist of ...' (a descriptive explanation). The production of a suitable explanation depends on the clarity of the signal given by the question. Clarity is manifest in two ways: the *explicitness* and the *scope* of the question.

Explicitness has to do with the physical obviousness of the question. A question can be posed either *overtly*, where it is actually stated (e.g. where a cause is asked for) or *covertly*, where it is implied by the structure of the discourse (e.g. where the need for a causal explanation is implied by the provision of an interpretative explanation followed directly by a predictive explanation). An overt question might be expected to more readily trigger an explanation of the corresponding type with, for example, a 'made of what?' question leading to an interpretative explanation. A covert question can produce a suitable explanation and can be the prelude to an overt question: thinking it just comes before saying it. However, too many unstated questions can lead to misunderstandings because explanations are not always supplied: the questions are not perceived by the explainer.

Scope has to do with the degree of specification of explanatory type that a question contains. Questions can be either of an '*open* ' or a '*closed*' type (Barnes, et al., 1969). A broad range of explanatory types, rather than just one, might be produced in response to the former because of the lack of precise cues as to what is expected. For example, the question 'what is an acid?' might produce answers corresponding to many of the types of explanation outlined earlier. However, the type of explanation expected in response to any one of a series of 'closed' questions would be easier to anticipate because the cuing provided is higher. Thus, 'why do we study acids?' should produce an intentional explanation (for example, 'because they play an important part in the digestion of food'), 'what happens when pieces of metals are dropped into hydrochloric acid?' should produce a descriptive explanation ('the metals will react at varying speeds, producing bubbles of a gas'), 'what is common to all aqueous solutions of acids?' should produce an interpretative explanation (for example, 'they contain solvated hydrogen ions'), 'why do metals react with acids?' should produce a causal explanation (for example, 'because one or more electrons are transferred from a metal atom to a hydrogen ion, producing a soluble metal ion and, initially, a hydrogen atom'), and 'what will happen when pieces of one type of metal are dropped into a series of different acids of similar concentration?' should produce a predictive explanation ('the rates of gas production will vary, dependent on the strengths of the several acids').

THE NOTION OF 'RELEVANCE' IN AN EXPLANATION

What needs might a questioner perceive him/herself to have? These might be classified as 'extrinsic' or 'intrinsic'. Extrinsic needs are those where the questioner perceives that some explanation, apparently valued by the explainer who is in a position of authority, has to be learnt. In short, there is a formal curriculum to be covered. This curriculum is nowadays centrally prescribed in an increasing number of countries. In an age where the accountability of education is manifest in the results obtained in public tests and examinations, a questioner will value those explanations which are likely to be assessed. These in turn will be governed by what can be validly and reliably assessed, leading indirectly to a emphasis in teaching on descriptive, interpretative, and causal explanations, and to the neglect of intentional and predictive explanations.

The relevance of an explanation to the 'external' needs of the questioner is, to a surprising degree, determined by the social context in which it is received. The contexts of a physics class and a chemistry class all too often seem to lead to an anticipation of different responses to apparently similar questions. For example, questions about colour in a 'physics' context produce causal explanations in terms of the wavelength of light, whilst a 'chemistry' context implies causal explanations involving the excitation of electrons in atoms or molecules. The nature of the entities discussed, the focuses of the assumed interpretative explanations, are different in the two cases. Producing a 'chemistry' explanation in a physics class is less relevant to the needs of questioners.

Intrinsic needs are those where the questioner wishes to *know* or *understand* something for its own sake. Learning is preferentially focused on phenomena in which the questioner has become genuinely interested. A 'typical' young person as a student will have a genuine desire to know how the natural world works and to be able to gauge the extent to which the environment can be managed and changed. Such a student will therefore judge intentional, causal, and predictive, explanations of valued phenomena to be relevant to personal needs. However, there are intrinsic needs which are even more compelling. These have to do with the 'affective engine' (Watts, 1998) which drives a wish to understand *whatever* is the focus of social attention at the time. The affective engine has been described in several different ways. Piaget (1954) saw it as the drive to achieve 'equilibration', the incorporation of new information either within existing mental schemata (assimilation) or within adapted schemata (accommodation). Ausubel (Novak, 1998) describes this as the achievement of 'meaningful' learning. Kelly (1955) saw it as the modification of core constructs until mental control and prediction of the environment is achieved.

The needs of a questioner may be perceived by an explainer in comparable terms to those above. In these circumstances, there is a good likelihood that a relevant explanation will be provided. However, an explainer can project a personal vision of 'a questioner's needs' onto that person. The explainer may know an explanation which is much valued and which it is thought a questioner 'ought' to know. This is what is supplied, rather than that which is relevant in the mind of the questioner. For example, an Einsteinian explanation may be provided to a question where a Newtonian explanation would suffice. The explainer may unwilling to supply a relevant explanation because it would take too much time or because it lies outside the 'subject' of the class. For example, a teacher of biology may not be willing to provide a 'physics' explanation, a less-relevant 'biology' alternative being provided instead. The treatment of questions about 'energy' seem particularly prone to this type of response. The explainer may be unable to supply a relevant explanation because it is not part of the repertoire of pedagogic subject knowledge (Shulman, 1986a) available to him/her. For example, a 'physics' teacher may not know enough about Darwinian evolution to be able to talk anthropomorphically about notions of chance. A mathematical treatment may be given instead. In such cases, a questioner will judge the explanation to be less relevant to her/his needs.

THE NOTION OF 'QUALITY' IN EXPLANATIONS

The quality of an explanation in science education is directly related to its status in science *per se*. The nature of possible relationships between the history of science and the science curriculum continues to receive attention (see Duschl, 1994, for a review). One form of relationship can be based on the historical models which are major landmarks in the development of any scientific research field (Justi and Gilbert, 1999a). The issue then becomes: how can the quality of the explanation provided by any model in a sequence of models in the history of science be judged? The ideas of Toulmin (1972) suggest four criteria:

- First, its 'plausibility'. To what extent: did it fit with what was already known about the field?; did it provide explanations for problems which were seen to be important at the time?
- Second, its 'parsimony'. To what extent did it use fewer and more wide ranging concepts than its predecessor?

- Third, its 'generalisability'. To what extend did it provide explanations for a broader range of phenomena than its predecessor in the field?
- Fourth, its 'fruitfulness'. To what extent did its use lead to a greater number of successful predictions than its predecessor? Rational curriculum design becomes possible when answers to these questions are available.

The designers of national curricula in a school science subject will produce a requirement which is governed by a number of factors: their own experience of science and of teaching, their views of the nature of the scientific enterprise, their judgement of what constitutes valuable scientific knowledge, and their perceptions of what it is desirable and possible for students of any given age to learn (Roberts, 1995). Curricular requirements often seem to be built around a listing of phenomena to be studied. For any phenomenon, the issue for teachers is then one of deciding on the quality of the explanations which are to be learnt. Which one of a series of models in an historical sequence should be studied? The choice often made is to pick the most recent model in a sequence (the consensus model), but such a model often implies more background knowledge than the students possess. The decision about which model to choose is limited by two additional issues: do teachers in general already understand it or can they readily come to understand it (Shulman, 1986b)? Is there a reasonable hope that it will be intelligible to the majority of the students to be taught it (Strike and Posner, 1992)?. Choosing a model for any given place in a science curriculum is a matter of striking a balance. Too 'early' a model in a sequence may not be able to carry the explanations which are required, whilst too 'late' a model in the sequence may not be taught and/or learned correctly.

THE NOTION OF 'APPROPRIATENESS' IN AN EXPLANATION

The interplay of suitability, relevance, and quality, in the explanation provided in response to any given question is subtle. Although the evaluation should be made on a continuum, it is likely that one of three simple judgements may be made by the questioner:

(a) The explanation is appropriate

The explanation is *suitable*. The meaning of the question was the same for both the questioner and the explainer. They both assume that a given type of explanation should be provided and that is what happens. For example, a 'why does that happen?' question produces a causal explanation. The mode of representation in which the core model is put forward facilitates the type of explanation involved. For example, the use diagrammatic modes may readily enable a causal explanation to be presented. The explanation was *relevant*. The explanation meets the extrinsic needs of the questioner to a high degree. The demands of the curriculum are met, so that the questioner can readily see how the explanation fits into assessment requirements. The explanation meets the intrinsic needs of the questioner to a high degree. The questioner finds the explanation interesting. The explanation can be incorporated, perhaps after some mental effort, into existing schemata: equilibration can be achieved (Miller, 1996, p.226). The explanation is of a suitable *quality*. The historical model used is such that it can provide the explanation required without needing unnecessary additional information to be learned.

(b) The explanation is inappropriate

This could have arisen for any one or combination of a number of reasons. The explanation may not be *suitable*. The meaning of the question might have been different for the questioner and the explainer. The form of the explanation chosen by the explainer did not permit the question, as understood by the questioner, to be addressed. The mode of representation of the core model used might not have been the most facilitative. The explanation may not have been *relevant*. The scope for failure is high here. The questioner may not: be able to see how the explanation relates to curriculum specifications; find the explanation interesting; be able to achieve equilibration after the explanation is given. Finally, the *quality* of the explanation provided might have been either too low or too high.

(c) A non-explanation is provided

This is the extreme case, with several wrong choices and failures of communication having occurred. The questioner may see no connection between the question that was asked and the explanation received. This may have been because the question was grossly imprecise in formulation, or because the explainer could not connect its expectations with her/his 'bank' of explanatory responses. A model may have been presented in a mode which was non-facilitative, e.g. the use of a symbolic mode when the question called for a descriptive explanation. The explanation may have been found irrelevant to anticipated assessment requirements, most uninteresting, or overly difficult to comprehend (or all three!). The historical model chosen to 'carry' the explanation may have been unable to do so.

Raising achievement levels in science education, in the broadest sense, requires an improvement in the appropriateness of the explanations that all students receive. This entails, for many teachers and students, the provision of explanations which are suitable, are more relevant, and which are of a more carefully judged quality.

IMPROVING THE SUITABILITY, RELEVANCY, QUALITY OF EXPLANATIONS

Achieving Suitability

This involves attention to the nature of the questions asked, to the nature of the explanations provided, and finally to the making of a match between them.

A prerequisite for good questions, i.e. ones for which suitable explanations might be provided, is that the questioner and the explainer have common purposes. All too often students have little idea of what is to be learnt, why something is to be learnt, how that learning will be assessed. The provision of basic information may help to overcome some of these problems. A copy of the syllabus, including the schedule of assessments and the scheme of work to be followed, in some form will help many students, especially if it is interpreted to them to allay their anxieties. However, it is essential that the issue of *why* they are learning something is taken up with them, that they receive intentional explanations of the curriculum.

There is a literature on questioning in science classes which emphasises the significance of the opportunity to ask questions (Graesser and Person, 1994) and of the context in which they are posed (Carlsen, 1991) for their value in enabling students to make their concerns known. There are practical guidelines available for teachers on how to produce questions whose meaning is evident, how to vary the types of questions asked, how to sequence these so as to achieve preconceived goals of understanding, and how to 'ring the changes' for students of different achievement in science (e.g. Brown and Edmondson, 1984). The prerequisites for an improvement in questioning thus seem to be already available.

The corresponding prerequisites in respect of explanation have recently become clearer, although the principles of explanation are not yet widely known. In one of the few research studies carried out so far, Dagher and Cossman (1992) analysed the explanations provided by 20 teachers in Grade 7/8 classrooms in the USA. A review of this work (Gilbert et al., 1998b) showed that, in addition to the use of explanations which can be mapped onto the typology outlined earlier in this Chapter, 'spurious' explanations (e.g. 'metaphysical', 'anthropomorphic') and 'atheoretical' explanations (e.g. 'practical', 'tautological') were also provided. These suggest specific gaps in what Shulman (1987) calls 'pedagogic subject knowledge', the ability to transform and present subject matter in such a form that it can be understood by students.

It is only when the principles of questioning and of explaining are widely known that strategies for achieving a match between them can be systematically produced.

Improving Relevance

Improving the extrinsic relevance of the science curriculum, beyond the simple measures of making it known, is no simple matter in any country. The introduction of National Curricula have 'frozen' curriculum development processes, with the substance of those curricula all too often being 'hereditary' a recapitulation of what has been done in the past. However, there are voices for change, e.g. Atkin and Helms (1993), Rutherford and Ahlgren (1990). It will perhaps only be when the flood of students away from the study of science is overwhelming, a situation which seems to be happening in physics, that something will be done.

Strategies for the general improvement of the intrinsic relevance of explanations cannot be rehearsed here. Suffice it to say that there are well-defined approaches available, each quite properly attached to a theory of the nature of learning, e.g. associated with the ideas of Piaget (Adey and Shayer, 1994), Ausubel (Novak, 1998), Kelly (Pope and Keen, 1981), and Lave (Rogoff, 1990).

Achieving Adequate Quality

The historical models in a given field of enquiry are often introduced to students in the order in which they were first devised. The curriculum is shaped so that the explanatory demands placed on students at a given point in their course are met by a particular historical model. Whilst this 'recapitulary' approach can be implemented in such a way that the circumstances of the invention and abandonment of a sequence of models is retained (as was done, for example, by Hoeve-Brouwer and de Vos, 1994), detailed knowledge of such sequences across a broad sweep of the science curriculum is not yet generally available. However, considerable progress has recently been made on approaches to the identification of the historical models in a range of different fields (Justi and Gilbert, 1999 a, b) whilst ways in which these can be incorporated in the science curriculum are being explored (Justi and Gilbert, in press, Justi, this volume).

Another way in which quality can be improved is through improved attention to the presentation of different models. Ogborn, et al., (1996) have recently published an analysis of a series of case studies strategies for *explaining* shown by a range of secondary science teachers in the UK. These teachers used an admirable range of tactics to generate the need for explanations, to construct the entities with which to provide these explanations, and then to use the notion of 'teaching models' in an attempt to make them fully intelligible, all within personal (but generalisable) 'explanatory styles'.

The notion of 'teaching model' as a way of raising the quality of explanations has come into prominence in recent years, e.g. Dagher (1985a). The basis for such work has been close attention to the selection of an historical model, from a sequence, for teaching. Mayer (1989) has put forward six criteria. A model selected should be:

- Complete, in that the entities of which it is composed and the relationships between them are already understood, or can be readily learned, by the students.
- Coherent, in that the level of detail of explanation it provides should match the needs of the students.
- Concrete, in that the operation of the model should be within the range of comprehension of the students.
- Conceptual, in that the model should form a clear bridge between the underlying theory and the phenomenon being explained.
- Correct, in that the scope and limitations of the model in representing the phenomenon are made clear.
- Considerate, in that the model is linguistically well presented.

The use of such criteria does enable a model of quality which matches the general attainment of the students to be chosen. They also enable a model of somewhat greater demand to be identified, so that the quality of explanation provided can be raised.

In the latter circumstance it is to be expected that students will value an intellectual 'bridge' between their existing explanation and that now required of them. This is where teaching models come in, whether devised and/or used by the teacher (Treagust et al., 1992; Dagher, 1995a,b) or by the students themselves (Wong, 1993; Cosgrove, 1995). A valuable teaching model is one which:

- contains entities and relationships which approximately match, both in number and in degree of similarity, those found in the historical model which is the focus of attention;
- serves as a bridge to an historical model which students both believe to be important and find difficult to understand;
- is based on a source with which the students are fully acquainted, preferably an a direct experiential level;
- can be used in combination with other teaching models in respect of a given historical model.

A broad reading of the literature suggests that, whilst such models can be very useful in both consolidating and in raising the quality of explanations, they do have a number of pitfalls. A teacher should only introduce a teaching model where there is evidence that it is needed (Abell and Roth, 1995) rather than in association with every historical model. Students must be familiar with the source of the teaching model, which is often a matter of socio-cultural background, otherwise it will be unintelligible. Finally, there is a risk, if they are not told its scope and limitations, that students assume too high a degree of similarity between a teaching and historical model: they learn incorrect information.

MODEL CHOICE AND EXPLANATORY APPROPRIATENESS

As was discussed in Chapters 1 and 2, a model is produced in science for a purpose. It is to forge a link between a theory and a phenomenon, to enable the phenomenon to be explained in terms of the theory. If the purpose for providing that same explanation in science education is the same as that which originally prevailed in science, then the same model or one of its successors may be used. In this Chapter we have discussed two problems that may arise. First, there may not be an sufficiently precise identification of the explanatory purpose being addressed to enable a model to be selected with any confidence. Second, even if the appropriate model is chosen, then it may not be used to give a suitable explanation.

The consequences of providing an appropriate explanation are fairly evident. The student will achieve a personal understanding, a condition which is a prerequisite for further learning with an efficient use of time and resources. The motivation to learn will be at worst maintained and at best improved. If an inappropriate explanation is provided, then the flow of a student's developing understanding will be halted or deflected. Further time and effort will be required on the part of both teacher and student if systematic learning is to recommence. There is a risk that motivation may decline, although individuals' responses to such adversity vary widely. Some will perceive it as a challenge to be faced, others the signal to abandon the attempt to learn. Non-explanations are the worst: the learner will have no landmarks against which to judge their intellectual position. A sharp decline in motivation seems much more likely. The selection of a relevant model and the provision of an appropriate explanation are central to the maintenance of active involvement in the learning of science. Efforts to improve appropriateness in respect of all the explanations provided must therefore, we suggest, be central to both the initial and in-service education of teachers.

Chapter 11

Teaching with Historical Models

Rosária S. Justi
Universidade Federal de Minas Gerais, Brazil

INTRODUCTION

As the first section of this book has shown, science is a special branch of knowledge that includes not only models and theories but also the processes through which scientific knowledge is produced. Assuming the relevance of scientific literacy to the citizens of the twenty-first century, it is important that their education aims at an understanding of (i) the structure of scientific knowledge, (ii) what the conduct of science involves and (iii) the limitations of scientific knowledge in a given context, as well as the development of the skills with which to develop reliable scientific knowledge. In Hodson's (1992) words, the purposes of science education should be learning science, learning about science and learning to do science.

In most science courses all over the world, the emphasis has been placed only on the conceptual nature of science. Moreover, such concepts are considered as absolute truths that were empirically 'discovered' in a linear and context independent manner. On account of this, students are generally asked to memorise the specific facts presented to them without questioning either their development or the relationships to other scientific or non-scientific knowledge.

In order to change the focus of science education so that it can produce a comprehensive understanding of science, many educators and philosophers of science have been advocating the introduction of the history and philosophy of science (HPS) in science education. Their main arguments are:

J.K. Gilbert and C.J. Boulter (eds.), Developing Models in Science Education, 209–226.

1. HPS would contribute to the motivation and engagement of students in science teaching (Golin, 1995; Matthews, 1992a; Monk and Osborne, 1997).

2. HPS would contribute to a better understanding of scientific knowledge and its development in specific contexts (Demeo, 1992; Matthews, 1992a, 1994; Monk and Osborne, 1997; Nersessian, 1992a; Stork, 1995).

3. By interesting students in historical and philosophical questions, they would exercise critical and reflecting thinking, thus making science more challenging (Bent, 1971; Cohen, 1994; Matthews, 1992a, 1994).

4. HPS would promote the understanding of the nature of science (Betts, 1992; Carson, 1992; Golin, 1995; Kipnis, 1996; Matthews, 1992a, 1994; Monk and Osborne, 1997).

5. HPS would dispel the myth about both the linear development of scientific knowledge and the status of 'genius' generally attributed to scientists (Brush, 1989; Betts, 1992; Feldman, 1995; Golin, 1995; Matthews, 1992a; Nielsen and Thomsen, 1990; Solomon, 1989; Weck, 1995).

6. HPS would promote an interplay between different areas of science and between science and other academic disciplines (Matthews, 1994; Nielsen and Thomsen, 1990).

7. HPS would contribute to making science teaching more humanised, to encouraging social perspectives and to making clear the interplay between science, social changes and technology (Carson, 1992; Golin, 1995; Matthews, 1992a).

8. Assuming the recapitulation perspective, HPS would both enable teachers to better understand students' naïve ideas and help students to change their ideas (Monk and Osborne, 1997; Nielsen and Thomsen, 1990; Stork, 1995).

By having such arguments (or some of them) in mind, many researchers have proposed ways through which HPS could be introduced into science education. Some of them consider HPS as an aid in the learning of science, whilst for others it can help in the learning about science (Matthews, 1992b).

Moreover, their proposals for science courses can be viewed as constituting a continuum with regard to the ways in which HPS is introduced. This continuum can be illustrated by looking at science courses in which:

1. Discoverers' names and dates of discoveries are reported.
   Nussbaum (1983) and Allchin (1995) recognise that this is a superficial approach because the most fundamental ideas, models, theories, and philosophical commitments are not discussed.

2. Brief biographies of scientists are introduced.
   Golin (1995) used to do this in his physics courses at Russian universities because he believed it was the way to promote a deep understanding of physics. On the other hand, Matthews (1994) relates the study of biographies of individual scientists to the humanisation of the subject matter of science. In the same perspective, Kauffman (1989) advocates the introduction of biographical material as a way to motivate students to study science.

3. Outlines of earlier ideas and some episodes in the history of science are presented.
   Many textbooks present them in boxes, i.e. excluding them from the main text. This may often mean that they are not read by students. Matthews (1992b, p.121) asserts that this means a 'minimalist interpretation' of the ways that the HPS should be introduced into science teaching.

4. Ideas expressed by students are compared to historical ones in an attempt to provide evidence that the former is not adequate and should be replaced.
   Wandersee (1985) made this proposal based on the result of a study of the parallels between students' conceptual difficulties and the historical evolution of the concept of photosynthesis.

5. A historical investigative approach is used, that is, experiments are conducted in their historical sequence, although with the use of modern apparatus that make them simpler, cheaper and less time consuming.
   Ellis (1989) and Kipnis (1996) describe their experiences in using historical experiments in chemistry and physics courses. For them, such a teaching methodology should be used because it favours students' understanding both the art of scientific investigations and how scientists produce knowledge.

6. Dramatisations of historical debates are performed.
   According to Solomon (1989), besides the motivation generated by the involvement in such an activity, it improves students' ability to deal with different arguments and to gauge different interpretations in specific contexts.

7. Historical and philosophical perspectives are conveyed in introductory science courses.

8. Arons (1988) provides examples of historical perspectives which he realised as intelligible to students. He claims that such a proposal both enhances students' scientific literacy and brings about a comprehensive understanding of scientific content itself.

9. Historical case studies are introduced, sometimes using original writings.
   For Allchin (1995) and Hodson (1988), the importance of the introduction of historical case studies is related to the learning of processes of science. This is because the historical context of a case would be highlighted. A more recent experience with the introduction of historical case studies in physics teaching is reported by Nielsen and Thomsen (1990). They produced textbooks for Danish upper secondary physics teaching in which selected excerpts from original materials are introduced together with texts providing the necessary historical and conceptual background.

10. A history line for the science content of a subject is created from the use of the history of science.
    According to Matthews (1992b), many science educators argue for the story-line approach because stories provide the coherence and linkage that both make students more interested in science and make science more understandable to them.

11. The overall curriculum or some of its topics are organised from a historical viewpoint.
    The Harvard Project Physics course, developed in the United States as part of a broader 'history of science' program, was the most well-known example of a serious and comprehensive application of history of science to science education (Duschl, 1994; Matthews 1992a).

As can be deduced from the analysis of these courses, some fit traditional ways of conceptualising the curriculum. Others are non-traditional in the sense that they require students to think about the process of construction of scientific knowledge. The different ways of introducing the HPS into science teaching imply different views of learning.

Independently of the approach adopted in any course, science teachers are an essential element in the introduction of HPS into science education. Therefore, it is of pivotal importance both that they understand and agree with the arguments in favour of the introduction of HPS into science education and that they have appropriate knowledge in order to be able to effect a successful introduction (Justi and Gilbert, 1999a).

By investigating the teaching of some topics in chemistry, the research presented in this Chapter suggests a new argument in favour of the introduction of HPS into science teaching and proposes a new approach to be adopted in such an introduction.

HISTORICAL MODELS

In this research, a *historical model* is considered as a consensus model developed in a specific context. Within this definition, a context is taken to mean a system of philosophical, scientific, technological and social beliefs. This implies that a historical model is not necessarily a model developed by an individual scientist, nor that its production and use is situated within a specific time period. The most important issue is that it achieved consensus status within a particular context (Justi, 1997).

In previous research (Justi, 1997; Justi and Gilbert, 1998b), historical models have been characterised by the use of a framework developed from Lakatos' notion of scientific research programmes (Lakatos, 1974). According to Lakatos (1978), a research programme can be characterised by the possession of a *hard core*, the major assumptions that identify it and that guide all who work within a given research programme. Each research programme would also have a *protective belt*, auxiliary hypotheses that protect the hard core from refutations, and a *positive heuristic*, a set of suggestions about how to modify the protective belt.

In the research discussed here, two elements play the role of the hard core: the *theoretical background* and the *main attributes* of the historical models. The theoretical background corresponds to the general scientific and philosophical ideas on which the model was/is based, as well as the analytical tools used in its construction. In this way, the theoretical background characterises the context in which the model was/is developed. The main attributes are the fundamental scientific ideas specific to the subject of the model. On the other hand, the protective belt is associated with the *secondary attributes*, ideas that complement the main attributes to permit a comprehensive characterisation of each model. Each secondary attribute may be discussed independently of the other, but all of them are directly related to the main attributes of a given model. In the definition of the historical models of chemical kinetics, for instance, the theoretical background corresponds to the concept of matter on which the model is based and the mathematical and statistical tools used in its construction; the main attributes comprise the meanings of chemical reaction, reaction rate and the determinants of them; and the secondary attributes are the ideas about catalysis, reaction path, and those related to the influence of energy on the rate of a chemical reaction.

Moreover, according to Lakatos (1974), a new research programme would be accepted when it supersedes a previous one. However, this would not be an immediate consequence of a 'crucial experiment'. The elimination of a research programme would be an evolutionary process in which the protective belt suffers defeat and the hard core must be changed. In other words, the overthrow of a given research programme results from a competition between the 'progressive problemshifts' of a new research programme and the 'degenerating problemshifts' of the original one. In characterising historical models some points were systematically investigated in order to probe such a competition. These were:

1. the deficiencies in the explanatory capability of a given model;
2. the features of that given model that were modified and incorporated into a new model;
3. the way by which the new model overcame the explanatory deficiencies of this antecedents;
4. the explanatory deficiencies of the new model (Justi, 1997).

By using such an approach, the historical development of two scientific topics, chemical kinetics and the atom, were analysed and the historical models of both topics were characterised (Justi and Gilbert, 1999b). The use of the same framework in the analysis of teaching situations produces very interesting conclusions about the use of historical models in science teaching.

ANALYSIS OF TEACHING SITUATIONS

The teaching of the atom was investigated through an analysis of Brazilian and English textbooks for 14 to 16-year-old students (Justi and Gilbert, 1999b, in press). On the other hand, the teaching of chemical kinetics was probed in a case study conducted in a specific Brazilian class. In this case study, the models expressed by both the teacher and the textbook were analysed, as well as students' ideas about such models (Justi, 1997). In both studies the models expressed in teaching were analysed in the belief that it is important to introduce HPS into science teaching in such a way that could promote a meaningful and comprehensive learning of science. Moreover, the models expressed in the teaching of each topic were related to the historical models characterised in each case. Some characteristics were found in common:

1. Neither the teacher nor the textbooks present scientific knowledge as consisting of provisional models which are developed and 'valid' in specific contexts. In general, chemical content is considered from an empiricist point of view and scientific knowledge is approached from an absolutist way, that is, it is presented as true and confirmed. There are no references to the development of scientific ideas as an evolutionary and continuous process. Even in the case of the teaching of the atom, of which some historical aspects are generally presented, a coherent discussion about how and why each of the presented models were provisional was not found.

2. The achievement of consensus status by each model is not discussed. In both cases there are no discussions about the contexts in which each model was/is valid. In the case of the teaching of the atom, an experiment whose analysis is based a model is sometimes presented. Notwithstanding, the authors do not discuss which philosophical or scientific ideas underpinned the analysis of such an experiment and make the model coherent. They do not discuss why the use of other philosophical or scientific ideas, which could exist at the same time, was not able to contradict a given model. There are thus no discussions concerning any aspect of scientific knowledge from different points of view.

3. The theoretical background on which each model is based is generally not clearly discussed. In the case of the teaching of chemical kinetics, for instance, the meaning of the Maxwell-Boltzmann distribution law, which is essential for the understanding of the influence of temperature on reaction rates in both the

Thermodynamics and the Transition State models (Justi and Gilbert, 1999b), was not explained to students.

An analysis of the models expressed in teaching using the framework developed to characterise historical models showed several interesting aspects:

1. A mix of different theoretical backgrounds is used as a basis for the main and secondary attributes of the models used in teaching.
   Initially it could be argued that this amalgam of theoretical backgrounds could contribute to the development of an overall view of the scientific themes by the students. However, this does not seem to be the case. As it was previously stated, scientific knowledge was not characterised as producing provisional models developed in an evolutionary and continuous process. Different theoretical backgrounds were presented without any discussion of the differences in the contexts in which they had been developed and in which they were valid. That is exactly the problem with the use of different theoretical backgrounds. Students thus cannot understand why a given aspect is analysed from distinct theoretical backgrounds. In the teaching of chemical kinetics, for instance, the Maxwell-Boltzmann distribution law is introduced, without any explanation about why it is needed, to underpin some secondary attributes related to the energy involved in a chemical reaction. As some students pointed out, it is not clear why it is introduced if they can already explain such attributes from the point of view of the collision theory (Justi, 1997).

2. The main attributes of the models expressed in teaching are associated with those of distinct historical models.
   This could be a very interesting contribution to the development of an overall view of the scientific themes by students. The problem is that in each of the historical models, the secondary attributes are closely and clearly related to the main attributes, so that there is a coherence between secondary attributes and the protective belt of a hard core. This cannot be seen in the textbooks analysed. This is because the authors discuss each specific attribute according to the approach they think is suitable for their immediate purposes.
   In the teaching of chemical kinetics, the authors, for instance, discuss the concept of activation energy from the point of view of the collision theory. However, when they say there is a species called 'the activated complex', they add elements of the transition state theory to the explanation. But 'activated complex' and 'transition

state' are different concepts, formulated from different theoretical backgrounds. The transition state was defined within the Transitional State Model as the configuration of atoms whose potential energy increases with deformation in all direction except that of the reaction path, whilst the activated complex, as defined by Arrhenius, is a real molecule in which the potential energy increases for all types of deformation.

3. Attributes of a given model are presented as if they were part of another model.
   In the teaching of chemical kinetics, there are two examples that show the disconnectedness between main and secondary attributes in the model used in the teaching.
   The first one is concerned with the utilisation of the energy profile diagrams, which were originally developed from potential-energy surfaces in the context of the transition state theory. In such diagrams, the energy of the system is plotted against the interatomic distances to give 'potential surfaces'. Thus, different positions on this surface mean that species have particular energies. Moreover, the activation energy can be obtained from the lowest point over these 'energy hills'. These potential-energy surfaces thus provide a valuable pictorial representation of the course of a chemical reaction. Maybe because of this utility, both the teacher and authors of textbooks present such diagrams as representations of the reaction path without explaining them or making any reference to their origin. On the contrary, they present them as if they were a diagrammatic representation of what is predicted by the collision theory. This is done by associating pictorial representations of molecular models with an energy profile diagram. Such a picture strengthens not only the idea that the energy profile diagram is a representation of the predictions of the collision theory, but also that a transition state is a concept within the scope of the collision theory. This means that elements of different models are merged as if they were within the same hard core.
   The second example is the teacher's use of the idea of affinity between substances within the discussion of collision theory. This was observed when he discussed affinity between substances as one of the conditions for the occurrence of a chemical reaction. He stated, as a matter of fact, that affinity between molecules is the original cause of the occurrence of their collisions. Assuming both that the teacher use 'affinity' meaning 'a tendency to attract' and that the collision theory deals with 'molecules as hard spheres colliding with sufficient energy and appropriate orientation' without discussing any

force of attraction, the model expressed by the teacher merges main attributes of different historical models, making one becomes the origin of the other.
In the teaching of the atom, the lack of coherence in the relationship of main and secondary attributes was found in three examples:

a. In one of the textbooks analysed it is asserted that, according to Thomson[1],

> matter would be composed of atoms made up of an equal number of (two) main particles: protons and electrons. As the mass of electrons is very small in comparison with the mass of protons, the mass of an atom is that of the protons. The electrons, homogeneously distributes among the protons, would keep the electric equilibrium.

When Thomson proposed his model, protons had not yet been characterised as a class of particles. Therefore, the relationship between the masses of protons and electrons was unknown at that time.

b. Four of the twelve textbooks analysed show pictorial representations for Rutherford's model in which there are circular, defined orbits. However, the movement of electrons in such orbits is part of the hard core of the later Bohr model.

c. Another textbook associates attributes of the Quantum Mechanics model (the existence of sublevels called s, p, d, and f, and the maximum number of electrons in each of them) with the Bohr model. These ideas are introduced after the presentation of Bohr's model in the text, but without saying that they constitute another atomic model. The aim of the introduction of such ideas is to show how to distribute electrons in sublevels. It seems that the author assumes 'atomic models' and 'distribution of electrons' as completely distinct subjects, that he does not realise that they are closely related, and that 'distribution of electrons' is an attribute of some atomic models. As the author asserts[2]:

> In each level of energy, electrons are distributed in sublevels of energy, represented by letters s, p, d, and f in an ascending order of energy.

This suggests to students that the existence of sublevels is an attribute of the model in which energy levels were defined.

4. There is no discussion about the attributes of a given model that continue in the one that succeeds it.
   In the teaching of the atom, none of the textbooks analysed recognise, for instance, the influence that the Greek ideas had on the work of many scientists (including Dalton) over many centuries. In a textbook[3], when models are explicitly compared, in a section entitled 'main ideas of the models', only the new attributes of each model are presented. The existence of the nucleus, for instance, is presented as an attribute of Rutherford's model, but not of Bohr's and the Quantum Mechanical models. This can contribute to students' lack of a comprehensive view about the processes of development of a model.

5. An attribute is correctly associated with a given model when it is presented in written form, but a (often physically close) pictorial representation of that same model shows attributes of another model or a mix of attributes of two distinct models.
   In the teaching of the atom it is possible to see this in the discussion of the Quantum Mechanics model. One of the textbooks[4] analysed, after saying that de Broglie proposed that electrons behave simultaneously as a particle and as a wave, shows a pictorial representation of an atom in which the orbit has the shape generally used to represent a wave[4] (Figure 11.1).

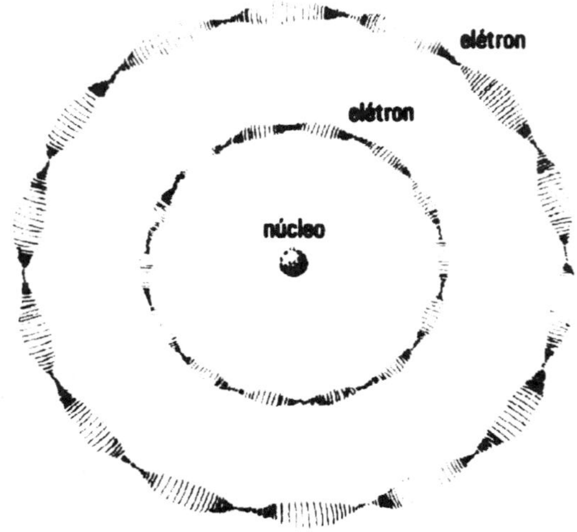

Figure 11.1. Pictorial Representation of an Atom

The same textbook asserts that the idea of an orbital is a basic characteristic of the Quantum Mechanics model. It represents an atom in two ways. First, as a nucleus surrounded by a circle called 'an orbit' in which there is a small ball (presumably the electron). Second, as a nucleus surrounded by a region containing a lot of points called 'an orbital' (presumably electron probabilities)[5] (Figure 11.2).

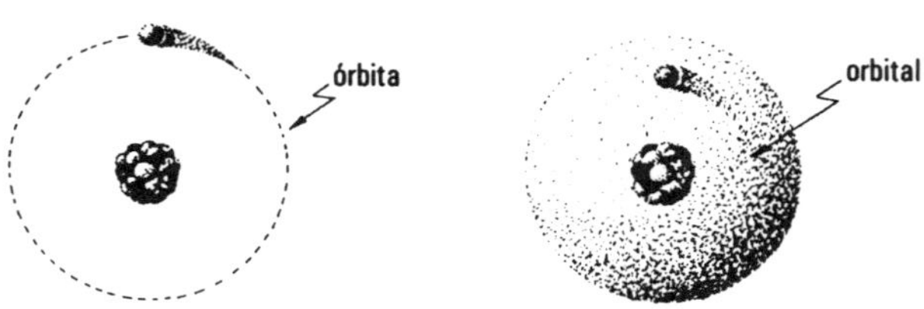

Figure 11.2. Difference between orbit and orbital

By so doing, the difference between 'orbit' and 'orbital' is clearly shown to students. However, after this, the textbook affirms that the identity of the electron is given by its quantum numbers and it then shows a representation in which a nucleus has semicircles cut into one of its sides. Each semicircle is identified by a letter (K, L, M, ...) and by its main quantum number (1, 2, 3, ...) respectively[6] (Figure 11.3).

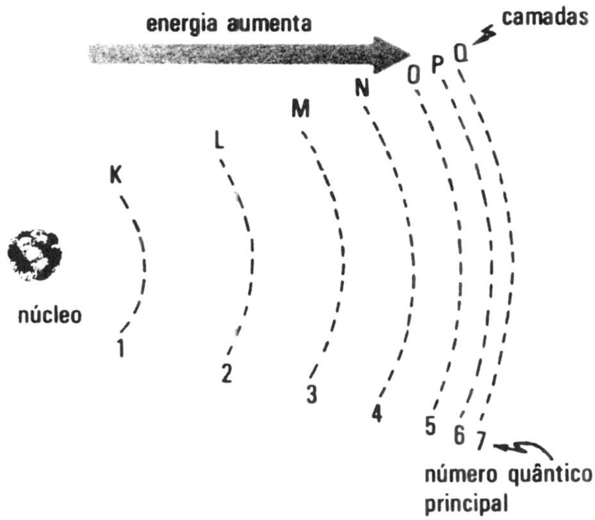

Figure 11.3. Identity of the Electron

The semicircles seem to strengthen the idea of orbit as a place in which electrons move. This is exactly contrary to the distinction made two pages earlier in the textbook. Such ambiguities result from an attempt to create a pictorial representation of a model that was essentially the result of a mathematical-statistical treatment. This seems to be evidence that the author does not understand what the phrase 'quantum mechanics as a theoretical background of a model' means.

Another textbook analysed may introduce the confusion as the result of an attempt to simplify matters[7]:

> The electrons move around the nucleus. They move at random and in a chaotic way, but to make the picture simpler we've shown them as if they travel in 'orbits' around the nucleus.

The points discussed above are evidence that the models of both the atom and chemical kinetics expressed in the teaching situations analysed represent no one single historical model of each of these topics. Moreover, such expressed models cannot be viewed as sequences of elements of different historical models, but as an aggregation of such elements. If elements of different models were sequentially presented, it would be possible to characterise the progressive and degenerating problemshift of each of these models as it was done in the characterisation of the historical models of each of the themes. This unexpected kind of relationship makes it impossible to assert that the models expressed in such teaching situations are based on the subsumption of different consensus models. The work described here proposes a new idea that these kinds of aggregation of elements from different historical models in expressed models are described as *hybrid models,* i.e. models which consist of elements of different historical models treated as if they constituted a coherent whole (Justi, 1997).

In teaching scientific themes it is evident that, depending on both the scientific theme and the level of the students, the latter should be presented with a model which is simpler than that of the current consensus. Notwithstanding, this does not mean that a new model should be produced by adding elements of different historical consensus models according to conveniences that converge on a simple outcome. Such an attitude is not consistent with serious science teaching and it does not contribute to students' understanding of either scientific reasoning or of how science evolves.

It is not being advocated that textbooks and teachers should present an expressed model that coincides with a given historical model, nor that they should present a linear progression of historical models of a given subject. What is important is that models expressed in teaching situations:

i. have their backgrounds clear, that is, the context in which they are valid should be explicit;
ii. do not present inconsistencies in the relationships between their hard core and protective belt.

Both these points mean that hybrid models do not enable scientific knowledge to be viewed as consisting of provisional models that are both developed and valid in specific contexts. Moreover, in a hybrid model the hybrid relationships are presented as natural ones, that is, as if there are no gaps between them, no questions requiring different ways of thinking about the phenomenon, no different approaches to interpret a phenomenon. This contributes to a view of scientific knowledge as something established, true, and that must be accepted. By presenting scientific knowledge in this way, hybrid models do not help students to understand the process of the development of such knowledge.

AVOIDING HYBRID MODELS

The identification of *hybrid models* provides a new insight through which teaching can be discussed. The existence of a hybrid model in teaching means that no history of science is possible because it implies that scientific knowledge grows linearly and is context independent. It leads students to have misconceptions in their mental models of the theme being discussed and/or to have difficulties in understanding the reasons for which hybrid relationships are introduced.

The existence of hybrid models in teaching also points to the need for some changes in teacher training courses. It leads to a discussion about whether such courses contribute to graduates' understanding of scientific knowledge and to their teaching with hybrid models. Three points should be considered:

1. The level at which basic chemical concepts are taught in university courses.
   When students enter universities they may hold misconceptions about some concepts or they may not comprehensively understand some basic chemical concepts. If university lecturers deal only with the current and most complex ideas, this would favour the construction (and consolidation) of many conceptual gaps and misunderstandings in future teachers' mental models.

2. The level at which philosophical discussions about the nature of scientific knowledge take place of university courses.
   This means a discussion about both the nature of models and theories and the relationships between them from different points of view. It is essential too that future teachers view scientific knowledge as provisional, that they understand how scientific knowledge changes, and that they develop a more critical view of scientific knowledge.

3. The level to which university courses attribute importance to the history of science.

   This may vary within the continuum previously proposed, that is, from the absence of discussion about this subject or the occasional mention of dates and names, at one extreme, to the introduction of the history of science as an intrinsic part of all subjects, i.e. permeating the discussion of all chemical themes, at the other extreme. However, independent of the position adopted in such a continuum, it seems that there is little sense in introducing the history of science into university courses if students are expected to just memorise facts or mathematical relationships, to uncritically accept what lecturers or textbooks present to them, rarely asking *why?* (Kipnis, 1996). The history of science needs to be part of chemistry courses in a way that effectively contributes to future teachers' understanding of either the process of construction of scientific knowledge or the identification of ideas produced in different backgrounds or in different contexts.

By critically and coherently considering these three points, universities could provide an *education* and not just a *training* in chemistry (Matthews, 1992a, 1994); they would not then provide conditions for the production of hybrid models; they would make chemistry more understandable to their students.

TEACHING WITH HISTORICAL MODELS

The approach adopted in the studies discussed here can also be used in providing an education in science. A change in the framework within which the history of science is commonly studied and presented in science teaching was advocated by Brush (1978), when he asserts that:

> I believe that the new style of history of science, which emphasises the dynamics of scientific change and its relation to the philosophical, technological and social background, is much more suitable for educational purposes than the older tradition that stresses the accumulation of facts and the assignment of credit for discoveries. (Brush, 1978, p.289)

However, as discussed in the Introduction to this chapter, the extent to which HPS really takes part in science courses varies a lot. With the proposal of historical consensus models underpinned by Lakatosian ideas about research programmes, it is possible to think in an alternative way about the introduction of HPS into science teaching. This is because the characterisation of historical consensus models implies the association of consensus models with the particular historical contexts in which they were developed. This calls for:

i. the explicit statement of such contexts;
ii. the definition of criteria for the characterisation of each model;
iii. the definition of criteria for the discussion of how each model reached consensus status among the scientific community in such a context;
iv. the definition of criteria for investigation into the competition between the 'progressive problemshifts' of a new research programme and the 'degenerating problemshifts' of the original one.

Such a framework proved valuable for the discussion of the historical development of the themes involved in the studies discussed here (Justi and Gilbert, 1999b, in press). This was because it emphasises not only the scientific ideas themselves (as it is often done when the initial approaches of the continuum discussed at the Introduction of this chapter are adopted), but also the process through which they were produced. The dynamic characteristic of such a process was stressed by the establishment of relationships both between different historical models, and between different ideas within the scope of a given model. It showed that scientists were normal individuals who communicated their ideas to others, dealt with technological limitations, and who constructed both good and not-so-good ideas. This contrasts with the linear and non-contextualised presentation of historical ideas in many approaches to the introduction of HPS into science teaching.

The approaches underpinning each one of the science courses that comprises the continuum presented at the Introduction of this chapter fit some of the arguments for the introduction of HPS into science teaching. The use of historical models as proposed from the studies discussed here not only avoid the existence of hybrid models in teaching, but also fits all such arguments. On account of this, its introduction in teacher training courses is advocated. This would enable teachers (i) to understand the importance of HPS into science teaching, (ii) to critically think about the approaches that could be used in doing so, (iii) to recognise the existence of hybrid models in textbooks and (iv) to not use models of scientific themes that may be characterised as hybrid models. This would enable teachers to understand that history of science should not be introduced as further and superfluous information (as it is often done in textbooks), but as an interesting way to explain the interpretation proposed by different scientists in specific contexts

(Sutton, 1993). Moreover, this could increase teachers' own understanding of scientific knowledge. As Matthews (1994) discussed, independently of a direct utilisation of historical and philosophical knowledge in their classes,

> A teacher ought to know more than just what he or she teaches. As educators, teachers need to know something about the body of knowledge they are teaching, something about how this knowledge has come about, how its claims are justified and what its limitations are. Teachers should have a feel for, or appreciation of, the tradition of inquiry into which they are initiating students. (Matthews, 1994, p.213).

In this way, an understanding of HPS could mean a comprehensible understanding of science. By understanding the process through which a research programme associated with a given historical model was succeeded by that of another model, teachers will be able to bring the dynamism of science to science teaching. The promotion of conditions for teachers comprehensively understand what they teach is, undoubtedly, a step toward the education of well-prepared teachers, who are essential to the complex task to educate citizens of the twenty-first century.

NOTES

1. Neto, G.C., 1995, Quimica: da teoria à realidade, v. 1. São Paulo: Scipione, p.91.
2. Carvalho, G.C., 1995, Quimica Moderna, v. 1. São Paulo: Scipione, p.31.
3. Costa, M.C. e Santos, G.O., 1995, Quimica: a visão do presente, v. 1. Belo Horizonte: Lê.
4. Sardella, A., 1997, Curso de Quimica – Quimica Geral, 23a. ed. São Paulo: Saraiva, p.89.
5. Sardella, A., 1997, Curso de Quimica – Quimica Geral, 23a. ed. São Paulo: Saraiva, p.91.
6. Sardella, A., 1997, Curso de Quimica – Quimica Geral, 23a. ed. São Paulo: Saraiva, p.93.
7. Holman, J., 1995, Chemistry. Victoria: Australia: Thomas Nelson, p.248.

Chapter 12

Models in Explanations of Chemistry: The Case of Acidity

John Oversby
The University of Reading, UK

INTRODUCTION

The discipline of chemistry occupies a special place in science since few of the macroscopic observations can be understood without recourse to sub-microscopic representation or models. Chemical models are constructed to provide a variety of explanations of parts of chemical phenomena. This chapter focuses on the development of modelling in chemistry, using the context of acidity for exemplification.

Acidity is one of the earliest and most well-developed concepts in the history of chemistry (Brock, 1992). The obvious sour and distinctive taste of acids marks them out as a class of substances both in everyday life and in the chemist's laboratory. Its taste is put to good effect in a variety of foods and its place in pickles and other food preserves is determined by the inability of microbes to reproduce in very acidic solutions. Ideas about the causes of acidity abounded during the time when chemistry developed from alchemy. The particular contribution of acidity to chemistry lies in its ubiquity in all areas of the discipline, e.g. kinetics in acid-catalysed hydrolysis, redox in the adoption of the standard zero for electrode potentials, in comprising the essential characteristics of phenols and carboxylic acids and in classifying the nature of oxides in inorganic chemistry.

At the heart of chemistry is an explanatory process that takes some points of interest, such as the behaviour of organic chemicals, and creates frameworks such as functional groups to make sense of the vast array of data available. Within this framework, specific functional groups are identified,

J.K. Gilbert and C.J. Boulter (eds.), Developing Models in Science Education, 227–251.

such as the hydroxy, OH, group. As the data are explored, the character of the hydroxy group is also developed, from being simply a pair of atoms attached to some structure which might be replaced in a substitution reaction, to the extent of its polarisation to understand why it is replaced more easily in one compound than another. Such a process is a modelling process. Chemistry seems to be composed of a whole variety of modelling processes for a variety of purposes, but all related in linking macroscopic behaviours with sub-microscopic explanations.

Chemistry texts provide valuable clues to the sorts of explanations in regular use in the classroom. They provide an element of consistency from one teacher to another as learners base their independent and guided study on the text produced by the author. They model carefully constructed and progressive explanations, at best. For teachers, they may well provide much of their understanding of what makes a good explanation for use in their own classes and may be a valuable source of personal professional development. This Chapter is, therefore, based on a survey of a selection of textbooks as exemplary material from what is available. The reasoning behind the formation of different models is rarely made explicit to learners in the textbooks surveyed and it is possible that this leads to confusion about which could be the best model. In terms of acidity, the idea of parallel models, each holding value in different contexts, is one which is more satisfying than a hierarchical progression. In place of a progression where the expert chooses the model with the greatest explanatory power, an alternative view of the expert choosing an appropriate model for a particular purpose from knowledge of the strengths and limitations of various models is proposed as fitting the evidence more satisfyingly.

EXPLANATIONS AND THE CHEMISTRY CURRICULUM

Explanations in science serve a number of purposes (Gilbert et al., 1998a, 1998b) which appear to progress in the following order:

- descriptive
- interpretative
- causal
- predictive

Science in English and Welsh schools is determined by a National Curriculum (DFEE, 1995a). The general format of this National Curriculum is similar to that found in many European education systems, the differences being in the examples required to be taught and the exact ages at which specific topics should be taught. The progression of concepts appears to follow a standard pattern. In the primary school, young learners regularly experience working with a wide range of materials, such as metals, clay, plastics, glass, and liquids such as water and cooking oil. Their experiences enable them to group these materials in a more or less coherent and useful way through their behaviours. The test of these groupings is their stability in the face of new evidence. The explanation for each grouping is descriptive, so that, for example, acids are grouped together because they taste sour and react similarly with carbonates, magnesium and metal oxides. Such an explanation appears to be reasonably satisfying at this level. The explanation is based on macroscopic properties.

At the next phase of education, the lower secondary school, the National Curriculum generally requires that learners adopt a particle model explanation to understanding the behaviours of materials. The particles used in such models are generally featureless, often represented simply by circles. Different particles are represented by different shading or colours in the circles, as a way of coding to distinguish between them. It is not intended that the viewer should perceive the exact shading or colour as that which the particles have. In the case of acidity, a particulate explanation is not given at this level. Rather, more detailed examples are provided, and more ways of identifying acids, including using indicators. Indicators are also used to classify acid solutions into finer categories, such as weakly acidic and strongly acidic. Further reactions of acids, with bases, are used as a vehicle for preparing and characterising salts, and to improve practical skills.

For the more able, at the upper secondary stage (14 to 16), the significance of water is often introduced as a prelude to the explanation of acidity based on the formation of hydrogen ions. 'A' level, post 16 years old, gives students a whole range of models and an insight into a quantitative treatment through pH calculations.

In general, progress in explanation at school level can be summarised in Table 12.1 below. In the case of acidity, development at lower secondary level is in terms of a richer descriptive form, rather than into a set of explanations based on particles.

Table 12.1.

Curricular development	*Phase of education*
Descriptive chemistry, phenomena based ↓	Primary ↓
Generalised particle model ↓	Lower secondary ↓
Detailed particle model ↓	Upper secondary ↓
More inclusive models, often quantitative	Pre-university and university

Table 12.2

Curricular development	***Phase***	*Phenomena*	***Explanation type***
Descriptive chemistry, Phenomena based ↓	Primary ↓	Simple descriptions based on taste ↓	Classification based on description ↓
Increasing detail of observed behaviour ↓	Lower secondary ↓	Extended behaviours, e.g. effect on metals, carbonates, indicators, metal oxides. Observe neutralisation in terms of behaviour. pH as an observation based on indicator colour ↓	Classification with a wider descriptive base. Multiple features used for classification. Greater discrimination between acids and non-acids ↓
Detailed particle model ↓	Upper secondary ↓	Acids conduct electricity. Some acids conduct strongly. Some acids conduct a little. Water is required for acidity. pH is lower for strongly acidic solutions. ↓	All acid formulae contain hydrogen. All acids form hydrogen ions in solution. Strong acids fully dissociate. Weak acids partially dissociate. Some non-metal oxides react with water to become acidic. pH depends on hydrogen ion concentration ↓
Variety of models, one quantitative model	Pre-university and university	Acidity and pH changes with dilution. Acid base reactions in the gaseous and non-aqueous phases. Base characteristic and extension to chemistry of complexes.	Acidity as an equilibrium. Calculations from dissociation constants. Acidity as hydrogen ion dissociation. Molecular structures of bases have lone pairs of electrons. Acidity is based on lone pair acceptance.

In terms of models of acids, there is a more detailed framework for progression of thinking about acids and this is provided in Table 12.2 above. Further explanation of the origin of this table is given later in this Chapter.

MODELLING

Modelling is the action of representing an idea, an object, a process, an event or a system. The principles of modelling have been treated in Chapter 1 of this book. The focus in this Chapter has two parts. The first is to highlight which parts of the phenomenon of acidity are being explained. The phenomenon of acidity simply has existence and has various significances for different people. For some, it is a recognition that the substance can form salts that is important, while for others the concern relates to the relative strength of the acidity in comparison with other acids. It is impossible to pay attention to the whole phenomenon of acidity and in modelling it is not required. It is inherent in the nature of explanation to limit interest to only some part of the phenomenon and to deliberately ignore other parts. This fits very well with the notion of modelling, since modelling is also only partial. The explanatory power of different models is one of the points of discussion in this paper.

MODELS OF ACIDITY

A variety of representations of acids have been created during the development of chemistry. In order of chronology, these are:

1. Behaviour model (antiquity to 1777)
   Acids are substances with particular attributes such as:
   - they taste sour;
   - they change the colour of certain dyes (indicators);
   - they react with reactive metals, producing hydrogen;
   - they react with carbonates, producing carbon dioxide;
   - they react with bases, losing their acidic properties through neutralisation

It was recognised that being in an aqueous solution is an important aspect of acids displaying these attributes.

It is likely that materials were first classified as acids on the basis of their taste (Brock, 1992) and that the development of mining led to the idea of acids producing effervescence with carbonate rocks. Alongside these clearly observed attributes of acids came the idea that acids were materials which could change the colour of various plant dyes and this eventually became part of the method of identifying the acidic nature of a substance. The reaction with carbonates, and later other bases, produced the notion that acids could lose their acidic properties, or become neutralised.

In this explanation of what is an acid, it is simply described as a material which possesses a set of attributes, or which operates in a certain way. This is an operational classification, and it is still widely used today by both novice and expert chemists. The value of such a descriptive model is that it enables chemists to predict the likely behaviour of a material once it is known to be an acid from some of its properties. As an example, if phenol is an acid, then it is expected to react with sodium hydroxide solution to form a salt, and this it does. It is also expected to react with sodium metal, to form the same salt, and this it does. However, the model is weakened in that there is no basis for defining how far its prediction will go. Again, with phenol, if it is a typical acid, then it should react with sodium carbonate to form a salt. It does not. It should turn pH indicator red, orange or yellow, and it does not. It should react with copper (II) oxide to form a copper salt, and it does not. A purely descriptive explanation gives us no idea of which of these features of an acid will be displayed and which will not.

Attempts had been made to explain the properties of acids in terms of something more fundamental than mere descriptions of these general properties such as that acidic materials taste sour because they contain particles with sharp edges which affect the tongue. In this sense, they made a relation between what they saw as obvious properties of crystals, sharp edges and the ability to slice through skin, and imaginary particles that make up the materials. Since there was no way of exploring this model further, it was not developed.

2. Lavoisier's model (1777 – 1787)
   Acids are substances that contain oxygen

The preoccupation with trying to understand burning and testing the behaviour of such products of burning gave rise to the idea that acids contained oxygen and that this caused acidity in some way. At this stage sulphur trioxide was understood to be sulphuric acid and sulphuric acid as we now know it was hydrated sulphur trioxide. The significance of the solvent was not understood at this time. A salt is made simply from the combination between an acidic oxide (or acid, as Lavoisier had it) and a basic oxide (the oxide of a metal). This was in line with the dualistic view of chemistry, which at the time ascribed many reactions to the neutralisation of opposite properties. In this model of an acid, attention is on the species involved, i.e. that an acid is an oxygen container.

This model attempted to make sense of acidity in terms of the nature of the chemical elements themselves. At a time when relatively little quantitative information was available about the chemical elements, the use of obvious features such as those that defined whether the element was a metal or not was an easy choice. However, some oxides of non-metals turned out not to be acidic, such as water itself, and this model had many drawbacks. In the absence of an understanding of atomic structure, there was not even a cause and effect relationship between being the oxide of a non-metal and being acidic in character.

3. Priestley's model (1772 – 1775)
   Acids are substances that contain hydrogen (Priestley)

This was devised after a considerable investigation of hydrochloric acid, variously named marine acid gas (since it was manufactured from common salt) and muriatic acid, thought to be an acid formed by a non-metal oxide. Interestingly, one of the early challenges to the oxygen model came about when Scheele, working in the Swedish mining industry, was asked to investigate pyrolusite (manganese dioxide) (Goodman and Russell, 1991) and he added hydrochloric acid to it as one of his first tests. The production of chlorine and its later identification as an element, after a fruitless exploration of its nature as an oxide of another element, gave rise to the idea that there were two sorts of acids, those containing oxygen and those containing hydrogen. It was Priestley who is credited with combining these two with the action of water on acidic oxides into the more inclusive model that all acids contain hydrogen. Again, attention is on the chemical identity of the fragment involved, i.e. that an acid is a hydrogen container. In this chapter, the term species will be used for the chemical part of the acid that is responsible for causing or explaining acidity, such as hydrogen or oxygen atom, or hydrogen ion.

The use of formulae soon demonstrated the worth of this model. It predicted, for example, why there was a stoichiometric ratio between the quantities of acid and base needed for neutralisation to form a salt. In the case of sulphuric acid, it even explained the formation of two salts, a normal salt and an acidic salt, formed by partial and complete replacement of hydrogen atoms by metals atoms. It also explained the formation of hydrogen when magnesium, for example, was added to an acid. The acidic character of some oxides was explained by the formation of hydrates when the oxides were dissolved in water. In the case of acids with two sorts of hydrogen, that which could be replaced and that which could not, the matter was resolved by labelling the replaceable hydrogen as acidic hydrogen. This practice still exists in organic chemistry today where, for example, the CH_2 hydrogen atoms in molecules are labelled as acidic hydrogen atoms because they react with hydroxide ions.

4. Arrhenius model (1884)
   Acids are substances that produce hydrogen ions in solution (Arrhenius)

This arose from two series of investigations, one by Raoult on measuring the depression of freezing point of glacial ethanoic acid and one by Arrhenius on the conductivity of a variety of solutions. It seemed that there was a blitz on using electricity as a way of inducing chemical change and this was coupled with the increasing trend towards being quantitative. The relative accuracy of both Arrhenius' and Raoult's measurements was remarkable for its time. Both of these investigations pointed to the model of partial dissociation into ions and the concept of a weak acid, as opposed to a strong acid. A weak acid was one that hardly dissociated into ions and a strong acid was one that almost completely dissociated. Interestingly, the evidence on sulphuric acid at this stage pointed to it dissociating into two particles, which we now know to be the hydronium ion and the hydrogensulphate ion, the second dissociation into sulphate hardly occurring in the experiments they carried out.

Further investigation of the nature of the hydrogen ion in aqueous solution has led to it being symbolised as H_3O^+. The significance of the solvent, universally water in this model, is embedded in the explanation of how the H_3O^+ ion is produced, since it can be considered to be a proton bonded to a water molecule. In an extension to this model of acidity, also provided by Arrhenius, a base is seen as a substance which produces a hydroxide ion in solution. Neutralisation is seen as the reaction between a hydronium ion and a hydroxide ion to produce water molecules. Despite the greater sophistication of this model, attention is still on species, i.e. hydronium (hydrogen) ions and hydroxide ions in acids and bases. The significance of the solvent is being noted more strongly at this stage in the development of ideas about acidity.

The Arrhenius model was profoundly satisfying since it explained a great deal of phenomena related to solutions of what we now know as electrolytes rather well. In particular, it explained the change in conductivity of strong electrolytes on dilution, although it failed at moderately concentrated levels. The addition of the Law of Mass Action, proposed by Guldberg and Waage, even explained the change in electrical conductivity of weak electrolytes on dilution. What is even more valuable, the measurement of the concentration of hydrogen ions, eventually using the pH scale, enabled chemists to compare the acidity of solutions in a quantitative way. Finally, a parameter that expressed the relative strength of the acid, the dissociation constant, was devised from the pH and the acid concentration. The restriction brought about by this model was that it only applied in aqueous solutions.

5. Brönsted-Lowry model (1923)
   Acids are proton donors

This model integrated the separate ideas of acid and base by promoting the concepts of conjugate acid and conjugate base. In this model, an acid is only acting as such when it can donate a proton to a base. It also uses the notion of equilibrium to consider the reverse reaction as an acid-base reaction. Brönsted and Lowry devised their model to explain neutralisations such as the reaction between gaseous hydrogen chloride and ammonia, which produces ammonium chloride as the reaction of aqueous solutions of these two materials does. In this sense, it is an attempt to seek a more fundamental understanding or explanation of how acids work by creating a model that is independent of solvent. This model is also used to extend the value of acid-base reactions to reactions in other solvents and to cases where no solvent is involved. Attention in this model is still on the species being transformed, rather than the nature of the bond.

The phenomena explained by this model involves a wider range of media, or no media at all, when compared with the Arrhenius model. This was its major strength, while it retains the notion of comparative strength by noting where the position of equilibrium lies after exchange of the protons.

6. Lewis model (1923)
   Acids are lone pair acceptors (Lewis)

Lewis was preoccupied with the lone pair of electrons and discussions in his research team were frequently focused on using this aspect of chemical change (Brock, 1992). In this model, attention shifts to the base rather than the acid. It attempts to extend thinking about acidity by concentrating on the bonds being formed and broken. Thus it represents a major change in that attention is no longer on an identified species. This is not surprising since this was an era in which there had been a great deal of thinking about bond formation, about different types of bonds, and attempts to generalise thinking about bonding. This model can provide a different dimension to reactions which might be otherwise considered as complex ion formation and the stretching of an explanation beyond what is normally considered an acid-base reaction can be both a strength and a weakness. The model can be generalised to such a level where it treats all reactions as the same and there is a point, sometimes, in differentiating between reactions in order to understand them better.

Lewis'contribution to explaining acidity lies in using the electronic structure of the base as the focus for defining basicity. Unfortunately, this extends the meaning of acid-base processes to such a wide range of reactions that the class becomes too great. Almost every reaction becomes an acid-base reaction and the phenomenon loses its distinctive significance. However, in organic chemistry, this explanation is of value in areas such as the reactions of amines.

7. Usanovitch model (1939)
   Usanovitch generalised solvent model

Between Lewis and Usanovitch, a great deal of work had been carried out in solvents other than water, such as liquid ammonia, liquid sulphur dioxide and liquid ethanoic acid. In these solvents the following autoionisations are assumed to take place:

$$2NH_3 = NH_4^+ + NH_2^-$$
$$2SO_2 = SO^{2+} + SO_3^{2-}$$
$$2CH_3COOH = CH_3COOH_2^+ + CH_3COO^-$$

Titrations between ammonium and amide compounds, thionyl and sulphate (IV) compounds, and strong acids, such as sulphuric acid and ethanoate compounds, had been explored using indicators, thermometry, and conductimetric methods, and these paralleled neutralisation reactions in aqueous solution. Usanovitch, devoting his attention to the reactions between the ions, appreciated that in these reactions the anion is simply a carrier for a lone pair. This explanation refers to a solvent model, considering an acid as that substance which increases the concentration of the cationic species in the auto-ionisation of the solvent. Its significance is in extending the ideas of acids to non-aqueous systems while still explaining reactions in water. Since many of the processes can be followed by the same indicators, or by conductivity, as in aqueous solutions, a model which brings together all these processes has intrinsic value.

MODELS OF ACIDITY AS EXPLANATORY TOOLS

Each model focuses on some parts of acidity, explaining a limited number of features of acidity. The hierarchy of explanations, given earlier (Gilbert and Boulter, 1998a, 1998b) has been used to characterise the set of acidity models used in this survey, and is given in Table 12.3.

Table 12.3.

Acid Base Models as Explanations				
Models	Originator	Chemical atom or ion defined?	Bond defined?	Type of explanation
Operational	Unknown	No	No	Descriptive
Oxygen	Lavoisier	Yes	No	Interpretative
Hydrogen	Priestley	Yes	No	Interpretative and causal
Hydrogen ions	Arrhenius	Yes	No	Interpretative, causal and predictive
Proton donor	Brönsted-Lowry	Yes	No	Interpretative, causal and predictive
Lone pair acceptor	Lewis	No	Yes	Interpretative, causal and predictive
Solvent cation	Usanovitch	Yes	No	Interpretative, causal and Predictive

How do these models explain the phenomenon of acidity? Table 12.4 identifies the strength of the explanation.

Table 12.4.

Model	*Type of Explanation*	*Parts of the Phenomenon*
Observable attributes	Descriptive. Establishment of a distinctive class	Observable characteristics in qualitative form
Acids contain hydrogen or oxygen	Recognition from formula or method of production. Cause and effect but not mechanistic	Formula and/or methods of production. Salt formation and formulae
Acids produce H^+ or H_3O^+	Cause and effect, with mechanism. Quantitative	pH, conductivity, colligative properties such as vapour pressure, depression of freezing point, Reactions with bases, carbonates, reactive metals, indicators
Acids donate H^+	Cause and effect. Mechanistic, relative strengths of acid/base pairs	Acid base reactions in all solvents, including non-aqueous, and the gaseous phase
Acids bond to bases with lone pairs	Cause and effect. Mechanistic	Acid-base reactions involving non-hydrogen acids. Focus on base rather than acid

It is possible to say that some explanations are more general than others, that they provide a deeper understanding than an earlier one. However, what is not clear is whether an expert will naturally choose a more fundamental explanation than a novice will. It seems to apply according to, at that stage of knowledge, the question being asked and to the person asking the question. In this way, the expert is demonstrating not only a greater understanding but also a greater facility in fitting the explanation to its perceived purpose.

A summary chart of these models is provided in Table 12.5.

Table 12.5.

Acid Base Models					
Models	Originator	Solvent-based	Operational attributes	Species defined	Bond defined
Operational	Unknown	Yes (water)	Yes	No	No
Oxygen	Lavoisier	Yes (water)	Yes	Yes	No
Hydrogen	Priestley	Yes (water)	Yes	Yes	No
Hydrogen ions	Arrhenius	Yes (water)	Unknown	Yes	No
Proton Donor	Brönsted-Lowry	No	No	Yes	No
Lone pair acceptor	Lewis	No	No	No	Yes
Solvent cation	Usanovitch	Yes (any)	No	Yes	No

IDEAS ABOUT ACIDS HELD BY LEARNERS

The development of understanding about acids in schools is a long process. Typically, in England and Wales, children become aware of significant attributes of acids, such as their sour taste and their ability to corrode ('burn') various materials, at primary school, i.e. by the age of 11. Carr (1984) has suggested that children's understandings about acids arise from everyday experiences such as tasting acidic foods and from stories and advertisements in the media. The notion that acids 'burn' or corrode materials was still a major understanding of pupils aged 15 interviewed by Hand and Treagust (1988). They discovered that these perceptions were remarkably stable for one third of these pupils, even after an active intervention teaching strategy designed to challenge and extend these ideas. Hand (1989), testing 17-year-olds, found that they still had problems in defining 'acid' and knowledge of ways of testing acids. Carr (1984) pointed out that there are various models of acids used in science teaching and that teachers do not make these explicit to learners as they deal with them. Ross and Munby (1991), in a test of high-achieving 17-year-olds, used concept maps to test their understandings of acids, discovering misconceptions about the produces of neutralisation, properties of acids, the nature of ions produced by acids (this was thought by some to be a hydroxide ion), and the relative pH of strong and weak acids. Cros et al. (1986), in a study of first-year university students, observed limited ability to give examples of acids and bases and of weak and strong acids. Schmidt (1991) noted that the word neutralisation was interpreted by German high school students as a position (the point) where a solution is neither acidic nor basic, rather than a process of moving towards, and often beyond, such a point. Schmidt's more recent work (1995) has further investigated the idea of conjugate acids and bases, noting significant confusion about this concept.

EVALUATON OF PREVIOUS RESEARCH EVIDENCE

Learners already have many ideas about chemistry before they come to the classroom. As regards acids, everyday ideas include terms such as 'burning' 'poisonous'. Although it is true that the meaning behind these terms does apply to the majority of acids, the use of specifically technical terms such as 'corrosive' in place of 'burns' establishes a special meaning, which excludes combustion. It is the induction into the chemist's refined and particular use of language that marks the focus on a chemical understanding. The term is an indicator of a special meaning, restricted to only part of knowledge. Many of the difficulties in chemistry appear to stem from a lack of explicit reason for using a restricted language. It may be that learners do not distinguish sufficiently their different roles in different circumstances. Within a chemistry lesson, they are expected to adopt the role of being a chemist by adopting the chemist's language, part of being enculturated into chemistry. Some teachers avoid the use of specialised language, arguing that it is better to use the accepted language of the student. Although this may help build a bridge between the language acquired by the student and that expected in the chemistry lesson, it does blur the new chemical meanings of the terms, which may be the cause of some of the misconceptions. The message is in the medium, so that the use of specialist terminology is part of the path to a specialist understanding.

The next set of misconceptions, that is, the conflict in meanings of neutralisation, may well be formed by the method of introduction of 'pH'. This is usually introduced as a refinement of the classification of weakly acidic and strongly acidic solutions based on pH indicator colour. This is best understood as mapping pH as the cardinal positions of a set of descriptions of solutions. Thus, 'weakly acidic' is given a more detailed description as pH values four, five or six. In this sense, pH is not to be seen as a continuous variable. The neutral point, pH seven, is the value that all neutral solutions of salts possess. The more advanced meaning of neutralisation, that of stoichiometric quantities of acid and base being added, conflicts with this, and involves appreciation of pH as a continuous variable. It is doubtful if teachers ever make this point explicit.

The idea of conjugate acids and bases belongs to a quite different way of looking at chemical reactions, where acidity is not seen as the production of a chemical species, but is an activity of a chemical series. In this case, acidity is explained in terms of proton transfer between two bases. It is probably this change from an entity to a process which is the main cause of misunderstandings of the conjugate acid-base model.

In this evaluation of the possible origins of misconceptions in acid-base chemistry, the focus has been on three potential causes:

- Challenges created by the adoption of new scientific language to formalise new and specific thinking, as part of the explanation.
- Challenges caused by an implicit and unexplained change in the quantitative character of the concept, in this case pH.
- Challenges caused by the introduction of a new way of thinking, without a detailed discussion of the purposes behind its adoption.

DETAILS OF THE TEXTBOOK SURVEY

'A' level chemistry represents the last stage in school where learners have the opportunity to use the chemist's models they have acquired earlier in an explanatory framework. In this regard, they and their teachers have role models in the form of the authors of textbooks. Since the textbook is such a significant resource for students and probably one of the most influential after the teacher, it was felt that an examination of textbook explanations would form a valid way of examining thoughtful approaches to teaching the subject. In addition, it may be that chemistry teachers also gain some of their understanding of chemistry, and of ways of teaching chemistry, from the textbooks they use with their students. In this way, textbooks may have a double influence on the learning that takes place in the classroom.

The textbooks chosen for the survey of their treatment of acidity are not unique in their explanations. There is a common core for the A-level chemistry syllabuses and so it would be expected that they would contain much of the same material. They were simply a sample of popular recent textbooks and there is no reason to believe that another selection of textbooks would produce different results. Early indications were that many pages were devoted to the idea of pH and pH calculations in many different contexts. Interest focussed on whether the space devoted to these calculations was related to their use in explanations in other chapters. It was quickly discovered that this was not the case. Similarly, note was made of the different models of acidity and whether these were used elsewhere. This was also found not to be the case. It was quickly becoming apparent that the textbook authors were providing descriptions of the models without considering their use. After collecting sufficient evidence to provide for an analysis of the ways that the authors discussed acidity, an attempt to provide a more soundly based approach was made and this is provided at the end of this chapter. It is hoped that these examples will indicate a new approach to a more coherent and rigorous view of acidity in chemistry.

Students studying A level have covered the whole of the operational classification of acids and many of them have been introduced to the Arrhenius model of an acid producing hydrogen ions. The Lavoisier model itself is not taught in schools in England and Wales, although pupils in lower secondary school (11 to 13 years old) are usually introduced to acids as aqueous solutions of non-metal oxides, and bases as metal oxides.

Explanations of acids are used in three areas of each of the books analysed. A chapter, or part of a chapter, on models of and calculations involving acids forms a major component of the books. These models are often employed in sections on both inorganic chemistry and organic chemistry, usually implicitly. Each of these sections will be considered in turn, identifying common patterns and noting differences.

WHAT IS EXPECTED OF A MODELLING APPROACH?

Essentially, acidity is a physical chemistry model devised to explain organic and inorganic reactions. The physical chemistry chapter should describe each model, and state the reasons for its invention. As part of this, its value in explaining chemical change should be asserted, as described earlier. However, its real significance comes in making sense of some reactions in the organic and inorganic processes. The choice of suitable examples would clarify the benefits and disadvantages of each model for different contexts. Given the restriction of book size for economic reasons, such detailed exemplification would be limited to a few in number, but some exploration of the reason for having such a variety of models should be made clear by using them in explaining some chemistry.

PHYSICAL CHEMISTRY EXPLANATIONS

All three books provide an overview of some of the models of acidity. Textbook A includes those of Arrhenius and Brönsted-Lowry. Textbook B describes the models of Arrhenius, Brönsted-Lowry and Lewis, as well as an introduction to the operational classification. Textbook C includes Lavoisier's oxygen model and Brönsted-Lowry by name but describes, in passing, the Arrhenius model. All three books include extensive sections on pH calculations of both strong and weak acids and in many other circumstances. In all the calculations on pH, there is very little justification of its use provided by the authors. It can only be assumed that it is an essential part of the study of chemistry at this stage. None of the books discuss why an acid is weak other than to produce a tautological statement that an acid is weak because it hardly dissociates into ions, or because it has a high pH value, or because its anion attracts protons.

In the calculation sections the models of acidity are not explicitly referred to and there is inconsistency in the use of the hydrogen and hydronium ions in equations. There is some discussion of the liquid ammonia system, but implicitly in terms of the Brönsted-Lowry model of proton transfer.

Textbook C provides a discussion of the evolution and significance of Lavoisier's model, leading on to the Arrhenius, model, and the Brönsted-

Lowry model. The author then asserts that an acid as a substance that reacts with a base to give a salt and water *only*, which rules out carbonates as bases. This is in contradistinction to the operational model provided by Brock (1992), which simply says that a base neutralises an acid, thus including carbonates as bases.

INORGANIC CHEMISTRY

Textbook A focuses most of its acid-base discussion in inorganic chemistry as the production of hydronium or hydroxide ions (Arrhenius model) and it describes acids as materials which react with hydroxide ions, providing ionic equations, in the main, to explain these properties. Very little attention is paid to operational features of acids in the explanations (e.g. effect on metals, indicator paper, carbonates), which would demonstrate value in the behavioural model. In discussing the acidity of the hydrogen halides, it is simply stated that these are acids. An explanation of acidity in terms of the ionic or covalent nature of the bond in oxides is provided for aluminium, group (IV) and transition metal oxides. In this section, there is no reference to any basis for calculation, or for comparison through acid dissociation constants.

Textbook B simply states that the alkali metal hydroxides are alkaline, with no explanation, not even in terms of production of hydroxide ions. The aluminium ion is considered to be acidic in terms of production of hydronium ions, but the relevant Arrhenius model is not mentioned explicitly. Reactions with alkalis are in terms of reaction with hydroxide ions, with some equations in ionic form and others in molecular form. There appears to be no consistency there. Acid strengths of the hydrogen halides are reflected in terms of acid dissociation constants but there are no details of the species formed, i.e. is it a hydrogen or a hydronium ion? Ammonia is discussed as a base by stating that it reacts with acids, but there is an implicit view given in a diagram that a lone pair of electrons is involved, i.e. that a Lewis model is being used. Carbon dioxide as an acidic oxide is explained as the production of molecular carbonic acid, referring to the idea that an acid contains hydrogen, Priestley's model although not clearly stated as such.

Textbook C uses the phrase 'proton grabbing' to explain the reaction of aluminium oxide, i.e. fundamentally the Arrhenius model. Liquid ammonia is discussed, with amide ions being identified as bases, but not in terms of one of the models of acids expounded elsewhere in the book. Interestingly, mention is made of the similarity of nucleophiles and bases behaving similarly because they are using electron pairs, although Lewis is not referred to. The acidity of carbon dioxide is discussed in terms of formation of hydrogen ions, back to the original Arrhenius model, although his name is not mentioned.

ORGANIC CHEMISTRY

Throughout the chapters on organic chemistry dealing with mechanisms, many proton transfer reactions are given, usually in equation form, as though they were facts and not considered to be based on the Arrhenius model. All three books provide a discussion of the acidity of carboxylic acids and a variety of hydroxy compounds such as alcohols in terms of the behavioural model. They refer to reactions with sodium metal (although the physical chemistry sections do not mention reaction with this metal as an example), sodium hydroxide and sodium carbonate to form salts. pH (Arrhenius model) is rarely mentioned. Acid dissociation constants are used to demonstrate that they are weak acids. Amines as bases are discussed in terms of basicity constants and lone pair donors, although, again, the relevant Arrhenius model is not mentioned. The Brönsted-Lowry model is not mentioned in terms of reactions of amines. All three books simply state that amino acids exist as zwitterions, with no reference to an internal Brönsted-Lowry proton transfer. The opportunity to describe an amino acid as its own conjugate, albeit in an isomeric(!) form, is missed.

There is an attempt in Textbook B to explain the acidity of ethanol in terms of the polarity of the oxygen-hydrogen bond. Differences in acidity are simply 'explained' in terms of the electron withdrawing nature of substituents.

DISCUSSION

Models provide explanations of phenomena (Gilbert and Boulter, 1998) and their inclusion in a chemistry syllabus can be judged by the use to which they are put. In the physical chemistry sections where the models are introduced, the reason for their inclusion is not given in any of the books reviewed. Progression is most frequently given in historical terms and not in terms of their power of explaining chemical phenomena. It might be thought that some review of their strengths and limitations as models would be given alongside a description of each model but this is not the case in the examples cited. In the 'application' sections of the books, i.e. inorganic and organic chemistry, there is no direct reference to any specific acid models at any stage and the models are simply used as though they were fact. In spite of this, it seems that the use of the models is eclectic, and readers are faced with trying to make sense of which model is being used, why it is appropriate, and why another model has not been chosen. Faced with such a challenge of trying to make sense of the plethora of models being used, it is not surprising that some learners adopt the strategy of learning each model and its use by rote, putting great strain on the mental capacity of the individual. Such a strategy militates against successful application of acidity models in new situations.

There is considerable variation in the treatment of models or explanations of acidity in the three books. Even Textbook C, which gives a much fuller description, does not exploit the opportunity to discuss the nature of science as a never ending quest for more inclusive and general models which remain useful.

It might be thought that the opportunities to use these models would come in the inorganic and organic chapters. However, the evidence provided here is that there is inconsistency in the use of the models and that there is rarely any explicit discussion of the value of these models in understanding the chemical reactions of these materials. The Brönsted-Lowry model is rarely referred to once it has been introduced in the physical chemistry section. The Arrhenius model is the one most commonly used, but its strength in explaining a great deal of chemistry is not developed. The Lewis model fails to be significant in any explanation in inorganic or organic chemistry. Acid dissociation constants are used to indicate relative strengths (often as pK_a) but that appears to be a different challenge to student understanding). The skills in performing the calculations in the physical chemistry chapters are not used in the organic or inorganic chapters. Apart from being a stipulated part of the syllabus, and therefore required to be presented in the textbook, it is difficult to justify the plethora of calculation that needs to be mastered.

The issue arising from this study of the three books as typical examples can be generalised:

- They provide a description of a variety of models of acidity without placing the development of each model in context, i.e. why was it developed and how did it help chemists to understand acidity better than the previous model?
- The development of quantitative use of acidity (based on the Arrhenius model) is not justified. The reasons for requiring students to come to terms with a wide range of pH calculations are not made

clear. Given the difficulties of measuring pH accurately at this level, it is difficult to see how many of the calculations could be checked in practice.

- There are very few examples of explicit reference to the models in the applications chapters, i.e. the inorganic and organic chapters.
- There is no exploration of the value of the models in context.
- The use of each model in the applications chapters appears to be eclectic.

There are legitimate reasons for choosing different models, for each model focuses on some specific features of acidity. In inorganic chemistry, reactions of alkali metal oxides with other species may well be best understood as simply the reactions of hydroxide ions. The basic character of the oxide ion in aqueous solutions of alkaline oxides could be explained as the:

- formation of hydroxide ions by reaction with water (descriptive and interpretative explanation based on the Arrhenius model), or
- the basic nature of the hydroxide ion in attracting a proton from a water molecule (interpretative and causal explanations, based on the Brönsted-Lowry model), or
- the ability of the lone pairs on the oxide ion to bond to a proton (interpretative and causal explanations based on the Lewis model).

A similar set of arguments applies to a discussion of the basicity of methylamine (produces hydroxide ions in solution, attracts protons from water molecules, has lone pair electrons to bond with a proton). The expert chemist is the one who has to have a repertoire of models and can choose an appropriate one for each circumstance.

IMPLICATIONS OF SYLLABUS REFORM AND TEACHING METHODS

The evidence provided earlier in this chapter indicated that many learners find it difficult to understand or apply models of acids. The books analysed give an indication of some of the reasons why this might be. Models are presented without reference to the development of the nature of science, i.e. the creation of explanations and models to promote understanding of experimental data, to generalise to novel situations. They are then not used elsewhere in chemistry in a consistent and coherent manner. Some of the models are not used at all by some authors. Calculations take up a large portion of the course without an understanding of the notion of mathematical modelling. At no point in any of the books does it say that the calculations based on the acid and equilibrium models are supported by experimental evidence. We should ask questions about whether we wish to see chemistry as an abstract component in the physical chemistry chapters, with experimental evidence in inorganic and organic chapters insufficiently related to the theoretical ideas.

The idea of separate physical, inorganic and organic chemistry sections may also be questioned. Such sub-disciplines in chemistry arose at the end of the last century and the first part of this century. It can be argued that such a classification is now obsolete. In the first instance, many recent developments in chemistry have taken place at the interface between chemistry and other disciplines. Molecular biology and materials science are two very good examples of such inter-disciplinary areas. Within the field of chemistry, most developments in both organic and inorganic chemistry now rely so much on high technology that it is difficult to separate these two areas from physical chemistry. A more appropriate approach might be to take a general topic, such as acidity, and explore its value in the contexts of inorganic and organic chemistry. This would produce an integrated approach that would link the theoretical parts directly with applications. Reflection on the practice would be referred back to the theoretical models produced, enabling them to be tested against data and appreciated in their contribution to understanding chemistry. Any deficiencies in explanatory power would then support the creation of better models and so the cycle of model production and evaluation against data from phenomena would continue.

In order to ground explanations in experimental chemistry, careful attention has to be paid to an appropriate selection of phenomena in the first instance. The author should be clear about which features of the phenomena are the focus for the explanations provided. It is helpful to be particularly clear about those features of the models that might be obviously distracting, as well as to those which might give rise to common misconceptions. Some examples could be selected for a comparison of the explanatory power of each model.

WHAT THIS COULD MEAN IN PRACTICE

The following examples are provided to demonstrate how the integrated cycle of model building and application might work in the case of acidity. It is not intended that this format would be implemented in all areas of physical chemistry. In practice, only some of the examples might be selected even for the concept of acid, so that the principle of integrated chemistry can be seen working. However, in order to give some indication of how the models of acidity might be explored in inorganic and organic applications chapters, examples in all four domains are given.

THE EXAMPLE OF HYDROCHLORIC ACID, IN THE CONTEXT OF INORGANIC CHEMISTRY

Hydrochloric acid, for example, can be characterised as an acid on the basis of its descriptive chemistry, such as its reactions with magnesium, sodium carbonate, sodium hydroxide and copper (II) oxide. In this, the purpose of the explanation is to demonstrate the fit with reactions of similar substances. No attempt at cause and effect explanation is made. The function of this exercise is classification by similar behaviour. By considering the process of making hydrochloric acid by passing hydrogen chloride into water, resulting in the formation of H^+, the Arrhenius model is invoked. This can be followed up by equations such as:

$$Mg + 2H^+ \rightarrow Mg^{2+} + H_2$$

to show how the existence of H^+ provides the basis for a mechanism for explaining the acidity of hydrochloric acid. The reaction of hydrogen chloride with water can also be interpreted as the transfer of a proton from a hydrogen chloride molecule to a water molecule:

$$HCl + H_2O \rightarrow H_3O^+ + Cl^-$$

This is tantamount to a Brönsted-Lowry model of explanation. In this case, a number of models of acidity can be used to explain why hydrochloric acid is considered to be an acid. Of course, the purpose of the explanation is closely related to the aspect of the phenomenon of acidity. As an example, the role of water in the formation of hydrochloric acid by dissolving hydrogen chloride in water is best explained by the Brönsted-Lowry model involving proton transfer. It is only in this model that the water is strongly considered to be an active chemical agent, accepting a proton from the hydrogen chloride. If the pH of the solution is measured, then the Arrhenius model of dissociation to give H^+ ions is the main attraction.

DISCUSSION OF PROPOSALS FOR CHANGE IN TEACHING CHEMISTRY

This chapter serves to provide some evidence for the following ideas within the context of acidity:

1. There is a hierarchy of explanations, which models serve, from descriptive to prescriptive.
2. There is a range of quantitative and qualitative models.
3. Quantitative models in acidity are not the most comprehensive, generalisable or applicable.
4. Later models are not reductionist versions of earlier models.
5. Models focus on different ways of looking at parts of phenomena.
6. No model can be said to be a best fit to a phenomenon.
7. Some models can be considered to be better suited to a specified aspect of a phenomenon than others.

In textbooks, models of acidity abound in all areas, and are reasonably well described in particular sections. Their status as models is grossly underplayed, leading to their ontological existence as real objects in themselves. Their relation to phenomena is frequently implicit and is sometimes not stated. Their explanatory power is not considered, nor are their explanatory weaknesses. Consequently, students using the textbooks have no rationale for independent selection of a model for a particular purpose. Without further guidance they must either accept the author's choice for the examples given, or proceed to construct their own rules for selection in novel contexts. In such circumstances, it is inevitable that errors in analogical reasoning will be made.

General principles of modelling acidity could constitute a basis for a physical chemical approach. An exploration of the value of this approach could then be carried out by application to examples from inorganic and organic chemistry. Organic and inorganic chemistry would then seen as test beds for the strength of the models chosen, rather than a collection of facts to be learnt.

A way forward for teaching chemistry might be one based on modelling as a central activity, with chemistry providing a particular and interesting context for this approach:

1. To include a discussion of the nature of science as it relates to the development of models of acidity, this being a clear and well-documented example of this process.

2. Using the concept of acidity to ask the question: what were the main imperatives for model development in history?
3. Are there any lessons to be learned from the answer to question 2, which can help us understand how learners in classes change their thinking?
4. Do chemistry teachers see some models of acidity as inferior to others and does this affect their attitudes towards teaching these models?
5. Are some models of acids too difficult to understand at certain stages?
6. How closely do established teachers and novice teachers follow the patterns adopted in textbooks?

Such a radical departure would require great changes in thinking of chemistry teachers and associated texts and other resources to support such changes. Further professional development to facilitate such changes would also be needed. It may be that it is the challenges of implementing any change rather than the desirability of change which limit the development of innovative approaches.

APPENDIX

The details of the textbooks surveyed are given below. The textbooks were chosen as typical examples, not to demonstrate idiosyncratic ways of explaining. Other textbooks adopt similar styles.

Two books most commonly used in A level chemistry teaching were analysed:

Textbook A: Chemistry in Context (Third edition) by Graham Hill and John Holman, published by Thomas Nelson (1989).

Textbook B: A-Level Chemistry by E.N. Ramsden, published by Stanley Thornes (Publishers) Ltd. (1985).

And a new book aimed at the same students:

Textbook C: Advanced Chemistry by James Maple, published by John Murray (Publishers) Ltd. (1996).

Chemistry in Context was written by two senior examiners of the Nuffield A level chemistry course, one of the most popular A level chemistry courses in England and Wales. The course is highly integrated in terms of physical chemistry with other areas and is based on a set of published books that form the basis of the syllabus. *A-Level Chemistry* is more specifically aimed at traditional chemistry syllabuses, emphasising the acquisition of factual knowledge more than the Nuffield course. *Advanced Chemistry* provides a much more lively approach, using a variety of novel analogies and explanations.

Chapter 13

Models in the Explanations of Physics: The Case of Light

Margaret Rutherford
The University of Reading, UK

INTRODUCTION

Over the centuries, the basic utility of physics explanations has laid a solid base for investigations, explanations and models in other subject areas. As the technology and instrumentation have improved, so the explanations and associated models have become more detailed, more predictive and a better fit to the phenomenon under investigation. In general, one consensus model has emerged which has held sway until some form of Kuhnian paradigm shift has occurred. However, there is one important phenomenon which has yet to develop a single consensus model, the phenomenon of light. This chapter therefore looks at the development of the models used to explain optical behaviour, how opinion has oscillated between the two major theories or explanations and the rather uneasy combination of them in the wave-particle duality explanation. This duality highlights the nature of models in that one is unable to say ‘light is ...’ but must always say ‘in this instance light behaves as if. .’ The phenomenon may therefore be a very useful tool in teaching about the nature of models. It could be argued that ‘the case of light’ is an ideal one for an examination of what is meant by a model, how models can be used and when they have been stretched too far. In the final sections of the chapter, the teaching of optical phenomena, specifically the phenomenon of colour, is briefly explored, the teaching models/textbook treatment examined and the possible links between the historical development of ideas about light and colour and classroom treatment of the ideas are tentatively developed.

J.K. Gilbert and C.J. Boulter (eds.), Developing Models in Science Education, 253–269.

MODELS IN EXPLANATIONS IN PHYSICS

In physics, which is frequently called the fundamental science, and which explains everyday phenomena as well as unpicking 'the way the world works', the use of models is commonplace. Models in physics have been developed for centuries ever since man became interested in more than basic survival. The first topics centred around things of practical interest such as how to navigate, how to determine the seasons, what governed the movement of the planets, what the stars were and why they seemed to move as they do. In other words many of the explanations and models used in these explanations were of utilitarian benefit. As the subject has developed so the immediate utilitarian aspect has diminished in importance and much of the current work and model development seems remote from any immediate relevance to everyday concerns. However, the metaphors used still reflect a concrete reality, despite the impossibility of ever directly perceiving the phenomenon for example, 'string' theory and the 'bag' model used in theoretical physics. What usually happens is that the idea or concept is reduced to its simplest form, assumptions are made and the model is constructed to relate to this. As the ideas develop so the model becomes more complicated and/or sophisticated and moves from concrete, through diagrammatic to symbolic. This final form will usually be able to explain more of the characteristics of the original than the earlier ones and makes fewer assumptions. The symbolic model, usually with mathematical equations predominating, is the most highly regarded in physics and possibly in other fields of science also (see Chapter 4). In this form more sophisticated predictions become possible and the field of study is expanded. Since a prediction may be termed a future oriented description, an explanation of what might or could happen in the future, this time dimension makes it probably the most highly valued and significant form of model. The power of a model is in its use to predict and this may lead to new understandings and modified or even radically changed models when the model is stretched so far that it becomes inappropriate. If we agree that prediction is the highest level of explanation (see Chapter 10), the use of models assumes an even greater importance.

Models may be represented as concrete or symbolic entities and these may well be mixed in presentation (see Chapter 3). Frequently they will have some verbal component. (Even a diagram needs labels unless the observer has a well-developed framework or 'expert' eye.) The more recent (and sophisticated) a model the more likely it is to be expressed in a mathematical mode. The case of light demonstrates a movement within a typology of models as we shall see later.

In the development of models to help explain the phenomenon under examination, it is usually the case that this development leads to one particular model being accepted as the best available and further development expands or refines this model. In Kuhnian type of revolution, a different model may be proposed and this new model may eventually come to be the accepted one. However, in the case of the nature of light this has not happened and, although there are attempts to unify the two major models, that using waves and that using particles, these attempts are not entirely satisfactory. It is usual to apply one model to explain certain phenomena and the other for different phenomena, although there are certain phenomena which may be explained using either model.

We cannot develop models or explanations without the use of language. If we accept that by 'stretching a model too far' we increase its power and develop our explanations into a predictive phase then language is being used on two levels, that of describing the process of modelling and that of using 'new' language to develop the model itself. This new language leads to a re-description of our model. By playing with the language we use we can revisit our model and explanation and look at them from a different perspective. The use of analogy and metaphor often widens our understanding of the models we are using but can also give rise to metaphorical explanations which students take to be literal descriptions There is some difficulty in separating the models from the explanations in which they are needed. In some instances the model is seen as the explanation for example, in the teaching of school level physics where sound is a wave. We will return to the language dimension later but first we must look at the two major models.

MODELS OF LIGHT

Electromagnetic radiation and its unique and peculiar characteristics has been the subject of debate and argument for centuries. It is only since the development of the quantum theory around a hundred years ago that some sort of consensus has been reached. And this is an agreement to disagree or rather to use competing models in different situations. It is interesting that because of the difficulty of explaining the phenomena and because of the explanatory power of both the extant models, this may be one of the few areas in science where students are confronted with a situation where they cannot mistake the model for the reality. Neither model explains everything and so we have to say 'Assuming that...' and the non-correspondence is highlighted.

One of the subsections of electromagnetic radiation is that of the specific wavelengths to which our eyes are sensitive, visible light (400 to 700 nm), and it is perhaps because of this that light and its properties have been the subject of interest for so long. Indeed, the study of light, its nature and properties has been of interest to natural philosophers since antiquity, (Von Laue, 1950). The explanatory models have oscillated between 'light as particles' (the original corpuscular theory) and 'light as waves' (originally in an elastic medium, the ether), to a duality model where both wave and particle models are used. However, there is another very simple, very powerful, diagrammatic model which seems almost to be taken for granted. This is the ray model and it is now used in conjunction with both wave and particle ideas to explain phenomena and in the development of the symbolic, mathematical, models constructed. There is no simple combined model which carries a useful information load since for the majority of cases, in the phenomena for which explanations are sought, duality is not appropriate. Each phenomenon is best explained by one or other model. The historical development of these two models, wave and particle, is interesting in that both models have compelling characteristics and opinion has wavered between the two for centuries. What is possibly of greater concern is that in the teaching of optical phenomena at school level, there is rarely any explicit link made between the two models and indeed a mechanical analogy or model (using rays), is frequently the most favoured.

One of the most appealing properties of light is the fact that it can be coloured and that objects appear differently coloured when they are all illuminated with white light (or light from the sun). The changes in the apparent colours of objects when the light itself is coloured is a continual source of amazement to many observers. Furthermore, the fact that shadows may also be coloured is a novel idea to most people. The phenomena of colour is however also one which most people take for granted, it is part of the background to one's life and is frequently only really noticed in its absence (for example, when a black and white film is watched on a colour television). It is however one phenomenon where the dual wave/particle is appropriate and thus explanations of colour are often used as exemplars in the rest of this chapter. The following section firstly describes the two major models and then looks at some of the historical features of the battle between the protagonists on each side.

The Wave Model

This model, although described here for light, is used for all electromagnetic radiation. It is the one which is evoked most frequently in explanations of interference and diffraction and in colour phenomena. The first two of these can only be satisfactorily explained using the idea that light sometimes behaves as if it were a wave (i.e. has wave properties) but the different colours of light can be explained using either model. Using the wave model, the different colours are said to be electromagnetic waves with different wavelengths and hence frequencies. The velocities of the different colours are the same in a vacuum but differ in a material medium. The rainbow effect and the production of a spectrum of colour using a prism are therefore explained in terms of the varying velocities of the 'coloured' components of white light in different media. This causes a differential refraction and the separating of the colours, Newton's 'differently refrangible' rays (Westfall, 1993). The addition of coloured lights to produce white or secondary colours is explained using ideas of the eye/brain combination. In the wave model the different coloured lights, electromagnetic waves of different frequencies and wavelengths, may be separated out from white light using a refracting medium which differentially refracts each colour. By choosing the angles of incidence of the light on the refracting surfaces a full range of colours may be produced and then recombined to make white light. By careful combination of colours we can produce any hue we wish. This rather simplistic description of the model leaves much to the imagination and can be expanded to answer more difficult questions. Using the wave theory, rays are drawn as straight lines perpendicular to the wave front and showing the direction of wave propagation (Figure 13.1).

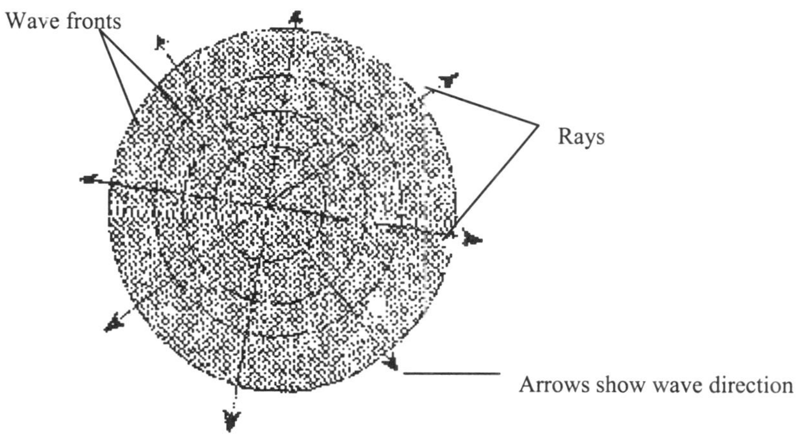

Figure 13.1. Wave Model Showing Rays

The Particle Model of Light

The particle model explains light in terms of quantum theory. A quantum of energy, called a photon, is emitted by an atom when one or more of the electrons in the atomic shells change energy level. This energy can only be emitted in discrete chunks (quanta) and the emitted photon will have an associated frequency and wavelength (the wave-particle duality). This model is rarely evoked in explanations of colour. It seems to be used only for the explanation of the photoelectric effect and in atomic spectra. The different colours are produced by the emission of quanta with different energies. The appearance of a coloured object is explained as follows: the molecules in the object have electron configurations such that electrons are in discrete energy levels. When an electron is excited by a photon of white light it jumps to a higher energy level. The electron then returns to its original energy state releasing once again a specific energy photon. Since this photon corresponds to energy with specific wavelength and frequency, this emission of energy causes the object to appear coloured.

One can see that whilst some phenomena may be explained using either model, there are some colour phenomena which can best be explained using one model and some using the other. For example, the spectrum produced from white light is best explained by the wave model whilst the particle model is more satisfactory when explaining atomic spectra and the appearance of objects as having different colours. (The usual explanation given in schools that some of the light is absorbed and some reflected only describes the effect and does not show why this happens.)

Using the particle model, rays are used to show the path of a particle. Figure 13.2 shows particles and rays used to explain reflection.

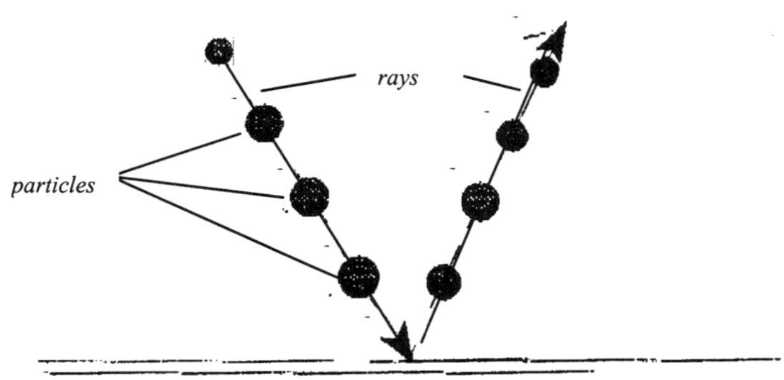

Figure 13.2. Using Particle and Rays to Explain Reflection

The historical development of these two ideas is interesting in that opinion has oscillated between them for centuries. The next section describes some of the highlights of this development

HISTORICAL DEVELOPMENT OF IDEAS ABOUT LIGHT AND COLOUR

An examination of the historical development of the (changing) consensus view about the nature of light is frequently useful in a teaching situation since students' views have been shown to mirror these changing opinions (Feher et al., 1992). Depending on the phenomenon under consideration, a physics teacher will usually use a wave or a particle model when explaining optical effects. The battle for supremacy between these two consensus models has been going on for centuries and, although many battles have been fought and won, the war is not over. There is now a somewhat uneasy truce between the models and the joint wave-particle explanation is considered by many to be a somewhat unsatisfactory consensus view. The terminology used above is not as fanciful as may be supposed. Consider the statement made by Espinet (1991):

> Controversy is at the core of the evolution of science. It is through controversy among practitioners that scientific knowledge is negotiated. Philosophers and historians of science have suggested interpretations of such controversies using different metaphors: Kuhn's shift from the old to the new paradigm ... or Lakatos's defence of the hard core beliefs of research programmes .... However, these struggles took place within the very specific community of scientists. (p.3)

Probably the first of these competing models of light was the one using particles (the original corpuscular theory) and this dates back at least as far as Aristotle (384 to 323 BC) (von Laue, 1950). However, shortly after this Euclid, in around 300 BC, studied optical phenomena and used a ray model, assuming rectilinear propagation of light, to explain perception, depth of vision and perspective. Rays, frequently starting from the eye, were used to explain reflection and refraction, and the explanations produced by Hero and Ptolemy, for example, used only Euclidian geometry to explain the phenomena (Cohen and Drabkin, 1948). Light was seen as emanating from the eye and returning to it. This notion that light comes from the eye to a perceived object, rather that being reflected to the eye from the object, is a naive conception, common in young children today (Feher et al., 1992). Some fifteen hundred years later, Snell developed a mathematical model to explain refraction, possibly the first mathematical model of optical behaviour. At about the same time Descartes explained the nature of light in terms of a mechanical model. Descartes used the idea of a stream of moving balls as an analogy for a ray or beam of light. In 1637 he published an account of the formation of a rainbow which had much in common with an earlier descriptive explanation given by Theodoric of Freiburg in the fourteenth century (Sabra, 1981). However, Descartes stated that the colours were caused by a tendency of the aether particles to rotate. Those particles which rotated fastest looked red and as the velocity decreased so the colours shifted through yellow and green to blue. He also thought that the colours of objects were modifications of white light caused by the surfaces of the objects. It was this theory that Newton questioned and which led to the acrimonious exchanges between the two (Sabra, 1981) Newton's Opticks, published in 1704, is probably one of the best known, though least read, of his major works. It brought together the ideas and experiments that he had been working on for at least thirty years. (His new theory of light and colours was published by the Royal Society in 1672.) Apart from his classical experiments with prisms to separate and recombine the colours in white light, he also experimented with water-filled globes, this being the initial factor in the development of his decision that white light is a mixture of colours, 'differently refrangible' (see, for example, Westfall, 1993). However, Newton was still a proponent of the particle model which might indicate again a notion of paint like grains.

Even Huygen's theory of light, using wavelets, was based on pulses and not on vibrations although it has subsequently been transformed into a wave theory. However, this could still not account for the different colours observed. Subsequently, when Newton used forces to explain interference he came to accept something of wave theory and so was forced to re-examine his ideas of colour in light. He had a real reason for wishing to understand the phenomenon since the aberration of lenses which produces coloured fringes around the edges of images was causing problems in the construction of telescope systems.

Newton therefore postulated two contrasting and incompatible theories of light. Firstly, that all particles of light were the same size. They were reflected or refracted according to their velocity. Colours in light were reduced to laws of elastic collision. Secondly, he thought that particles of light were of different masses and that these excited condensations and rare factions in the medium, that is, wave phenomena. The medium then alternately reflected and transmitted the particles. He claimed that white light was an 'aggregate of different homogeneous rays'. It seems that

Newton was trying to reconcile the ideas of particles and waves (Westfall, 1993).

Nicolas Malebranche in 1699 also propounded a medium or wave theory of light. He thought that light was caused by pressure waves in the aether and that the colours were related to the frequency of these light waves (similar to the relationship of the frequency of sound waves and musical notes). He claimed that white had the highest frequency then a reduction in brightness from yellow through red and blue was caused by a reduction in frequency, hence to black which had no frequency (no vibrations). This notion was in contrast to those who faithfully followed the order of colours observed in a rainbow and with hindsight, seems to have been based on intensity rather than on frequency. However, Malebranche changed his view when he read Newton's Opticks and accepted Newton's ideas that white light is not homogeneous. He was also seduced by Newton's attempts to relate colour on a harmony scale similar to a musical scale (Dampier, 1966).

In this latter theory, Newton developed a relationship between the perception of different colours based on an analogy with musical sound, light particles excite vibrations of different 'bigness' (wavelength or amplitude?) in the aether. These caused corresponding vibrations in the optic nerve. Thin film effects were explained by saying that at the surface of the film the aether was condensed or rarefied by the wave motion excited by the light particles and this caused either reflection or refraction. Next Newton tried to divide the colours of the spectrum up to correspond with the musical scale and to make a match between seven colours and seven notes. We now consider that visible light has a continuous spectrum which goes from a red sensation through green to a purple hue; however, most texts and teachers still teach the seven colours of the spectrum which is maybe more a tribute to Newton than a result of careful observation.

Hooke criticised Newton's theory because he thought that light was a motion propagated through 'aether' and that colours were produced by a disturbance of vibrations in the aether, not by the action of different light particles. He backed this by claiming that his theory corresponded more closely to observations since it could account for both diffraction and the periodicity of light. However, he also thought that there was an analogy between musical notes (harmony) and the pleasing mixing of colours.

Whilst the seventeenth century was very fruitful in advancing theories of light and colour, there was practically no progress in the eighteenth century possibly because the practical problem of lens aberration had been by-passed by Newton's reflecting telescope. However, at the end of the eighteenth century, Young's experiments on diffraction and interference led to the idea that each colour has its own wavelength and frequency (Bynum et al., 1981). Least it be thought that the discussions were carried on between a select group of natural philosophers, we should reflect that Goethe was also interested in trying to explain colour phenomena. Although the actual phenomena under consideration were known and accepted, the explanations provided by Goethe were very different to those of Newton which have come to be accepted by scientists. Espinet (1991) argues that it was the process of enquiry used by the two protagonists which led to the alternative views about light and colour held by Goethe. Goethe's observations led him to a view that colours were once again, as in the ideas of Theodoric, caused by the existence of bounded and unbounded surfaces. His methods were qualitative and humanistic and might well be similar to the methods used by the majority of people (non scientists) today. The fact that a Newtonian perspective is the one generally held may say more for the influence of Newton as perhaps the premier scientists of his day than for a logical arguing of the methods used (Ribe, 1985).

The work of Young, together with that of Fresnel, Arago and Fraunhofer, finally established that all the then known phenomena associated with light could be explained if light is considered to be a transverse wave. Originally these light vibrations were considered to be analogous to elastic mechanical waves in solid bodies. It was therefore necessary to continue with the postulation of the existence of an aether filling empty space. There were however difficulties in explaining why longitudinal waves were not apparently propagated in this ether. It was not until Maxwell's experiments at the end of the nineteenth century that the possibilities of electro-magnetic waves (with light as an example) were able, at least partially, to explain this. However, the transverse wave theory was by now generally accepted by scientists.

This wave model explains how light travels through space, but it cannot explain the interaction of light with matter, that is, it cannot explain absorption and emission of light. For this the discoveries of Plank, Lorentz, Einstein and others in the early twentieth century were needed. Their works marked the beginning of the re-instatement of the particle model of light. The difficulty was that, in the intervening time, the wave model had become firmly entrenched. The discovery of particle-like properties shown in the photoelectric effect could not be explained using this wave model. It was Einstein, becoming more and more concerned about the increasing complexity of the explanations and models used (Einstein and Infeld, 1938), who proposed that light itself might be quantised within the wave, with short wavelength light containing energetic quanta and the long wavelength light containing less energetic quanta. The connection between the colours of light and the frequency of the photon emitted during electronic transition had to wait until Planck in the early 1900's. He developed the theory of the relationship between energy level of the electron and the energy and frequency of an emitted photon. This finding was substantiated by physicists working on the emission and absorption spectra of gases and hence an acceptance of the wave-particle duality of light and of the colours of light enabled an explanation to be found for all the observed effects.

In summary, therefore, the predominating model is represented by a symbolic-diagrammatic using directed arrows to indicate the rectilinearity and direction of the light. This model is then transformed using Euclidean geometry into a mathematical mode which enables predictions to be made. Alongside these two models are the wave and particle models used to explain the nature of all electromagnetic radiation, not just the visible portion of the spectrum, but both of these also use rays to develop the mathematical models which enable prediction. I asked an eminent physicist, researching in the field of optical phenomena, how he would explain such phenomena and what models and diagrams he would use. In common with almost all the people interviewed on this topic (Gilbert, 1997), he said it would depend on who he was explaining to. At first year university level he would start with waves and draw rays to show the direction of propagation of the wave. He would then also introduce ideas of electron/quantum mechanics to explain in a simplified manner the behaviour of the materials in the phenomenon of, for example, refraction. However, when using Huygens' principle he said he would also introduce particles since the traditional explanation of the lack of backward propagation of the wavelets is unsatisfactory.

> 'I think if you take something like Huygens' principle and the way it is presented very much in the words of the originator, it sounds mechanical, it sounds contrived, it doesn't sound specially convincing quite frankly! And anybody who thinks a little bit more deeply about it probably wonders about things like backward travelling waves, and if you can simply point out that this is a useful concept but it is really a concept that's a part of a much more comprehensive theory that actually explains a tremendous amount of different optical phenomena, for example, diffraction, then at least that provides them with a little bit more confidence...' (physicist interviewed).

He was the only person who I have interviewed who seemed to carry both wave and particle ideas concurrently and indeed stated that they were not competing but complementary models. He also stated that the usual textbook drawings of ray diagrams for multiple lens instruments were in general unsatisfactory since they did not show what happened phenomenologically at each lens made it difficult for students to understand what was happening. However, he stated that to explain many of the phenomena in a totally satisfactory way it was necessary to use mathematical models even for an adequate description.

> '...if you want to interpret the spectrum fully and account for the full form or even to extract the maximum information from it, then I'm not aware of any other way of doing it apart from fairly advanced mathematics ... the sort of simple Doppler shift model is really totally inadequate..' (physicist interviewed).

He was totally comfortable with the joint model but pointed out that all our understanding is but a step on the way to an explanation of why and how things happen.

> 'Physics is not a cut and dried subject, it's dynamical, the product of peoples' minds, subject to test all the time and these are not hard and fast ideas.' (physicist interviewed).

When talking about his approach to teaching, as a research scientist, this physicist also stated

> 'I think that goes for almost any form of teaching, if someone is either researching or is more expert in a particular area, you can bring some of those concepts, I think you have a richer knowledge of, you know, about the intricate subtleties and I think you can inject these at appropriate intervals even at an elementary level, pointing out to the students that what appears to be a rather superficial model actually is part of a much deeper more comprehensive one …' (physicist interviewed).

And indeed this emphasises the physicists' use of models rather nicely!

However, many teachers, certainly of elementary physics, have not performed research in any particular field of physics and so cannot claim to be 'experts' in the sense meant in the above quotation. How do these generalists teach light and colour? What models do they use? The examples in the next section are drawn from three sources; firstly, a lecture given by such a generalist; secondly, textbooks and, thirdly, very briefly, teacher educators.

AN INQUIRY INTO THE TEACHING OF LIGHT

The next part of this chapter is based on several inquiries carried out from 1993, some aspects of which have been reported elsewhere (Rutherford, 1997). The intention was to investigate an underdeveloped concept area in the school syllabus, that of colour. Colour was chosen since although it would appear to be important and of relevance in all science subjects, very little has been written about children's ideas of colour (see Feher and Meyer, 1992, as one example). One has only to watch an old black and white film on television to realise that colour plays a very important role in our lives. Apart from the discriminatory aspects of the variety of colours, we often associate colours with moods and symbolism; for example, red is hot or angry, blue is cold, yellow is cheerful, black is for mourning and white indicates purity. We use colour in decorating to change the moods of a room or to visually change the shape (e.g. a dark ceiling reduces the apparent height of the room). Animals and birds use colour for camouflage and for attracting mates. Whilst it is difficult to establish exactly what the visual ranges of animals, birds and insects are, there seems to be some evidence that they may well have a range which extends wider than the human range and encompasses frequencies in the infra red/ultra violet ranges. Even plants seem to need colour (to attract pollinating insects) so that it might be said that colour variation is important for all living things. This being the case, it is surprising that the mechanisms and models of colour production and colour perception are given very cursory treatment in science education. It is only in the technical fields of printing and stage lighting that these seem to be considered. It may well be that colour is so important that it is taken for granted (as will be illustrated later). Nonetheless, as with other everyday phenomena which are amenable to a rational scientific explanation, it would seem reasonable that colour should receive attention as a topic in many if not all scientific disciplines taught in schools. One of the findings of the research reported in this chapter is that this is not so, only in physics is it given any 'space', and this as a subsection of light. In other subjects it seems to be a phenomenon which is almost totally neglected. The investigation into colour concepts inevitably included light phenomena and the major light models and this investigation forms the basis of the following section. The data were collected from interviews with six teacher educators, examination of twelve textbooks and is fully reported elsewhere (Rutherford, 1997). Summaries of the findings are given here and links drawn between these and the historical development of ideas about light.

THE TEACHING OF LIGHT AND COLOUR

The single most used model in teaching light is neither the wave nor the particle model. It is the original diagrammatic model where a straight line with an arrow on one end is said to represent a 'ray' of light. The arrow indicates the direction from source to object and also direction of the light after it is reflected, refracted, undergoes interference or is diffracted. A group of such lines is said to be a beam of light. If we relate this to the wave theory, as said before, then the line represents the direction of the movement of the wavefront and is drawn perpendicular to the lines representing crests or troughs. Using the particle model, the line represents the path of the particles. The use of rays dates back at least as far as Newton with his 'differently refrangible rays' used to explain the colours when white light is passed through a prism. This use of straight lines as rays has given rise to the term 'geometric optics' when considering the reflection, refraction and dispersion of light. No mention is needed of either waves or particles, although reflection is often described as light 'bouncing' off a surface (a mechanical model). However, once interference and diffraction are studied, it becomes necessary to use the wave model and talk about wavelength and (sometimes) frequency. In geometric optics, the usual methodology is to show some phenomenon such as reflection and then to draw a diagram using rays to explain what seems to be happening, the 3D phenomenon has been converted to a 2D symbolic model. The laws of reflection are deduced by drawing ray diagrams. This activity is then extended to look at refraction and the change in direction which frequently accompanies the change in speed of the light as it passes from one medium to another is again represented by straight lines. The next stage is to look at the geometry of the diagrams so constructed and to develop mathematical equations which model the phenomenon. By this stage the actual physical event has become more and more remote from the activity of the students. However, the mathematical model is more powerful in that it is more accurate than the ray diagrams and enables some prediction to be done. To take a first-year university class as an example:

> 'If I start off with an object, object is one metre high, on the other side (of the lens) it forms an image ten metres high, then is seems to me logical to say that the magnification has been ten because…, because it made it ten times bigger'.

Then:

> 'However, I can also express this in terms of the object distance and to see that we have to go back to that drawing of

> ours.... so just look at the triangle AHP, this one on my drawing.....'

And finally, after looking at similar triangles:

> 'Mathematics can show you much more accurately what is actually happening' and 'So you don't have to do that elaborate way of finding images any more. You just need the mathematics for that'.

This seems to be the standard way used by textbooks observe or describe the phenomenon, draw a diagram, relate this to mathematics and try to find some simple equation relating some of the quantities to each other and how they affect each other. Then use this mathematical relationship to find out what should/might happen in other situations. So far this is a fair analogy with some of the activities carried out by 'real' scientists but the next step of testing these predictions rarely seems to happen, in other words we have a model of scientific enquiry which is content to stop at finding a model which will explain most of the empirical data. This is insufficient. To quote Giere (1988):

> The rule is, roughly, that a model that can correctly predict part of the data is preferable to one that is constructed by empirically fitting all the data. (p.199)

Looking at the textbook treatment of light and colour, light as a subject is only in physics textbooks. The topic of colour, which might be assumed to be of importance in most science subjects, is treated in differing ways. From the twelve textbooks examined, the analysis was in terms of the models used and the coherence of the treatment. In the biology texts, and in the majority of the chemistry texts, the most common use of colour was as labels. The phenomenon of seeing colours was not mentioned. Colour was not used as means of classification except for colour changes in chemical reactions (red for acidic and blue for basic). The physics textbooks, on the other hand, divided light into geometric and physical optics with atomic spectra included under a section on quantum theory. Light as waves was the main model used with the ray model being almost taken for granted. Some texts explicitly made the connection between wave fronts and rays and others used the bouncing ball analogy. However, none of them produced a coherent picture of the models and their interrelation (Rutherford, 1997). The textbooks seemed to echo the historical divide between the corpuscular and the wave theories and the ray model was used extensively, frequently with no link to either of the explanatory frameworks. The level of explanation was almost exclusively descriptive with an occasional interpretative aside or causal link provided. Discrete isolated explanations for the various phenomena are provide and the link between particle and wave theories considered to be essential by the physics researcher mentioned earlier were non existent. Colour was explained with wave ideas when looking at a Newtonian spectrum and with particle ideas when looking at individual atomic spectra. There seemed to be no link between the various phenomena and their explanations.

Since textbooks are usually written by teachers, another investigation of relevance is that into the views of teacher educators. Interviews with several of these led to the conclusion that the model most widely expected to be used was the wave model (Rutherford, 1997). Although the particle model was used in selected contexts, the two were never used simultaneously. The interesting finding here is that the majority of the teacher educators interviewed would expect most explanations to be given in terms of the ray model, i.e. geometric optics. Only one of the interviewees (a physicist) said that he thought that even quite young children could cope with a 'wiggly waggly wave' model for light. Indeed some respondents had reservations about introducing wave phenomena before the age of about seventeen years. There is little surprise therefore in the findings that textbooks also divorce the two major models and revert to a simplistic ray diagrammatic model and thence to a simple mathematical equation to explain light phenomena.

CONCLUSION

In this Chapter, the development of the two extant models of light has been explored and the evidence for the ray model possibly being a unifying idea has been presented. Historically, the two major competing models, that of waves and that using particles, both have compelling characteristics and the evidence used by the proponents of each side was well documented and presented. Consensus opinion therefore oscillated between the two models with good reason as explanations for different phenomena developed. With increasingly sophisticated measuring techniques and developments in mathematics, the phenomena could be more and more closely examined and models of greater complexity produced. This inevitably led to the invention of explanations for one phenomenon which produced difficulties in another, an example being the aether which was postulated to help explain many phenomena but could not be identified. The combination of models and explanations for the phenomena associated with light is comparatively recent. This may well explain the teaching approaches which were found. Several sources have been reported and it would seem from the evidence presented that the majority of resources (teacher, textbooks and teacher educators) still keep the two models in isolation. The explanations given use either the wave or the particle model and convert these to diagrammatic models using rays. This in turn leads to symbolic models which at school level revert to the mathematics used by Snell. The predictions made with these models are simply 'more of the same', for example different focal lengths in lens combinations, different types of mirrors and so on. The history of the development of ideas of optical phenomena is restricted to a brief look at some of Newton's experiments and the exciting controversy which could be used to illuminate ideas about the nature of science are not in general presented to students. Indeed a missed opportunity. Although the literature on children's ideas of light has only been touched on briefly there is some justification for claiming that, as in other areas of physics, these ideas mirror the historical development of physicists' understanding of the phenomena, which is another reason for incorporating the 'battle for a consensus model' into the teaching of light.

In summary what can we say? That modelling in explanations in physics is as old as man's wish to explore and explain his surroundings: that the models used in teaching have a tendency to revert to the simplest of those used historically and only move to more and more comprehensive and predictive models in a sophisticated research; that the early models are extremely robust and have a tendency to occur in studies of children's naive ideas; non-specialists are content with these simplistic models and, in the case of light, only one model is used at a time with the complicated dual model ignored. It would therefore seem that at school and early tertiary level, students are not expected to engage with the current ideas and models used by 'real' physicists in explaining light and electromagnetic phenomena, but to be content with simplified models and descriptive explanations.

Chapter 14

The Role of Models in Biotechnology Education: An Analysis of Teaching Models

Bev France
Auckland College of Education, New Zealand

INTRODUCTION

Time magazine has nominated the 21st century 'The Biotech Century' and predicts that the 20th century revolution in infotechnology will merge with the 21st century's revolution in biotechnology (Isaacson, 1999). These days biotechnology has a high profile and strong opinions are generated when the dilemmas of genetic engineering and its products are placed before the public (Van Brunt, 1991). There is a consensus that there is a need to educate citizens living in this new century about biotechnological innovations. In New Zealand the need for public education about biotechnology has been identified (Macer, 1990) not only because the public feels insecure about biotechnological products and processes but also because this technology makes a significant contribution to the production of value-added primary products (Kennedy and Davis, 1994). This challenge has been met in New Zealand education through the avenues of science and technology education. Biotechnology education is claimed by biology teachers to illustrate biological concepts (Farmer, 1994) and its classroom expression has a 'Technology as Applied Science' (TAS) focus (Gardner, 1995). A modern form of technology education which includes biotechnology has been introduced into the New Zealand curriculum in response to the recent world-wide curriculum reform (Layton, 1994). When biotechnology is given a technological focus there is the assumption that biological concepts contribute to a technological outcome. This technological viewpoint broadens the educational perspective for the learner and teacher and provides opportunities for an exploration of the perplexing questions that this technology will pose in the new Millennium.

J.K. Gilbert and C.J. Boulter (eds.), Developing Models in Science Education, 271–287.

The introduction of a new technology education curriculum in New Zealand (Ministry of Education, 1995) provided an opportunity for professional development for science and generalist teachers to explore biotechnology not only from a biological perspective but also from this broader focus. The public perception of biotechnology tends to limit biotechnology to genetic engineering which can restrict a practical classroom approach. The definition in *Technology in the New Zealand Curriculum*:

> Biotechnology involves the use of living systems, organisms or parts of organisms to manipulate natural processes in order to develop products, systems, or environments to benefit people (Ministry of Education, 1995).

signals that biotechnology can encompass a range of technologies employing biological processes and provides a broader viewpoint for exploration (France Farmer, 1998, p.12).

During 1992-94 I researched and developed a professional development model for biotechnology education that enabled teachers to plan and teach technology education programmes with a biotechnological focus. The potential for model development in biotechnology education became apparent in this research project. Teachers realised that their teaching programmes required students to develop an understanding of biological concepts as well as the dimensions of the problem-solving situation as experienced by the biotechnologist within their community of practitioners (Lave, 1991). Because the biotechnological agent was microbial, teachers were faced with exploring and explaining biotechnological processes carried out by 'invisible' organisms. In addition there was an expectation that students would manage and monitor their growth during the production of biotechnological products. Faced with this dilemma, these teachers used models to make these objects visible and their metabolism comprehensible.

This Chapter will explore the role of teaching models that were developed during this research programme by:

- describing how the teaching models were identified
- justifying the choice of these teaching models
- exploring and analysing the nature of these teaching models
- examining problems encountered with model use
- identifying some implications for teaching.

IDENTIFICATION OF TEACHING MODELS WITHIN THE RESEARCH PROJECT

This research project chose an interpretivist methodology because it accommodated the complexity of the classroom (Lather, 1992; Merriam, 1988). This methodology provided space for teachers to develop appropriate strategies to enhance learning and the range of data collection methods enabled the researcher to explore the reasoning behind their decisions. Classroom data were collected from a total of 19 teachers in the research period (1992-94) who developed and taught biotechnology-focussed teaching programmes. The data included written reports (diaries, lesson plans and flow charts of lesson sequences), interviews (semi-structured interviews with teachers and students, unstructured with teachers alone and in groups) and classroom observations. This analysis is concerned with teaching models that were employed during the project.

The definition of a model as 'a representation of an idea, object, event, process, or system' was the starting point for this analysis (see Chapter 1) and is based on model formation as a process of analogy drawn between a source (something perceived to be somewhat like the phenomenon under study) and the phenomenon itself, which may be called the target.

Biotechnology involves biological processes and biotechnologists have an understanding of the biological concepts on which they are based. Teachers needed to explain these principles and used teaching models in their explanations. As Gilbert (1997) and Gilbert and Osborne (1980) assert, science models take up an intermediate position between the observed reality of phenomena and the theory explaining it. It can therefore be assumed that the teacher's role in this process is to develop a student's mental model of a phenomenon towards a scientist's mental model (Gilbert et al., 1998a).

Gilbert et al. (1998a) have identified teaching models as those used by teachers to explain scientific phenomena and comment that consensus models are often adapted for this function. In this research project the biological consensus models used by teachers included:

- representations of microbial morphology
- flow diagrams using words or chemical formula that describe the metabolic pathways
- diagrammatic representations of microbial techniques.

These teaching models were employed to make the invisible agents visible. Microbial representations used were textbook diagrams. The metabolic pathways selected were simplified word or chemical equations. The descriptions for making microscopic slides or manipulating microbes were represented by a series of pictures identifying the important aspects of the process.

Because microbes are invisible they need techniques to make them visible. This can occur through magnification under the microscope or from amplification when they are grown in large quantities either as a colony on a substrate or in a liquid broth. These procedures require the learner to develop safe and effective handling techniques and these teachers used models to demonstrate these procedures as well as evaluate their students' competence when carrying them out.

Biotechnological processes require biotechnologists to have an understanding of metabolic pathways. Such an understanding can be obtained by theoretical study. However, at a junior level teachers may identify the critical stages by showing students the chemical changes that have occurred. These may involve colour changes via pH tests, identification of gaseous by-products for example, carbon dioxide or the teacher may provide an example of the endpoint of this metabolic pathway. These physical manifestations of the stages can provide the basis of a technological model or be demonstrated using a model that mimics the process.

However, model use in technology education is problematic. The situated nature of each technological learning means that there is less emphasis on using consensus models (Gilbert et al., 1998a). Sparkes (1992a) observes that in technology education models are used to 'create simplified versions of reality for a particular purpose'; however, when modelling is given a cognitive function and technological activity is perceived as 'thought in action' (Kimbell et al., 1996) then models provide a focus for developing the learner's technological understanding.

Access to the biotechnological community of practice (Lave, 1991) provided teachers with a range of consensus models. These provided procedural knowledge (McCormick, 1997) that ranged from:

- models used in product standardisation
- working models used to predict and evaluate biotechnological processes
- organisational models.

The research data demonstrated that understanding of biological concepts and the acquisition of biological technical skills was just a small part of the procedural and conceptual knowledge needed to realise biotechnological solutions (McCormick, 1997). Access to the community of biotechnological practitioners provided the framework for teachers and students to develop their teaching models that enhanced their understanding and enabled students to realise biotechnological solutions. These models provided the data for the following analysis.

JUSTIFICATION FOR THE SELECTION OF THESE TEACHING MODELS

Teachers involved in this research project were aware that they had to bridge the gap between the expert (biotechnologist) and the novice (student). The professional development model proposed in this research supported them in their role as mediator between the community of the classroom and the community of biotechnological practice (France, 1997).

All of these teachers had experience in employing constructivist teaching strategies and identified the level of prior biological knowledge and prior technological capability (Ministry of Education, 1995) that their students possessed. Such gaps in understanding included: the inability to perceive how small microbes were and therefore that they were invisible to the naked eye; ignorance of safe handling procedures; a lack of knowledge of their metabolism; an unawareness of their wider role than as disease-causing organisms. In response to the technical demands required to carry out these technical processes teachers used models as a tool to access technological procedural knowledge (McCormick, 1997). When teachers realised that the learning potential was much wider than biological knowledge and skill development they identified a need to access information from the biotechnological community and used their models of organisation to give authenticity to their classroom programmes.

DESCRIPTION AND ANALYSIS OF TEACHING MODELS

Teaching models (see also Chapter 10) developed in this research project have been summarised in Table 14.1. The table identifies the class level and age of student; model (by name); a descriptive summary; characterisation that includes model type, mode of representation and its function; an analysis in terms of model components and the learning outcomes that occurred in these teaching situations.

The literature on teaching analogies has proposed a range of features that enhance the analogical efficacy of their employment: Abell and Roth (1995); Clement (1998); Dagher (1994, 1995); Glynn and Takahashi (1998); Thagard (1992); Treagust (1993, 1995); Wong (1993). These features have been translated into a list of requirements for effective teaching models and will contribute to this analysis.

Teaching models need to:

- be a reasonable fit to enable the learner to draw close comparisons between the source and the target
- use a mode of representation that is accessible and appropriate to the culture of the learner
- indicate a broader field than the phenomena under explanation
- be self-generated by the learner.

Table 14.1. Teaching models developed for the biotechnology education research programme 1992/94

Level and age of students	*Model name*	*Description*	*• Model type • Mode of representation • Function*	*Analysis*
Year 11 (15-16 years)	The mighty microbe	Text book diagrams of microbes.	• teaching model • visual representation • to enhance meaningful observation	Source: diagrammatic representation. Target: bacterial morphology. Outcome: Colony mistaken for microbes.
Year 11 (15-16 years)	The magic circle	Protocol for transferring microbes to a new culture. Red dye represents microbes.	• teaching model • material representation. • practising aseptic culturing techniques	Source: red dye. Target: bacterial culture. Outcome: hot bacterial loop was mistaken for presence of red microbes.
Year 7+8 (11-12 years)	Microbial plating	Gelatine model of an agar plate. Blue dye represents yeast.	• teaching model • material representation. • developing microbial plating skills • evaluation of skill development	Source: gelatine plate + blue dye. Target: agar plate + yeast colonies. Outcome: students compared distribution of blue dye with distribution of yeast colonies on agar plate.
Year 7+8 (11-12 years)	Yoghurt product	Yoghurt product that set standard specifications for further production.	• teaching model expressed model • material representation. • for standardisation	Source: yoghurt. Target: yoghurt (future product). Outcome: students found it difficult to produce a consistent evaluation checklist.
Year 12 (16-17 years)	Pop bottle fermenter	Fermenter construction. Monitoring growth rates of yeast in different substrates.	• teaching model • material representation of fermenter • expressed model • prototype used to develop problem-solving solution	Source: pop bottle fermenter. Target: fermentation technology (bioprocess). Outcome: concentration on technical skill development to make and use a fermenter.
Year 9 (12-13 years)	Quality control process model	Class simulation exercise to produce soft cheese. Protocol developed by students to monitor + co-ordinate cheese production.	• teaching model (company organisation + delegation of roles) • visual and verbal representation • expressed student model (for planning + organisation) • visual and verbal representation • flow chart for monitoring cheese production	Source: Organisational brief modelled on company requirements. Target: organisation and monitoring cottage cheese production. Outcome: development of a sequence of testing procedures to monitor production of cottage cheese. Collaborative problem-solving.

Model: The mighty microbe

Katherine (the teacher) was aware of the need to identify her students' prior knowledge of microbial morphology and used a range of teaching strategies to help students interpret a microscopic examination of microbes. She asked students to make a stained slide of a microbe that they had grown on an agar plate and compare it to a generalised diagrammatic textbook representation. The source was the diagram that illustrated the shape and the internal structure of a generalised bacterium. The target was the stained bacterium seen under the microscope where the shape was discernible to the student but no internal structure was visible.

Such microscopic examination is notoriously difficult. It is easy to miss these small organisms, let alone identify any details of their morphology, as school microscopes are incapable of providing magnification that portrays bacteria any larger than a dot. When these students were asked to draw the bacteria as it appeared under the microscope, they drew the bacterial colony that was present on the agar plate, because the dot under the microscope was an unconvincing analogy to the text book model. When questioned about this anomaly, students pointed to a bacterial colony as an example of a bacterial cell and used a book illustration to justify their accuracy.

In this situation these students had mixed the model with the reality. Katherine had used the model to help her students visualise the abstract 'invisible' (Duit, 1991). However, the fit between the source and the target did not provide a close enough comparison for these students and this model became a barrier to their understanding of bacterial morphology.

Model: The Magic Circle

In the same teaching programme Katherine explored the manipulation of microbes via plating and sub-culturing via a teaching model called 'The Magic Circle'. The teacher modelled a sub-culturing and plating technique using red ink instead of microbes and then required her students to carry out this procedure before translating this procedural knowledge to a situation where bacteria were used. The ink would solve the problem of invisibility and illustrate the importance of safe manipulation techniques to prevent contamination. This procedure showed how easily bacteria can be dispersed away from the working area with the formation of aerial droplets. Such contamination would be seen by spots of red on a sheet of paper set on the working bench. In addition, this model would illustrate serial dilution as the solution representing the bacterial culture would loose its redness as the dye was transferred from test tube to test tube. This teaching model set the source as the red dye and the target as the bacterial culture.

When asked to describe and explain their plating of bacteria, students commented that when the bacterial loop became red during flaming it indicated that bacteria were present. Although this teaching model appeared to provide close comparisons of source with target, the learners were unable to distinguish these differences. For this group of students, the source became the target for they had not separated the simulated protocol from reality (Abell and Roth, 1995). They gave imaginary features to the model because they could not reconcile the target with the source.

This model appeared to foster misconceptions, as the mode of representation introduced another step away from the original process. The red colour was given other significance instead of just being seen as a method of making the invisible visible.

Model of Microbial Plating

In this programme, Elizabeth (the teacher) wanted her students to be able to grow dispersed yeast colonies on an agar plate. She was aware that microbial plating was a specialised technical skill that involved carrying out a procedure with a substance that was invisible. Normally a yeast solution is opaque but this solution had to be diluted and then spread out thinly on the agar plate so that individual yeast colonies could be seen, otherwise the culture would be seen as a 'lawn' of yeast growing on the agar plate. To add to the problem, yeast cells are invisible and the dilute solution was clear, therefore the spreading out and plating of these individual cells on agar plates became an imaginary process.

To solve this problem Elizabeth developed a gelatine model of an agar plate which used gelatine spread on a tinfoil circle. The gelatine surface provided a medium for her students to practise their plating techniques using blue food colouring. In this exercise the gelatine and the blue colouring were the source. Students were able to evaluate their plating techniques by looking at the distribution and thickness of blue dye on the gelatine plates. When the students were given the real plating episode (target) they produced agar plates that had spread out yeast colonies that were relatively contaminant free. These results indicate that these students had mastered a complicated plating technique.

Elizabeth had used a material model to visualise the abstract unseen 'microbes' (Duit, 1991) to develop her students' procedural knowledge. Because she had modelled the process and emphasised the role of the blue dye in the modelling stage, the students were able to transfer this activity when plating out a yeast culture.

In this situation the teaching model provided a close fit between the source and the target. The teacher was able to emphasise the similarities and differences as the students worked through the plating techniques. She added to this close fit by making the gelatine model approximately the same size as the agar plate which the students used in the 'real' plating situation.

Model: Yoghurt product

Sparkes (1992a) observes that technological models are used for development and evaluation. This teaching programme provided an opportunity for the teacher (Lance) and his students to explore the validity of setting up a biotechnological model for on-going evaluation. Yoghurt is a product formed by the fermentation of milk in controlled situations. In this teaching programme, yoghurt was adapted to fulfil a particular niche market with the addition of additives and the development of appropriate packaging and marketing. This model is a material representation of the final product and provides the prototype (proposed endpoint for testing) against which further production can be standardised. However, because it is a copy of reality as well as being the final product, both the target and the source are the same.

Lance encountered problems when using the yoghurt product as a prototype. Biotechnological products have a limited shelf life, so how can this standard be maintained in succeeding batches? What are the standards for evaluation? How can this product be produced on a larger scale? Issues of scale and standards of evaluation are problematic when using living material that is liable to deteriorate. In this situation Lance found it difficult to develop a common understanding of the qualities of the product in order for the group to establish a quality standard for further production. The role of yoghurt as a model will continue to be problematic when it is used as a model of reality for evaluative purposes.

Model: Pop Bottle Fermenter

Fermentation has an important role in biotechnology (Chisti and Moo-Young, 1991; Primrose, 1991) and can be demonstrated in a classroom using a range of fermentation techniques and apparatus (Olejnik and Farmer, 1989). An important role of models in technology is to explore problematic aspects of a proposed solution (Sparkes, 1992a) as it allows the participants to evaluate potential solutions through a concrete representation of reality.

The teacher (Bud) posed the biotechnological problem to identify the optimum carbohydrate substrate in which to growth yeast cultures. The working model (a 'pop' bottle fermenter) was produced from a plan and detailed directions (Olejnik and Farmer, 1989). Students were expected to build and adapt the apparatus and monitor the growth rate of yeast over a 24-hour period in order to determine the substrate that supported the highest growth rate of yeast.

This teaching model was presented to students in an effort to bring the knowledge of the expert and the novice closer (Dagher, 1994). This model enabled the students to acknowledge similarities between the source (1 litre plastic soft drink bottle) and the target (an industrial fermenter). The model employed the same microbe and the substrates were similar to those used in the industrial process. However this is where the similarities ended, for variables and rates of fermentation identified in a small fermenter cannot be extrapolated to large industrial processes as microbial metabolism is affected by large volumes (Chisti and Moo-Young, 1991). At best this model provided students with a means of manipulating a working model and using it to identify variables in the fermentation process.

In this situation any attempt to bring the expert knowledge of the biotechnology community to bear on the learning of the novice classroom community was superficial. Although the students were given a flow chart showing the major stages in a fermentation process, the focus of model-use remained technical rather than indicating the wider dimensions of technological knowledge. These would be employed in carrying out this bioprocess, for example in organism selection and culturing, as well as disposal of the effluent (Chisti & Moo-Young, 1991). In this prescriptive teaching situation, the directions for assembling the model fermenter were given to the students and there was little opportunity for them to adapt and generate their own versions (Wong, 1993).

Model: Quality Control Process Model.

This teaching model was developed and implemented by Katherine (the teacher). It fulfilled more of the criteria for successful teaching models identified at the beginning of this analysis. The biotechnological problem was set firmly within the biotechnological community of practice when she developed and presented to her class a teaching model that represented the organisation of a cheese company (the source). This model was an example of a teacher using a model to enhance communication (Ministry of Education, 1995) and simulate how a biotechnology company organises and carries out its biotechnological activities.

The class project was to carry out a production simulation (Jamieson et al., 1992) to produce a soft cheese. The class was organised into groups that were responsible for a specific part of the process; that is marketing, product development, quality control and packaging. The realisation of this organisational structure was this teacher's target and the groups were required to co-ordinate their activities via a student managing director, write reports on their progress, and present the finished product to a manager of a local cheese company on which the model had been based.

Katherine's goal was to enable students to work through a biotechnological process where the model of organisation was the source and the organisation and production of cottage cheese was the target. Katherine presented an organisational model that did not overtly demand that her students developed biological knowledge of cheese-forming microbes, or details of the metabolic pathway that results in cheese formation, or even a microbial skill base that would enable them to grow microbes and monitor for contamination. Instead, she introduced a teaching model that assumed that such knowledge would be accessed when the students had carried out the technological process. In fact this assumption was correct. The student-expressed model (the source) and the successful production of a delicious cottage cheese product demonstrated that these students had accessed appropriate information to make links between the source and the target. They were able to access information from a range of sources and utilise this knowledge to realise this biotechnological solution (Staudenmaier, 1989).

One of the components that contributed to the solution was a student-generated expressed model which provided a system for monitoring the cheese-making process. This procedure, called a 'quality control process model', was developed by 6 students in year 9 (12-13 years) who were responsible for determining hygiene requirements, monitoring the production process and determining the shelf life of the product. Figure 14.1 has the researcher's analysis (in type) superimposed over the students' work. The idea for a process flow monitoring sheet originated from the group's visit to the cheese factory where they noted that each batch of cheeses were accompanied by a data sheet. After the group had interviewed experts at the factory they identified the key stages in the cheese-making process and decided on feasible testing procedures for a classroom laboratory situation. This quality control process model became the organising framework for the production and quality control groups and enabled these groups to co-ordinate their activities. Such activities included the identification of optimum pH for curd formation, the establishment of temperature parameters for the fermentation process, and the optimum time to add rennet to the mixture. In addition the quality control group were able to inform the production team of the necessity of establishing and maintaining aseptic conditions during cheese production.

This self-generated expressed model illustrated a microbial metabolic pathway and this conceptual understanding was employed by these students when setting up their quality control protocol.

In summary, Katherine's teaching model enabled students to co-ordinate their activities and the students' expressed model fulfilled the role of being a communication conduit during this technological activity (Ministry of Education, 1995). The teaching model enabled the learner to draw comparisons between their classroom situation and the company organisation. In addition this teaching model enabled her students to perceive the wider picture of the biotechnological process and allowed them to generate their own models that co-ordinated specific activities within the cheese-making process. Another factor that contributed to the successful employment of this model was that it was accessible to her students. It had a level of authenticity which resulted from her close links and research at this cheese-making factory.

Although this analysis has identified the specific problems encountered when developing and using models in these contexts, I believe it is possible to make some generalisations about model use in order to provide a basis for further research of the use of models in biology and biotechnology education.

PROBLEMS ENCOUNTERED WITH MODEL USE

The research analysis given above demonstrates the problems encountered when these teachers used models to explain biological phenomena and make aspects of the biotechnological process clear. These can be summarised under the following points:

(1) Teachers often had insufficient pedagogical content knowledge to use these models effectively.
(2) Although familiar with the biological concepts, these teachers were new to biotechnology education and were unfamiliar with the knowledge base that this technology utilised.
(3) Generally these teaching models were used to tell or direct students rather than providing a framework for discovery.
(4) In some cases students gave imaginary characteristics to the model.
(5) The models used to explain microbe morphology did not incorporate a sense of scale.
(6) Teachers were not aware of the differing focus of model use in a biological, as compared with a biotechnological, context.

These problems have significance for teachers using models in biotechnology education.

IMPLICATIONS FOR TEACHING

This research has identified the following features that have implications for classroom practice. When biotechnology education is a focus:

- the purpose for model development needs to be understood by both teachers and students
- the pedagogical content knowledge of model use must be given attention by the teacher
- the manipulation of microscopic biotechnological agents is problematic
- biotechnological outcomes may pose a problem when models are used for evaluation and assessment.

Purposes for Model Development Need to be Discussed in Classroom Programmes

Grosslight et al., (1991) analysis of student and experts' perception of the notion of models in relation to their epistemological view of science knowledge has resonance in biotechnology education. They noted that students' views of models were largely limited to copies of reality or that they were produced for problem solving. Teaching models developed in this research tended to echo this perception. If model use in biotechnology education is to be developed for the purpose of analysing the developing ideas that Grosslight et al. (1991) perceive to be at an 'expert' level, it will be important to demonstrate that model building can provide a means of approaching complex intellectual problems.

One such intellectual problem is the co-ordination of the direction and pace of a biotechnological outcome. Flow charts provide such a function. The development of such models needs attention so that students are aware of their power in providing communication channels and a focus for analysis. As well as providing a planning focus, models may provide a means for students to explore the scientific concepts that provide a foundation for biological application as well as testing these understandings through a physical expression of reality. Conversely the view that the model becomes the 'reality'(Jones and Carr, 1993) also needs to be explored by students so they can identify the function of the model and recognise that the planning process should identify a need and purpose for model building.

Pedagogical Content Knowledge of Models Needs Attention.

Pedagogical content knowledge of model use is important, for not only does the teacher need to know how to use models to establish a meaningful relationship between the prior knowledge of the learner and the new concepts being introduced, but he/she must also be able to critique these models in terms of their classroom effectiveness (Glynn and Takahashi, 1998; Van Driel, 1998).

This research showed that 'The Magic Circle' model provided an obstacle to these students developing an understanding of plating techniques. Although the teacher had identified the gaps in their prior knowledge, the model was too removed from their experience and the comparisons were not made clear. As Glynn and Takahashi (1998) observed, younger students appeared to appreciate the opportunity to set their thinking in a concrete intuitive framework before moving to a more abstract situation. In this case the material source of the gelatine plate and blue dye provided the bridge between the need for developing a technique and the abstract activity where children were expected to manipulate 'invisible' organisms (Clement, 1998).

Experience of model building will give both teachers and students an appreciation that model development requires careful analysis of both the target and the source in order to closely map the source analogy with the target problem (Thagard, 1992). If such analysis occurs the teacher will be able to identify the essential parts of the model analogy that will give the model relevance to the student. Clement (1998) notes that such analysis can provide a link for the learner and this 'bridging analogy' as he calls it could provide the starting point for teachers to identify effective starting points in the development of teaching models for their students.

In this research, model development as such was unfamiliar to both the teacher and the learners, so there was a tendency to impose the model on the learner. The efficacy of self-generated analogies is well documented (Abell and Roth, 1995; Dagher, 1994; Middleton, 1991) and this aspect of professional development needs attention so that the teacher can create opportunities for learners to develop their own models.

Another problem that was apparent was that children had little conception of scale (Boulter and Marsh, 1997). This was very apparent with Katherine's class when they were asked to draw and identify a bacterium. When students do not have an understanding of the dimensions of microbes it is understandable that they will have problems reconciling a dot seen under the microscope with a generalised drawing. Issues of scale need attention in professional development and will provide another facet of the pedagogical repertoire of the teacher using models in their teaching.

Finally, teaching models and their implementation in classroom programmes will only be effective when teachers are aware of the role of models in the communities of experts that use them (France, 1997). This requires teachers to have some access to these communities as well as to acknowledge that model use in biology may have a different expression from that in the biotechnological world. Teachers will need to acknowledge the different forms that models can assume and when their use is appropriate.

The Manipulation of Microscopic Biotechnological Agents is Problematic

Many biotechnological agents are microscopic and this poses problems for their manipulation for there are hygiene and safety considerations as well as the problem of using organisms that are 'invisible'. The problem of invisibility is compounded when the learner is required to develop mental models of these invisible microbes and employ them in practical problem-solving situations.

During this research project the difficulty in manipulating microbes led teachers to be pre-occupied with developing their students' microbial skills. The teaching models that were developed demonstrate that there was a need to practise with something 'larger than life'. The gelatine and blue dye model was effective in allowing Elizabeth's students to develop plating skills and translate this to the distribution of yeast cells on the agar plate. The success of this teaching model could be due to choosing an appropriate mode of representation which Abell and Roth (1995) contend is a fundamental consideration for intelligibility for the learner. Glynn and Takahashi's findings (1998) that younger children (10 to 12 years old) were more comfortable with concrete intuitive thinking than more abstract reflective thinking points to similarities with the students in this research.

Further research needs to be done to investigate the efficacy of such models not only in the development of students' technical skills in microbial manipulation but also to see if these models enhanced their conceptual understanding of the nature and form of microbes.

Biotechnological Outcomes May Pose Problem for Model Development

The transient nature of biotechnological prototypes pose a problem when models are used for evaluation. Often the model and endpoint (product) are the same when they become a standard for quality assurance (Sparkes, 1992b). There is a need to develop a common technological language for describing such prototypes that will provide a means of communicating such information when the prototype standard is so transitory.

The use of living material also poses ethical and conservation problems which reinforce the importance of exploring all the dimensions of technological literacy (Ministry of Education, 1995).

The predictive power of models are well established in technology education with working models developed to establish parameters for the process to be investigated (Elmer, 1996). However, these parameters can be suspect in biotechnological situations as living material does not respond proportionally when kept in large volumes. This aspect of modelling introduces areas of complexity that is yet to be explored in biotechnology education. The analysis by teachers and students of complex process models used by experts will provide further insights into the development of biotechnological knowledge in classrooms.

MODEL USE IN BIOLOGY AND BIOTECHNOLOGY

Biotechnology education with its close links to science education provides an opportunity to study the role of teaching models in both curriculum areas. In this research the role of models changed as the focus changed from biology to biotechnology. Whether one can make a distinction between the disciplines is debatable. However, this research did show that the models can assume a diversity of roles. This experience demonstrated to teachers and students that models are not just a representation of reality but also as means of approaching intellectual problems. The links between biology and biotechnology are strong and further research on model development in teaching situations will enrich both curriculum areas.

Chapter 15

Language, Models and Modelling in the Primary Science Classroom

Carolyn J. Boulter
The University of Reading UK

INTRODUCTION

The perception that pupils should become scientifically literate and act as authentic scientists has fuelled the rise of science education for primary (5 to 11 years) pupils. The contemporary development of methods for analysing whole class talk between teachers and pupils has shown three main scenarios in the relationships in the control of language and knowledge: informing, questioning and collaborating. The collaborative problem-solving scenario uniquely allows for pupils' understandings to be voiced and deliberated. In this scenario patterns of persuasion which allow challenging by pupils in dialogic argumentation are possible. It is probable that there are interacting patterns of questioning, explanation and narrative. The model building that is possible in the collaborative scenario allows the pupils to explain their own ephemeral explanatory models in the example of the eclipse using structural, behavioural and mechanism aspects of the phenomenon, representing them in verbal, graphic and gestural forms. To become authentic in the knowledge and use of models and modelling pupils probably need to experience a range of scenarios, including the collaborative.

SCIENCE IN THE PRIMARY CLASSROOM

There has been a rapid development of the significance of science learning for primary schools and as this proceeds the connections between science learning and modelling are becoming apparent.

J.K. Gilbert and C.J. Boulter (eds.), Developing Models in Science Education, 289–305.

The Developing Importance of Science in the Primary Classroom

The last fifty years has seen large changes in primary (5 to 11 years) school education, especially in science. This is not just a local western phenomenon but there is evidence of worldwide trends (Meyer et al., 1992) and reforms of the curriculum for this age. In England and Wales the mandatory National Curriculum puts Science for the first time as a core subject with equal status to Maths and English. In this process of concentrating on the curriculum and making it explicit, it is the science content that has become more central rather than the processes of science, as method and philosophy, or its social implications (White, 1994). Within this development, the acquisition of 'scientific literacy' is being seen by governments as an entitlement for all pupils who will need it in an in internationally competitive world where societies are increasingly technocratic and where the findings of science are applied to become profitable for corporate industry (Lemke, 1993).

Differing Views of Scientific Literacy.

The nature of this scientific literacy is debated within science education. Is science to be a subject-driven theoretical specialism for training scientists or a problem-orientated thematic popularist endeavour (Fensham, 1985)? Is school science to mirror what real scientists do? If so, what is the nature of that process and is it possible and effective to imitate it in the classroom (Millar and Driver, 1987)? There is debate too over the way science comes into being: is this an inductivist, a hypothetico-deductivist or a contextual consensus process (Lakin and Wellington, 1991)? The discussion around these questions, between the sets of ideas involved in scientific literacy and authentic science education (Roth, 1995), is far from settled. It will be a suggestion of this chapter that different viewpoints on these issues influence the way science is approached in the classroom and thus the sort of talk that takes place and the resulting models and modelling that are present.

Models and Modelling Are Seen as Critical in Primary Science Learning.

As curriculum development and revision proceeds, various important processes with content outcomes which are agreed to be part of science are being identified and worked upon. Amongst these is models and modelling (Gilbert, 1993). Others are observation and evidence (Lubben and Millar, 1994), reading and representing in graphs , tables and charts (Lowe, 1993).

In other papers in this volume the significance of models and modelling in the practice and learning of science have been explored. It is sufficient here to emphasise that models are essential as both content products of science and in the process of coming to understand the world scientifically, however that may be described. So, in primary classrooms it is essential that pupils are introduced to the end-product consensus models of science which form a substantial part of the school content framework. Scientists also engage in critiquing existing models and in constructing new ones no matter how they see the relation of those models to a pre-existent world and no matter how they envision the process of being authentically scientific. So, it is also essential in classrooms that pupils come to appreciate the scope and limitations of the agreed models of science which they learn about and that they themselves engage in using models and making their own (Gilbert and Boulter, 1995a). In these ways the demands for an authentic science education and the development of scientific literacy, however that might be constructed, may begin to be met.

LANGUAGE AND THE PRIMARY CLASSROOM

As science has developed in its significance as a subject in the curriculum and as an important skill for primary pupils, so too has the significance placed upon the language through which that learning is mediated.

The Developing Importance of Language in the Learning of Primary Pupils.

For primary education in England and Wales the publication of the Plowden Report (1967) was a turning point. It is looked back upon as the crystallisation of what came to be known as the child-centred approach to education. Group work was advocated and children were to be encouraged to talk to each other and to the teacher as a means of learning. This concern with language as a means of building understanding through interaction with others shows the strong influence of Piaget (1924/69) and Vygotsky (1978) at this time. The report also highlighted another aspect of language, that of providing access to education, a vital theme of the time. The work of Bernstein (1971) had suggested that pupils from some social groupings were language deficient and that this needed to be addressed by additional school provision so that all could claim their entitlement to education. In the 1960/1970s, language had therefore become closely linked to learning for primary pupils and was seen as the means by which to give all pupils access education. This included science education which was not yet then seen as a vital core subject.

The preferred means of research in language then was the process-product methodology involving pre-designed observational systematic schedules (Flanders, 1970). The ORACLE project (Galton et al., 1980) in a large sample of schools in England looked at styles of teaching, grouping, specific aspects of language such as questioning and who was speaking to whom. It showed that, in the primary classes of that time (mid 1970s), 78% of the teacher's time was spent in addressing and questioning the class as a whole and that, for 65% of the time that the pupil spent on a task, he/she was not interacting with anyone. So, far from demonstrating the Plowden ideal of the teacher interacting through talk and assisting individual children to new understanding, it revealed a rather formal structure of whole-class exposition and generally silent pupils who responded to teachers' questions. Studies today frequently show the same patterns. It is what Janda (1990) refers to as the 'default pattern' of primary teachers and what Boulter (1992) calls the 'questioning classroom scenario'. It allows primary teachers to get the attention of all the pupils and still address each one whilst the class listens. Thus the demands of large numbers of children and the teacher's child-centred aspirations are both met.

The Analysis of Child-centred Questioning Whole Class Talk.

The child-centred questioning of much whole-class talk does not reveal the competence of children to talk about science to the teacher or other pupils. Only a very limited performance is possible: to try to answer the teacher's question within a framework that the teacher will recognise. The recognition of how this process of control of language worked in the classroom was enhanced by the development of the tape recorder and the introduction of ethnomethodological approaches from sociology. It meant that the roles of teacher and learner could be scrutinised. At the same time the discipline of linguistics provided ideas about how to analyse the structure and patterns in the dialogue in the transcribed tapes of the real classroom events. The synthesis of these two strands into the sociolinguistic research tradition to study language in the classroom came of age with the publication of Sinclair and Coulthard's (1975) hierarchical system of analysis into acts, moves, exchanges and transactions. From the 1970s onwards, the sociolinguistic analysis of language in the classroom has dominated research into classroom language, becoming known as discourse analysis. Reports describing the ways in which teachers controlled classroom talk through the rules, rights and obligations they imposed showed how the child-centred questioning discourse of the primary classroom operated in whole class settings with the teacher present (Fig.15.1).

- Asymmetry of control. The planning, introduction, in fact every decision that affected what happened was taken by the teacher (Edwards and Furlong, 1978).
- The process of knowledge building was one of cultural induction into the teacher's world. The teacher was scaffolding access by the pupils into her/his world as a representation of the official understanding of science (Edwards and Mercer, 1987).
- The talk was disembedded from the real life of the pupils. If it didn't arise from the context the teacher knew, it was discounted. Lessons are about what happens at school (Donaldson, 1978).
- The process was implicit. The goals and aims of the process of the discourse were rarely spoken about. The planning of the sequence of experiences was the property of the teacher (Edwards and Mercer, 1987).
- The questioning pattern had a distinctive Initiation-Response-Feedback structure. Children's initiations were generally ignored (Sinclair and Coulthard, 1975).
- The teacher's question was often reformulated until she/he obtained the 'correct' response. The process was one of 'guessing what's on teacher's mind' (French and Maclure, 1980).
- The rules were very different from either conversation with peers or at home (Barnes and Todd, 1977; Wells, 1986).
- Children learnt the rules by participation, they were seldom made explicit (Willes, 1983).

Figure 15.1. Characteristics of Questioning Classroom Discourse

Teachers recognise the importance of pupils speaking, but in the constraints of the normal primary classroom, teacher language is generally used to inform and question about an area of the curriculum that has been identified by the teacher and to control the shared 'floor' of the whole class discussions. It is possible for pupils to be informed and questioned about consensus models of science. It is difficult for pupils to talk about their conceptions of the scope and limitations of models, to struggle through talk to make their own models, or apply accepted scientific models in new situations. The scales are weighed against them acting as authentic scientists and gaining practical scientific literacy.

Identifying Collaborative, Problem-solving Classroom Language

The crucial aspects of this questioning classroom scenario are the teacher control of the language and of the knowledge that is allowed. Boulter (1992) developed an analysis framework along three axes for analysing transcribed classroom texts (Figure 15.2) which enabled the control of the language to be described using discourse analysis. The control of knowledge could be described by scrutinising the text of that discourse with questions about whose knowledge was included and what it was about. The nature of the control on both these axes shaped the relationship between the language and the knowledge on the third apex, the construction of meaning.

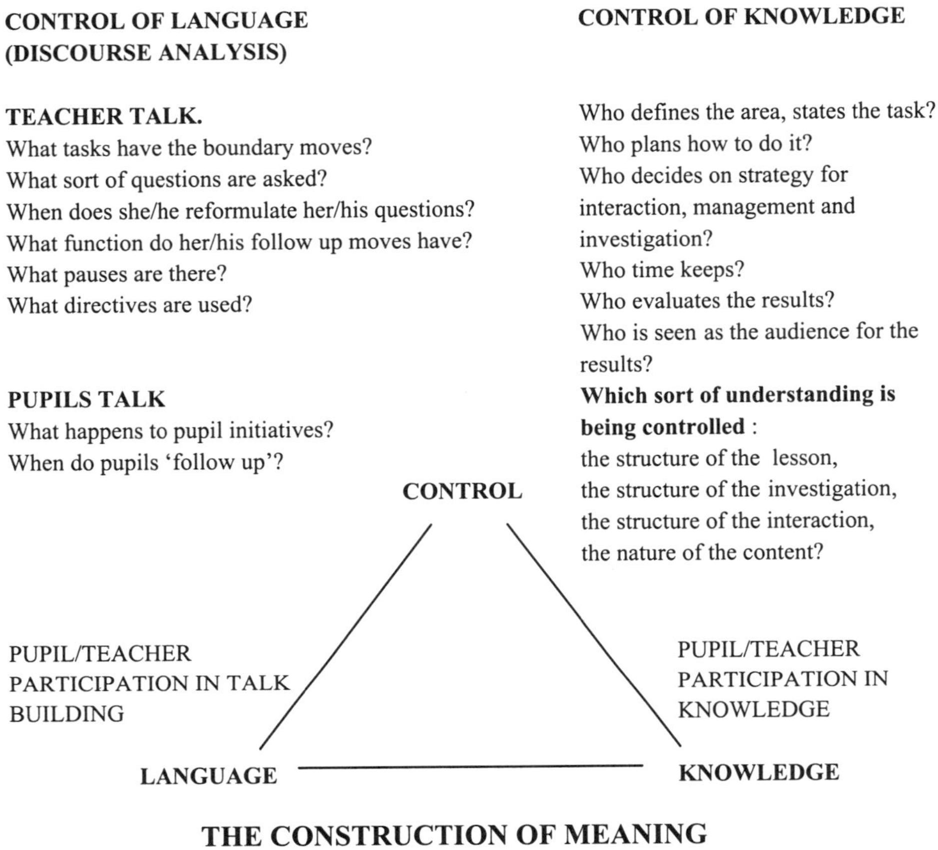

Figure 15.2. Analysing the Control of Language and Knowledge

The identification of an unusual whole class scenario where the pupils were actively encouraged to find answers to their own questions and problems rather than those of the teacher, provided texts with which it was possible to examine where and when the teacher consciously placed control with the pupil. In a typical transcribed text pupils engaged in deliberating with their teacher for 36 minutes on a pupil's question 'What's going to happen in the eclipse tonight?' This text was analysed for the control of the language by discourse analysis and initiating, directing and informing exchanges identified. This whole class discussion breaks down into five sections or transactions: Alex's eclipse model; The colour of the moon; The two theories; Is anyone going to look? and Simon's daytime eclipse. The same text was scrutinised using the questions for analysing the control of knowledge. The synthesis of the analysis of this and several such texts from the same class showed a distinctly different pattern to that of the questioning scenario and this was termed a 'collaborative scenario' (Figure 15.3)

(Boulter, 1992). In this collaborative scenario pupils were allowed to voice their own models and ideas and to bring in their outside school experience to support what they said. Here pupils were being able to engage in using models from their experience, criticising how well they worked and building new ones as they sought through language to find an explanation for a phenomenon (a lunar eclipse) which was to happen that evening. The questioning scenario allows the transmission of consensus models to be examined but the collaborative scenario allows the process of pupils' own models being developed to be open to analysis. This is the language of the construction of understanding through modelling which can now be explored.

- Some symmetry in the control. Children take part in some of the planning, timing, and evaluation often in an iterative fashion, e.g. children's journals provide questions to investigate.
- The process of knowledge building involves the children and allows their own ideas and models to be deliberated. The teacher does not reveal her scientific knowledge.
- The experiences of everyday life of the children often gave rise to the talk.
- There are strong explicit interactional rules. The sequence of what is going to be done is brought into the discussion by the teacher.
- Teacher's questions are almost all to do with investigative and interactional process and clarification of something a child has said. The teacher's follow-up move is seldom there to evaluate what a pupil replies.
- Reformulations of questions only occur when the pupil fails to reply or there is ambiguity. The teacher pauses a lot.
- Strong interactional rules of turn taking, listening for understanding, speaking to be understood, and keeping to the subject, apply.
- The rules for participation are made explicit to the pupils. Pupils are allowed to initiate and reply to other pupils.
- The boundary moves are marked and set the frame. The teacher makes it clear what is to be talked about and why.

Figure 15.3. Characteristics of Collaborative Classroom Discourse

The Nature of the Language Used to Build Understandings

Although the sociolinguistic approach to language by discourse analysis of texts reveals the elements of control of both the talk and the knowledge that is allowed, it does not have anything to say about how understandings are built through language. The language-knowledge axis is left rather bare. It becomes important to attempt to understand the some of the processes involved that affect and relate to modelling.

There appear to be a number of potential avenues to explore, all of which relate and interweave with each other. When the pupils and teacher talk together, the teacher may, as has been suggested, choose to have differing patterns of control over the rules of who talks and what they talk about. The talk in the scenarios identified as informing, questioning and collaborating affects the knowledge that is built. Certain processes will always take place in any scenario.

The first of these is questioning (see also Chapter 10). Questioning is not only used to control, it is also a means of building understanding through language. The analysis of the sorts of questions and their outcomes in understanding is a significant area. van Zee and Minstrell (1996) illuminate the changes from the traditional questioning style and particularly the 'reflective toss', in which the teacher, committed to elucidating the pupils' understandings, returns the pupils statement as a clarifying question.

The nature and processes of explanations is another avenue (Gilbert et al., 1998 a,b) and it is the substance of a major study by Ogborn et al., (1996). They see the process of explanation in the classroom as one of opening up difference, constructing entities, transforming knowledge and making matter meaningful. This is set within an understanding that, to explain scientifically, entities are identified which act together to produce phenomena.

A third avenue is the telling of stories and the application of narrative theories to the texts of classrooms (Gilbert and Boulter, 1995a). This approach to the building of understanding is taken up by Sutton (1995). He stresses the importance of pupils understanding the narrative nature of science in which the experiments that they may do are embedded and the turning of science into stories written by persons who can be imagined.

The last is argumentation and the ways in which persuasion operates as understandings are built (D. Kuhn, 1992). It is this avenue which has been taken up and developed by analysis of the texts of the eclipse discussion by Boulter and Gilbert (1995). In the collaborative scenario the pupils debate their particular explanatory models. The patterns of argumentation might illuminate the relationship of argumentation to modelling. The five transactions in the text 'What's going to happen in the eclipse tonight?' were taken. Alexander and Judy's (1988) categories of knowledge: declarative, procedural and conditional, were used and an analysis system built. A declarative move always starts off the exchange and then there are three patterns of argumentation that may follow: didactic, socratic or dialogic. It is only in the dialogic pattern that the declaration can be challenged and reformulated. This dialogic pattern is only likely to be found in a collaborative problem-solving classroom scenario. It appears to be signalled as permitted by the teacher in her preceding boundary moves (Figure 15.4). It is possible that each type of argumentation is signalled by the teacher's boundary moves. These have the function of allowing certain events to happen, such as particular forms of interaction and shape the lesson structure. These were called procedural boundary moves. Conditional boundary moves describe those that allow access to out-of-school experience.

59. You've got a question, David?	Teacher uses procedure to nominate child	T.PROCEDURAL	
60. David: I just wouldn't have thought the moon was red.	Child challenges the concept of the red moon. Frames a new focus but makes a statement which is allowed	C.CHALLENGES C.DECLARES	
61. I wouldn't have thought it, because that would shine on there - it would cause a shadow over here so it would make the moon completely invisible.	Child demonstrates on diagram and reformulates in terms of shadows making the moon invisible. Child changes the type of representation without invitation	C.EXPANDS /COMPARES	DIALOGIC EXCHANGE
62. So you wouldn't be able to see the moon at all.	Child explains the effect	C.SUMMARISES	
63. T: Right, would you like to draw your theory of what's happening?	Teacher directs a change in procedure, a change in type of representation. Another theory is acknowledged	T.PROCEDURAL /CHANGE	PROCEDURAL MOVE
64. David: It's the same really.	Child considers the diagram to be the same and challenges the idea of difference	C.CHALLENGES	
65. It's the same...but all right...with a different explanation.	Child draws Compares with previous diagram	C. DECLARES	
67 The light would go on this half of the earth and the shadow that half and shadow the moon as well.	Child focuses on difference in explanation Child explains in terms of shadow being thrown	C. EXPANDS	DIDACTIC EXCHANGE
68. So it would completely shadow the moon	Explanation is repeated	C. EXPLAINS	

Figure 15.4. An Example of the Dialogic Pattern

It does seem from this study that natural texts of classroom talk, where the pupils' voice is allowed, can usefully be analysed for their argumentation style as pupils try to persuade each other of their explanatory models using rhetoric within the boundaries allowed by the teacher. It is also probable that it is only in collaborative talk, where pupils' own models can be expressed and the official model of science is as yet not voiced, that they will be free to challenge each other in a dialogical form of argumentation.

The language used to build models is multifaceted, involving questioning, explanation, narrative and argumentation. Each of these probably interacts in complex patterns in particular contexts. The language used to build understanding can therefore probably only be ultimately analysed in a meaningful way by using these multiple perspectives (Prain et al., 1998).

MODELS AND MODELLING IN THE PRIMARY CLASSROOM

Studies of the nature of the language used in building understanding such as have been outlined tell us a great deal about how meaning is produced and communicated. They do not deal with specific content, i.e. the meanings themselves of the eclipse that are built through the argument, explanation, questioning and narrative which are allowed. It is in this realm that research in modelling links to that in language, enabling the unpicking of the constructions of meaning in the models of the eclipse that the pupils voiced in the text taken as an example.

The Nature of Models in Collaborative Classroom Language

A theoretical base for the modelling of agreed understandings has been set out earlier in this book (see especially Chapters 1, 3 and 6). It revolves around agreed answers to the following five questions:

> *What is a model?*
> The representation of a phenomenon: an object, event, process, system or idea.
> *How do models work?*
> By transfer or mapping of similarities between the model and the phenomenon.
> *What different types of model are there?*
> There are mental and expressed models. Expressed models may be ephemeral or consensus. Consensus models may be agreed classroom models, teaching models, or agreed scientific models.

What aspects do all phenomena have that can be mapped onto models?
Structural aspects, behavioural aspects and mechanism (causal) aspects.

What modes of representation can models have?
Concrete and symbolic. Symbolic modes can be textual or gestural. Textual modes can be verbal, visible or mathematical.

So, we can look at the text of 'Alex's model' and the 'Colour of the moon' transactions and ask these questions of the model which Alex was so persuasively defending.

The phenomenon involved is a lunar eclipse, identified by Alex,

'There's going to be an eclipse of the moon' line 6. Video VD2

This is an event (a time-bounded interaction between objects). The original model of the event is likely to have been a elementary orrery of sports balls and a mounted light bulb provided by the teacher (a teaching model) with which Alex and his partner have experimented. Alex may well be recalling this original model.

This orrery model has similarities with the eclipse phenomenon which are transferred by analogy as the model is used. The possible structural similarities are that balls have roundness, opaqueness, reflective surfaces, relative size. They can move and align relative to each other in various ways and thus have behavioural similarities. The bulb can produce light, a structural similarity, which travels in all directions, a behavioural similarity. The similar causes underlying the similarities of behaviour, the mechanism aspects, are for Alex the orbits and static nature of the sun.

'It's because the earth moves round the sun and the moon moves round the earth' line 27 Video VD2.

The mechanism aspects may be different for different levels of understanding and explanation. For an astronomer the causal mechanism would probably involve gravitational and rotational forces.

At the start of the talk Alex's model is hidden from the class and the teacher, it is his own mental model based on his previous experience. The teacher asks him to explain his statement about the eclipse.

'Could you explain exactly what's happening....in order for me to understand' line 12 and 13.

This allows Alex to begin to express his model and he starts with the aspects as suggested by the teacher's question.

> *'... there's sun and the earth and the moon all in a row and the sun's light doesn't go round the earth,.... It doesn't reach the moon.' line 14*

Alex's mental model is being expressed. There are four entities or parts of the eclipse: sun, earth, moon and the sun's light. The sun, earth and moon have behaved in a certain way to line up in relation to each other. The sun's light behaves so that it will not go round the earth (*pace* Einstein). The structural and behavioural aspects are expressed, the same aspects as he had probably transferred using the orrery.

Alex is encouraged to express his ephemeral model for the class and he hopes that it will be agreed as the class consensus model. The teacher withholds her understanding of the consensus science model so that both Alex and the other pupils will be empowered to voice theirs. The teacher might have produced an official teaching model in the form of a manufactured orrery but she does not.

At this point the teacher asks Alex to represent his model on paper on a flip chart. She invites a visible textual mode of representation to enhance his previous verbal mode, his model is becoming more stable by this move as it now has permanence as ink on paper. He draws it as a pictorial image of the three bodies as spheres represented as two dimensional circles, concentrating on the structural round aspects of them and then names them.

> *'This is the sun ... .and that's the earth and that's the moon' line 20 and 21*

He goes on to give the behaviour of the light from the sun, tracing its path with his finger in a gestural mode as he does so.

> *'Light from the sun is going to the earth, but it won't reach the moon because it's in the earth's shadow'.*

The teacher asks him to write letters for earth, sun and moon and goes on to elicit a mechanism.

> *'Why is it happening tonight and not tomorrow night or last night?' line 26*

Alex gives the orbiting of the earth and moon as the cause showing that this is the mechanism aspect he is transferring by analogy.

The teacher picks up on his previous gestural mode and asks him,

> *'Show me where the light is going' line 34*

Alex explains the behaviour of the sun's light, using his hand.

> *'Its going from the sun to the earth and it doesn't turn round the earth to go to the moon so it doesn't reach the moon'.*

The teacher seems not to be content that the model expressed by Alex is sufficient for the other pupils to relate to their likely experience of the eclipse that night, and she asks,

> *'So what shall we see tonight?' line 36*

It is at this point that the teacher receives a surprise when Alex tells her the moon will be 'a reddy colour'. This is an aspect of the phenomenon, which hadn't been mentioned, before relating to their life outside the classroom and what they may see in the sky tonight. It may well be that Alex is recalling an item from the television pictures of a previous lunar eclipse, shown the previous night, in which the moon glowed red. Alex is envisaging an image of the moon which is unexpected, to her at least. She challenges him to relate this image to his expressed model and he replies,

> *'I think it's because the white light won't go round the earth but the red light will' line 50*
> *'Red light will go on the moon and the moon will be red' line 52*

She allows David to challenge Alex and David discounts this extra expressed aspect of the eclipse phenomenon, the redness of the moon. He uses only the behaviour of white light in causing shadows and gestures over the drawing,

> *'(The light) would shine on here (the earth) and it would cause a shadow over here (space between earth and moon) and would make the moon completely invisible' line 61*

These are the first two transactions. The talk then proceeds to a discussion, focused by the teacher, of: the two possible phenomena and their related explanations; the earth stopping the white sunlight but the red light getting round it to give a red moon; and the earth casting a shadow on the moon and the moon being invisible.

In these first two transactions we see two pupils' expressed models as they struggle to question, explain and argue what they will see of the lunar eclipse event to happen that evening. They use their mental models and visualisations from experience of teaching models and pictures to focus on the structural, behavioural and mechanism aspects of the phenomena in relation to their models as they express them for the class to hear. They represent them texturally (verbally and visually in graphics and written letters) and through gestures. In this collaborative whole class talk we are able to catch a glimpse of pupils constructing their own models of phenomena in a setting which allows the debate of alternative models and the coming to an agreed consensus model for the class. The range of aspects and modes of representations used is no doubt significant, as are the sequence in which the aspects are elicited and the modes of representation used.

A FRAMEWORK FOR ANALYSING CLASSROOM LANGUAGE

What has already been described can allow models in the making to be analysed in any classroom talk. The example given is of collaborative talk where the pupils' own models are valued. The emphasis is on the authentic voice of the child as ephemeral models are being constructed and criticised. In informing classroom scenarios or settings such as lectures the emphasis may be on the transmission of the consensus models of science and on accepted teaching models. The same categories can be used to analyse the modelling involved in the transmission of aspects of phenomenon and modes of representation as the entities which make up the consensus model are expressed. This can then sit within the analysis of the same text for participation and deliberation patterns (through discourse analysis and control of knowledge analysis questions) described in the first part of this Chapter. In a similar way the socratic questioning scenario can be analysed for the teaching models and the consensus models of science.

The exercise with the collaborative talk provides a framework for the analysis of any text of classroom discourse (Figure 15.5). As with any texts, they are set in particular cultures. A primary school in rural England in the late 20th century where this particular talk was collected has a particular cultural frame. Cultural frames influence the stories that people tell about themselves and about science. Although this Chapter deals essentially with texts of talk in whole class groups in primary classrooms doing science, it is important to remember that the texts sit within cultural influences and their related stories (Boulter and Gilbert, 1995).

Issues Arising from the Analysis of Modelling and Whole Class Talk

This Chapter shows how we can move forward in analysing the relationship of modelling and whole class talk. As yet it is not clear what the relationship of classroom language to the different types of models (pupils transitory models, teaching models, teacher's models and consensus scientific models) might be. This is an area where important research might be done. What sorts of models do we find being used in different sorts of talk (informing, questioning, collaborative) in specific content areas?

Related to this is the major issue for science education, that of conceptual change and how the relationship of modelling and language might illuminate how children's understanding develops. What modelling processes and types models help children to develop their conceptual understandings and what sorts of talk will support effective conceptual change?

From the studies so far it is clear that pupils are readily cued by the teacher into producing what they consider appropriate models for certain situations and are also cued for arguing in particular ways, or not arguing at all. No doubt they may be cued to use certain sorts of question and explanations and to introduce different narratives too. How this cuing operates is a major task of research in language and modelling.

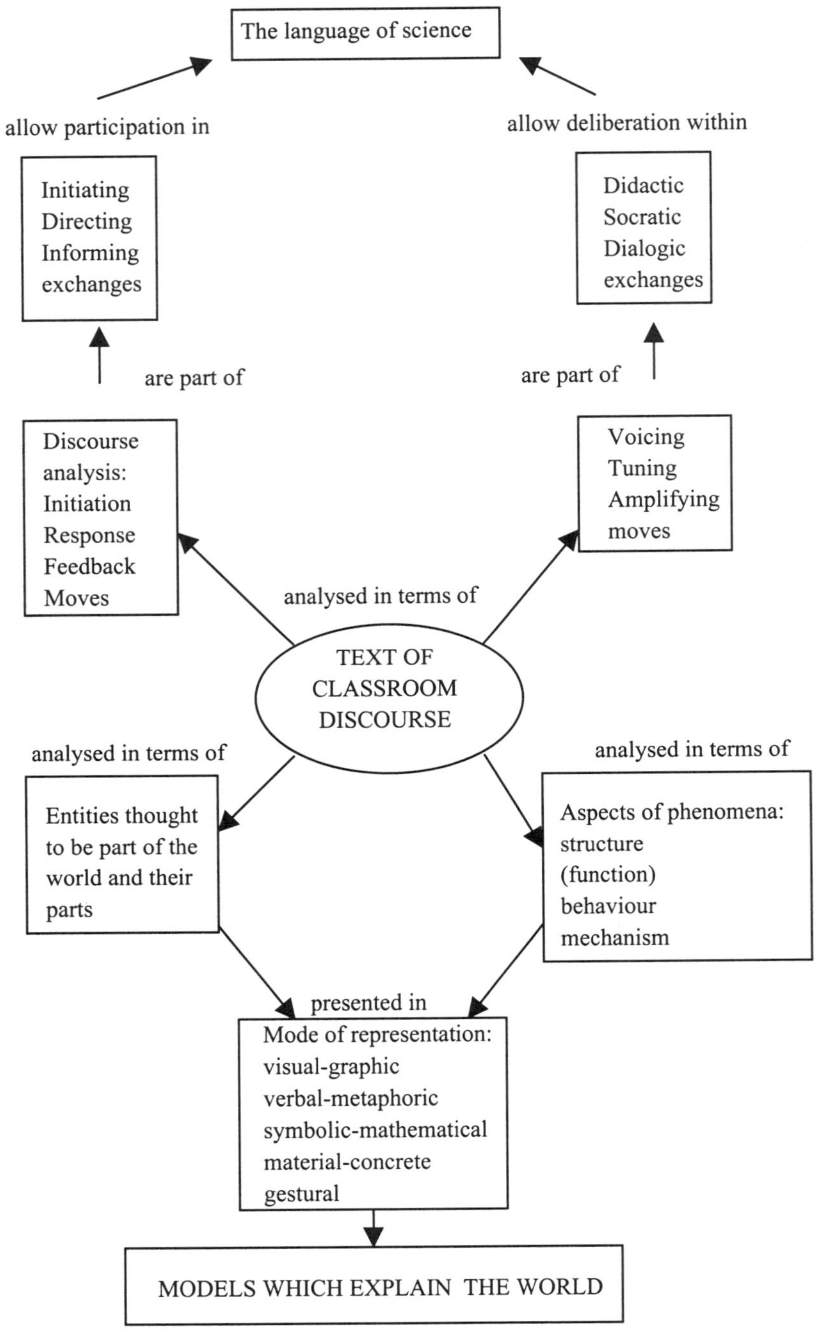

Figure 15.5. Analysing the Text of Classroom Discourse

CONCLUDING REMARKS

The present rapid development of primary science education has set the scene for research into the processes involved in pupils becoming authentically scientific. One of the most important of these is modelling: that pupils should become adept with accepted scientific models, knowing when and how to use them and also that they should model themselves as a way of acting as scientists to come to understand the world. For all this to happen pupils need to experience a variety of classroom talk scenarios which both allow them to meet accepted scientific models and to make their own. It seems that whole class collaborative talk is particularly effective for allowing pupils to construct and argue with their own models and find, in practice, the strengths and weaknesses as models are used to explain phenomena. Other talk in which the teacher has more control over the content may be more effective in presenting existing scientific models to children. The sustained and developing interest in research into classroom and other discourse, taken with the developing analysis of the modelling process, provides a means of beginning to understand the intimate connections between words and meanings, pictures and their comprehension in ordinary classrooms. It may also give us something upon which to hang our assessment of children's learning in process.

Chapter 16

Teaching and Learning about Chemistry and Modelling with a Computer managed Modelling System

Nitza Barnea
The University of Reading UK; Technion, Israel Institute of Technology, Israel

INTRODUCTION

The advantage of using computer managed modelling systems (CMMS) to illustrate and explore phenomena in chemistry teaching and learning stems from the convenience and simplicity they provide for building molecules of any size and colour in a number of presentations. This Chapter summarises aspects and thoughts about CMMS that were collected from four different populations: Chemistry and microbiology professors, graduate and undergraduate students at university and high-school teachers and students. The data comes from semi-structured interviews, written reports, and questionnaires. Academic members who use CMMS for research and teaching, and schoolteachers, who use it for teaching, all see the advantages and importance of using this tool. Students' feedback on the use of CMMS both in high school and university was found to be positive. Most of the students enjoyed using CMMS and indicated that it had helped them understand concepts in molecular geometry and bonding, through the improvement of their visualisation skills.

DIFFICULTIES IN CHEMISTRY LEARNING

Students who study chemistry seem to experience difficulties visualising the structure of matter in terms of a particulate model (Garnet et al., 1995). These difficulties are due to the abstract, unobservable, particulate basis of

J.K. Gilbert and C.J. Boulter (eds.), Developing Models in Science Education, 307–323.

chemistry and the need for rapid transfer among the three levels of thought – the macroscopic, sub-microscopic and symbolic (Johnstone, 1991; Gabel, 1992). The macroscopic is the phenomenon level, where properties can be seen and measured; the sub-microscopic is the molecular structures of the particles, which cannot be seen; and the symbolic level is the way a substance is represented by its chemical formula. While chemists and chemical educators operate across the various levels quickly and easily, students have difficulty creating links across these levels, and must therefore use models, analogies or computer graphics. Gable (1992) found that high-school students exposed to worksheets connecting the molecular level to macroscopic and symbolic levels alongside transparencies portraying the particulate nature of matter improved on all three levels.

SPATIAL ABILITY AND VISUALISATION SKILL

Another problem that many students face while studying structural chemistry is the need to perform a variety of spatial tasks and visualise the three dimensionality of molecules that are represented by two-dimensional diagrams in books. In chemistry, as in many other scientific areas, three-dimensional visualisation is an important skill. Students who lack this ability usually have problems in developing their understanding of topics which require it. This skill, referred to as 'spatial ability', involves representing, rotating and inverting objects in 3D when they are presented graphically in 2D. There are different levels of difficulty associated with visualisation skills as defined by several authors (McGee, 1979; Richmond, 1980; Eliot and Hauptman, 1981; Newcomb, 1982; Lohman, 1986). These levels in ascending order of importance are:

1. Spatial visualisation, the ability to understand accurately three dimensional objects from their two dimensional representation.
2. Spatial orientation, the ability to imagine what a representation will look like from a different perspective.
3. Spatial relations, the ability to visualise the effects of operations such as rotation, reflection and inversion, or to mentally manipulate objects.

Spatial ability is influenced by several factors: genetics, differences between cultures which concerning learning opportunities, and the everyday environment (MacCoby and Jacklin, 1974; Waber, 1976). In addition to those factors, a gender difference was found in some studies (MacCoby and Jacklin, 1974; McGee, 1979), while others claim these differences are usually small or not found (Mayers et al., 1988; Tuckey et al., 1991). The explanation given for why boys obtained better results was the intensive practice they got in building models from diagrams while playing with their toys, which certainly could improve their spatial perception. This is in contrast to girls who did not get the same opportunities to develop their own spatial perception (Tracy, 1987). Another explanation is that many of the tests used to determine spatial ability are based on speed of performance and not on reflective and careful thought. It was found that, while solving spatial problems, both boys and girls were capable of finding solutions, but boys were significantly faster. When the effect of time was minimised, the male advantage decreased (Kail et al., 1984; Gallagher and Johnson, 1992). If enough training and practice is given to females, the gap will vanish (Koslow, 1987; Linn and Hyde, 1989). Spatial ability requires the ability to organise the various aspects in the representation of information and the ability to generate and manipulate visual images. Kosslyn (1987) explained the importance of imagery in producing the experience of seeing in the absence of appropriate sensory input. Having a visual mental image produces the conscious experience of 'seeing' with the 'mind's eye' rather than with the 'optical' eye.

Early studies suggested that visualisation skills cannot be taught and that this ability is innate (Smith, 1964; Witkin, 1969; McFie, 1973). Many later studies, however, indicated that these skills could be improved via learning tasks and proper exercise (Lord, 1985, 1987; Baeninger and Newcomb, 1989; Kiser, 1990). Teaching aids such as models, stereo-diagrams, mirrors, shadows and dynamic pictures have been used in remedial instruction programs (Tuckey and Selvaratnan, 1993), and have proved to be useful in improving spatial ability skills. The suggestion of these authors is to use instruction methods that guide students to visualise three-dimensional structures from their two-dimensional representations. Imagery is also used as a thinking tool to retrieve tacit knowledge from memory and as a way to perform mental simulations. In learning, memory may be improved by having an imaged object stand in for the actual object, allowing one to associate a perceptual representation of the object with the relevant content.

A strong correlation has been found between spatial ability and achievement in science. Generally, students who studied science improved their visualisation skills in comparison to control groups who did not study scientific topics (Baker and Talley, 1974; Bishop, 1978; Lord, 1987; Pallrand and Seeber, 1984). These inquiries have also shown that students who had difficulties in spatial ability tests were not able to solve problems in subjects in chemistry, biology and physics.

THE USE OF ANALOG MODELS AS A LEARNING TOOL

One of the ways of improving both spatial ability and achievement in science generally and in chemistry particularly is the use of analogue models (Lazarowitz and Naim, 1985; Talley, 1973).

A visual base for learning may be a real object, a theoretical model or a graphic image of an individual object. According to Gilbert and Boulter (1998), models can be differentiated between; mental models, expressed models, consensus models and teaching models. The value of models is in enabling visualisation of complex ideas, processes and systems. In chemistry the most used models are in the symbolic and the scale area: formulae and equations or concrete analog models of ball-and-stick type that represent the theory of molecular structure. Many concepts that were described only verbally and therefore were vague and unclear can be fully understood with pictures and drawings. One way of performing a simulation with different model types quickly and efficiently is by a computerised environment: an animated simulation creates a vivid model and enhances understanding.

Many teachers who use models do not emphasise the fact that they are simulations of theory and that no molecule looks exactly like any one of the models. Teachers also often use only one type of model. Consequently, students' perceptions of the model are sometimes partial and wrong (Gilbert, 1993; Ingham and Gilbert, 1991). However, many teachers do not emphasise the corresponding and non-corresponding features between the analogy and the new concept (the target). If the use of analogies is not appropriate it can cause misunderstanding and confusion. The best ways to overcome the effects of 'dis-analogies' are firstly to point out to students the places where the analogy/model breaks down and secondly to use multiple analogs, comparing the target to sources with different characteristics (Thagard, 1992).

THE USE OF COMPUTERS IN DEVELOPING SPATIAL ABILITY

A curriculum that integrates computer graphics, which give a three-dimensional representation of objects, is a very efficient way of improving the visualisation skills of users. This is true even if the main purpose of the studyware is to teach certain scientific subject content (Kiser, 1990; Rodriguez, 1990; Wiley, 1990; Bezzi, 1991). The use of computer technology in mathematics teaching improved achievements in both mathematics and in spatial ability (Teles, 1990; Vazquez, 1991; Shoaff-Grubbs, 1992). The advantage of integrating dynamic animation while developing spatial ability was known before the personal computer became common. However, with the computer technology now available, the provision of animation is a lot simpler. Kiser (1990) claims that the advantage of computer graphics lies in the enhanced learning pace and in the flexibility while operating translation, rotation, and drawing functions. The control of the animation by the learner that can be achieved on the computer helps in the development of spatial skills (Wiley, 1990).

THE HISTORY OF COMPUTERISED MOLECULAR MODELLING

Computational chemistry was born as a consequence of the development of digital computers in the early 1950s. The 1960s were a decade when the application of digital computers to chemistry was expanded to include a wide range of subjects. The value of the computer, as a tool, was applied in statistical mechanics, crystallography, and organic synthesis and structure determinations (Bolcer and Hermann, 1994). Until 1978 computational chemistry was limited only to those who had access to mainframe computers. With the introduction of minicomputers, and later personal computers, more institutions and individuals could use computational facilities. In 1986 computer programs for molecular orbital and molecular mechanics were adapted to PCs, thus spreading their use even to small colleges and small research groups. So computational chemistry has flourished during and since the 1980s.

An important research area in this field is molecular modelling, which is a key skill in the chemical and microbial sciences. It enables one to explore the three-dimensional shape of known molecules, to verify the possibility of existence of molecules with desired properties prior to attempts to synthesise them, and to simulate the mechanism of complex reactions. In describing molecular structures, reactions and interactions, chemists use 3D images and models of atoms and molecules.

Most chemists and chemistry educators working in field today were educated when a variety of wooden, plastic and metal models were used to illustrate how a molecule ‘looked’, to speculate on the feasibility of its existence and even predict some of its physical properties. While organic and inorganic chemists were using this approach, theoretical chemists were developing theories which involved mathematical models, quantitative calculations and computer graphics to simulate and predict more accurately molecular properties (Gorson, 1988). To overcome the fear of changes because of the uncertainty element involved, it was suggested that chemists should be trained to use computer graphics and simulations, and that they must be integrated into the curriculum, both of chemists and of chemistry educators. It was found that an in-service training course for chemistry teachers, in which they were exposed to new technologies through active participation, eliminated computer anxiety (Dori and Barnea, 1997). The research reported here follows up ways of introducing the use of this new exciting technique and its fruitful effects.

THE EFFECT OF COMPUTERISED MOLECULAR MODELLING ON VISUALISATION AND UNDERSTANDING

The advantage of computerised molecular modelling (CMM) over the use of rigid models lies in the fact that, through the use of software, molecules of any size and number can be conveniently constructed. Each molecule can be presented in a model of any type and each representation highlights different properties. The ball and stick model, with each atom type having a different colour pattern and a precise size, shows the proportional bond lengths and whether the bond is single double or triple. The space filling model gives indication of the van der Waals radii and the overall volume the molecule captures in space (Hardwicke, 1995). As well as drawing pictures of molecules, computers can perform calculations and find the optimal conformation even for very large molecules. The study described here contains various aspects of using CMMS in different environments with diverse populations. The populations involved were three university professors from the chemistry and microbiology departments, undergraduate and graduate university students in the UK and high-school chemistry students in Israel. The main purpose was to provide quantitative, qualitative, and cognitive explanations, for the influence of using computerised molecular modelling on students' visualisation skills and their perceptions in chemistry and in modelling. Both populations of high-school and university students used a PC-compatible package 'Desktop Molecular Modeler[1] (DTMM)' (Crabbe et al., 1994). This visualised three-dimensional molecules through the use of various representation options and completed several tasks. Though the teaching purposes, strategies and styles were different, the aim of using the software was the same.

THE CMM SOFTWARE

The software tool used in the study was the DTMM, a PC-compatible package which enables three-dimensional molecule visualisation through the use of red/green eyeglasses to view stereo paired lines display. The package also provides coloured lines, space filling, quick filling, and ball and stick, representation options

The most powerful features of the software are molecular synthesis and energy minimisation (Gulinska et al., 1991). The energy minimisation routine optimises the geometry of the newly created molecule by using an algorithm that calculates bond, angle, Van der Waals radius, and charge energies, to create a viable three-dimensional conformation. The minimisation process is shown both graphically, as the display changes, and in a box at the bottom of the screen, which counts the number of iterations done. The software is controlled through pull-down menus which are easy to master. A special school version of DTMM has been designed and is available for a nominal fee, and was used at the school level, whereas in the university DTMM together with other software alike were being used.

Although this software is a simulation and by using it students are exploring a model that has already been created, using it contributes also to the modelling process (Boohan, 1997). The opportunity to envision various representation modes of the same molecule, combined with the computational accurate capabilities help students create their own mental model of molecules.

LECTURERS' PERCEPTION ABOUT THE IMPORTANCE AND USES OF COMPUTATIONAL CHEMISTRY

Three staff members from the University of Reading with expertise in molecular modelling were interviewed for this study. Two were from the Chemistry Department and one from the Microbiology Department. When interviewed about the role of computational chemistry in the University teaching, Prof. C. (Microbiology), who is involved in developing Molecular Modelling software (DTMM) for education and research, highlighted three aspects. First, using computers enables the molecular visualisation of the quantitative basis of the modelling to take place. He mentioned that some students have a high 'activation energy' barrier when it comes to using computers and having programs in this field that are easy to use will certainly help them get over that barrier. Second, it will help people visualise molecules so that they don't actually need to build them with 'kits', and they can see them on the screen and manipulate them in real time. The third element is that it enables to get some quantitative understanding of molecules and interactions.

Dr. R. (Chemistry) emphasised the fact that chemistry is much more theoretical, and the computer is used to test the theory. Therefore his requirements from students are to have mathematical skills and a wide background in chemistry, so that they will use the software to examine mathematical models and their prediction ability. He sees CMMS as a link between theory and experiment as it can interact with theory and can interact with experiment and tells you a bit about both (Figure 16.1).

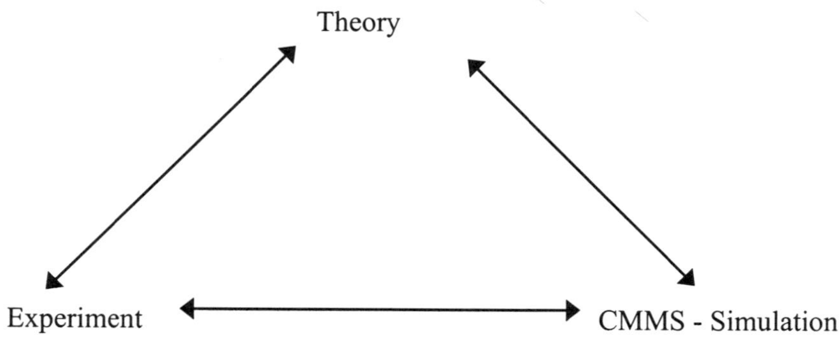

Figure 16.1.

Dr. D., from the same department is less interested in the mathematical side of the software, but more in its uses to plan the creation of new products. Therefore he thinks that the most important requirements of a research student are enthusiasm, motivation and commitment to the research.

Professor C. (Microbiology) pointed out the fact that the borders between different areas in the sciences do not exist any more. Therefore, every scientist needs to acquire the modelling technique.

> People who work in biology, biochemistry, pharmacology, etc. using molecular modelling tools, can compare what they model with what they do in the laboratory, and enhance their knowledge. In the next century we are going to be looking at targets for drug design; we are becoming pharmacist computational chemists. Software like DTMM are involved in this training and are therefore so important.

Prof. C. is aware of the strong linkage between experiments and the modelling system.

> It is an iterative process between what goes on in the computer and what you design experimentally in the laboratory. This is

> why programs like DTMM, where the same person can do both, are absolutely vital. It shortens that iterative time and quickens the mental processes. You do everything on your own and combine the *thought experiment* with the *real experiment*.

One of the main questions addressed in the interviews was whether the use of the molecular modelling package helps to improve visualisation skills. Although there was consent about the usefulness of the package the answers varied. Dr. R. was sure that visualisation skills can be learned, but only when students learned to use it properly can the CMM be of help in this respect.

> This is a skill that one learns. We find that with students who study a course in symmetry, only then do they learn to visualise three-dimensional shapes from paper. I have found out the computer does not actually help them very much until they are much more familiar with it. When we first started using DTMM we thought about tetrahedral shapes and we kept finding that students were stumbling, so we kept on using the frame models and explained the relation to what was seen on the screen. Although the computerised models can be moved around on the screen, you have to teach students how to interpret the image as a 3D image. Once they have learned that it is much easier to show the more complicated molecules. I use plastic models as a stage in the learning, but when they understood it we use only the computers.

Dr. D. believes the software improves visualisation, but it takes time till he sees the results. At first he uses both computer and plastic models and when his students understand it they use just the computers.

From these interviews it would appear that both lecturers underestimated the contribution of this program in improving spatial ability. They did not try to use it as a tool when teaching symmetry and only did it later. On the other hand, students' responses were much more favourable.

Professor C. thinks the use of the software helps in terms of the number of display options that are offered and the quick change from one visualisation type to another:

> From wire frame to ball and stick, then to the overall charge distribution of the molecule. All the time it's underpinning the

> visualisation with basic physical chemistry, basic organic chemistry and chemical principles.

He also mentioned the contribution of peer learning that comes from working in pairs at the computer laboratory and the need for verbal explanations among students that enhance understanding.

PERCEPTION OF GRADUATE STUDENTS ABOUT THE NEED FOR COMPUTATIONAL CHEMISTRY

Four graduate students who are working on their Ph.D. thesis in molecular modelling in the chemistry department were interviewed. All the interviewees emphasised the advantages of CMM. The main and most important property that was mentioned was its complicated calculation ability that gives really accurate predictions in a very short time. But students were aware of the fact that it was not used much. The explanation for that given by G. (a Ph.D. student) was:

> Many people are very unsure of the computer predictions. They would rather not believe it. It's because many have not used it and do not know it. If you haven't seen its capabilities you do not think about using it. I do think that in the future a lot more people are going to use it and appreciate its useful predictions.

Graduate students identified a range of prerequisites for computational chemists

- Liking computers combined with a knowledge of and experience with computers.
- A sound knowledge of chemistry.
- Curiosity, the will to know why things are happening.
- Good literature skills, the ability to derive information from the literature.
- The ability to relate experimental values to the software predictions, and to correct the model accordingly.
- The ability to analyse the structures, questioning and thinking whether they are reasonable.

All the graduate students interviewed thought that the best way to train someone to become a computational chemist was by using the program.

> Using the program provides a good start, looking at the models, training to see different shapes. Just using the DTMM software itself is good practice. You can do simple molecules, simple shape exercise, looking at two molecules fitting together. Just start using the computer.

In order to introduce CMM to more people so that they will appreciate it, believe its predictions, and use it more, the students suggested starting using those programs at an earlier stage.

> Like they exposed us during our learning of the elements of the lab, they have to expose us to the computer to give us the opportunity to plant the seed in our mind. If you start when you are younger and you have always known it, it's easier than when you introduce it to some Doctor (member of staff) who has been working in the lab for forty years. It will be a lot more difficult to start working on the computer.

M. thinks that it is not known enough because:

> In undergraduate studies you really do not come across CMMS. You do not use computers, CMMS is out of territory for a lot of people who think that it will be too hard. If you come across it once a week in school, for say 4-5 years, it certainly will help a lot. It's like with everything, the more you do it the more you become happy with the idea and the more you find it a useful technique. The reason why not so many people are doing computational chemistry is because they are not really introduced to the idea, not familiar with it. It is unknown territory, and a lot people will rather stick with what they know and keep away from the unknown.

Concerning the advantages of CMMS to improve spatial ability there is agreement among graduate students. They all see the contribution of animated molecular programs to visualisation. However, they express it differently. J. mentioned the contribution of the program to bridge between the macro and the micro levels without really naming it:

> In chemistry you always get colours and physical changes but you never really see what you are actually dealing with. When you get a huge protein enzyme, and something is docking, you can actually see the interactions on the screen, it's the visual

> stimulation that can help you to think. That how I find it helpful for understanding.

G. said that when working with the molecules on the screen a certain level of spatial awareness is needed. He expressed his thoughts about the advantages and problems:

> All the molecules are presented in 2D, but shading and depth cues are used to achieve a 3D view. It's very easy to look at 3D structures, because almost all packages have some way of rotating the molecule. You can rotate the molecule and see how the atoms are bonded. Having said that, I still have problems sometimes, especially when looking at the chiral systems. I need to sit down and look at the stick model, because with it you can look at everything at once. I use both. I generally use computers, but there are times when plastic models are the way to go.

It seems that the graduate students adapted the approach of their supervisors. Dr. R.'s students (M.and E.) believe like him that it is important to understand the mathematical approach and be able to check if it gives logical results, whereas Dr. D.'s students (G. and J.) do not bother about the functioning and accuracy of the software they use. They are concerned with the chemical aspects of the results they receive.

As G. says:

> You have to understand how the computer is doing it, but you do not have to understand the high level mathematics. I really do not understand it. I take it for granted, that they did it right. It still depends on which technique you are using when solving wave equations.

He stated that at the most basic level CMMS is a general aid for understanding that can be used as a visualisation tool. It should be introduced to high-school students to get them more interested in chemistry, help them understand what is happening and help them become more interested in computing.

M. found CMMS very helpful in improving his visualisation skills and in relating 2D to 3D representations. From his explanation, it is very clear that the use of the software helped him construct his mental models, and enhanced his ability to move easily between two- and three-dimensional models.

> When you draw a molecule and look at it on the screen, it offers many options to rotate it easily, different methods of looking at the molecule, so that you can get 3D structures. If you get a 2D structure it's very hard to look at and visualise it, but when you get it on the screen in a 3D style then it helps to see more things. When I see a flat 2D structure, I do not think of it as 2D, I try to look at it in my mind as 3D. I constantly strive to make it 3D. When you are using the computer it's always 3D, that why it's easier with the computer. It sort of the 'translation of your brain' turned into 3D structure, it does it so much quicker and more easily. That's how it helps.

UNDERGRADUATE STUDENTS' PERCEPTION ABOUT THE NEED FOR COMPUTATIONAL CHEMISTRY

Undergraduate students of the Chemistry Department learn to use CMMS and are given several tasks. The data reported here comes from semi-structured interviews conducted with 13 students who used the software as well as from their written reports on their results and assessment on the software.

The students interviewed were in their second year of study and had used the software just once in an introduction course on computers. The semi-structured interviewees were tape-recorded in the computer lab during a training session. Students were interviewed individually in a separate part of the lab. Although the sessions were open ended and flexible in nature, they usually followed a general format. The participants were asked to answer some or all of the questions that focused on:

- Students' background with computers.
- Students' experience with the software.
- The advantages and disadvantages of the software in comparison to ball-and-stick models.
- Visualisation skills before and after the use of the software.
- Attitudes towards further use of the software.

Analysis of the data collected from the interviews revealed that, although the students had diverse computer backgrounds (one third were complete novices, one third had a little experience and the rest were very computer experienced) none of them had experienced molecular modelling before. Despite the wide differentiation among the students, most of them found the software friendly and easy to use. 11 out of the 13 thought it helped them to visualise the molecules, understand and remember their structures. Four students complained that it was difficult at the beginning to get used to it, and one student said the instructions were not clear enough. All the interviewees enjoyed this experience, even those who were critical.

When asked to compare plastic to computerised models, most of the students preferred the use of the computer. Although at first it seemed that the two kinds of models are very similar, students found several explanations for their preference. They mentioned the wider range of models the computer presented and its availability for every student in the class, as opposed to plastic models, which were sometimes only for the use of the teacher. Students commented on the quicker and more precise way of molecule structuring as well as the ability to create molecules of any size, to calculate their most stable conformation and to find the charge distribution, activities that can not be done with the plastic models.

Most of the interviewees thought their spatial perception before the DTMM experience was quite good, but they were sure this software helped them to visualise molecules in 3D. Two students declared their spatial ability was excellent and said that they did not need the aid from the computer, and two others admitted their spatial ability was poor, and they did not feel it improved at all. (One of them had problems with computers in general, which might have caused her hostile attitude.)

The more interesting gain from the interviews was the insight they offer into how this software facilitates visualisation and chemical understanding, when it does it. Students pointed out the most useful features of the software for them. Mention was made of its 3D rotation, the provision of calculations like those of measuring bond angles and bond length, and the minimisation process. The graduate students raised similar ideas and it explains how this technology when used properly can be of great help. Students' explanations of the way these properties improved their comprehension were placed into five categories:

Category 1 - Rotation in all directions.
Category 2 - Quick shift between two- to three-dimensional representation.
Category 3 - Great help in remembering and memorising.
Category 4 - Relationship between molecular shape and energy.
Category 5 - Representation of different model types.

HIGH SCHOOL IMPLEMENTATION

In Israel, CMM was used in an urban high school. It involved five heterogeneous classes (N=169) of tenth graders (age 15) who were studying chemistry for the first time. The experimental group, three classes (N=97), worked on the subject of geometric shapes of molecules with the molecular modelling software, while two other classes, which served as a control group (N=72), studied the subject in the traditional way (Barnea, 1997).

When measured, spatial ability, the understanding of new concepts related to geometric and symbolic representations, and the perception of the model concept, were higher among the experimental group students than the control group. Gender differences were found neither in spatial ability, nor in the model concept and chemistry understanding.

In two representative questions related to three-dimensional molecule structure and properties, the experimental group scored significantly higher than the control group (Barnea and Dori, 1996). Students' spatial ability in both groups improved, but students from the experimental group scored higher, and the difference was larger, when the comparison was restricted to the students of average attainment in the two groups. These results match the findings of the university undergraduate qualitative study, in which the most significant improvement in spatial ability was reported in the group of the average students. Students with high spatial abilities could not improve more. The very low spatial ability students did not improve so much either because they were not motivated enough or because they could not use the computer efficiently. But the average ability users in both groups found it most useful.

This innovative learning approach improved students' perception of different geometric structures and their relation to chemical formula. The results also indicate that, overall, the new approach improved many aspects of the trainees' model perception and visualisation skills. The importance of integrating the modelling software into the curriculum is that, through their experience, students could see various ways of representing molecules, each emphasising another aspect. They could realise the role of models and their limitations, so their initial perception of models was expanded and a notable difference between experimental and control groups was found.

AN INTERVIEW WITH A TEACHER OF THE EXPERIMENTAL GROUP

An interview with one experimental group teacher highlights the teachers' side of the CMM learning environment. The main factors that were examined were the atmosphere, the effectiveness of the new tool as a motivator and as a learning aid.

The teacher described the atmosphere as very pleasant. Students entered the class with curiosity, after they had waited for the lesson to begin, and their expectations were very high. When they started to use the computers their enthusiasm grew even more, especially while using the spatial glasses. Students worked individually or in pairs. Most of the students worked on their own, and asked for help only when a problem arose. The teacher expressed her surprise that the time it took to get used to the software was very short. The 'computer hackers' among the students did not need the explanations in the booklet. They discovered all the features of the software on their own. The novices, who never had experience with computers before, needed no more than 15 minutes to get used to the new environment. The teachers described some interesting events during classes:

> There were some exciting moments for all of us. When the students discovered new phenomena, such as the facts that a bond length between the same atoms was constant, and the tetrahedral bond angle is 109 degrees, they were excited. 109 was an unfamiliar number that kept on popping all the time. Students were mostly enthusiastic to view the three dimensional representation of diamond and graphite, and they got confirmations for their predictions when they later saw plastic models of these two substances.

The teacher thought that the discovery learning with the computer made the learning exciting and interesting. After the students had seen all the molecules they had better mental models and back in class when they were discussing those structures they were all talking about the same things, which were very realistic and not abstract as they usually are. She thought that the use of molecular modelling has definite advantages over the use of blackboard and plastic models.

DISCUSSION

The findings indicate that overall the use of computerised models has improved many aspects of the students' model perception and made them realise the role of models. This study advocates the use of various models to represent the same phenomenon. By doing this, students and teachers appreciate the value of models as well as their limitations, they then build in their mind a more scientific and reliable representation of the studied phenomenon. The dynamic rotation of models provided by the computer and the users' ability to manipulate and dominate this movement is substantial and central in achieving the most efficient usage.

It has been established that CMM has considerable advantages:

- simulations of all major modelling systems are available simultaneously and to every student;
- students are able to make quantitatively accurate changes in all the causal determinants of molecular shape;
- ability of the software to illustrate concepts studied at the lectures

Improved understanding is reported by students in respect of the relation between 2D and 3D representations and in respect of the shape of complex molecules.

MM is an important tool which can enhance visualisation skills and understanding of core concepts in structural chemistry. It should be used more often in undergraduate studies and it is not too early to introduce it in high school chemistry classes. Molecular modelling should not be used just as a separate exercise, but as an integral part of basic courses like: symmetry, organic chemistry, biochemistry, etc. In this way it is hoped that the next generation of scientists will use it properly and benefit from it.

A greater emphasis on the development of integrated projects for undergraduates, which combines various courses and the increasing use of molecular modelling, both as a modelling and experimenting system is advocated.

NOTES

1. This package is published by: Oxford University Press, Walton St., Oxford OX2 6DP, Great Britain.
2. The names of the teachers have been changed.

Chapter 17

The Structure and Development of Science Teachers' Pedagogical Models: Implications for Teacher Education

Erika Zimmermann
Universidade Federal de Santa Catarina, Brazil

INTRODUCTION

This Chapter discusses the structure and development of science teachers' frameworks of pedagogical practice, their models of pedagogy. It is suggested that the explanations that teachers give of their classroom practice contain embedded models of the nature of science, the nature of science teaching, and of the nature of science learning. Case studies of a range of secondary school teachers show the scope of these models and how they interact to form models of pedagogy. Drawing on Toulmin (1972), Chinn and Brewer (1993) and Zimmermann (1997), the Chapter provides insights into the ways that these models of pedagogy are developed. Finally, the implications of the findings for teacher education are discussed.

THE RATIONALE FOR THE NOTION OF A MODEL OF PEDAGOGY

> Where do teachers' explanations come from? How do teachers decide what to teach, how to represent it, how to question students about it, and how to deal with problems of misunderstanding? (Shulman, 1986a)

The above questions initiated a research agenda (Wilson et al., 1987; Shulman, 1986a, 1986b). In-depth case studies have been carried out by Shulman and his associates to provide an overview of 'a framework for a knowledge base for teaching', to examine 'the *sources* of that knowledge

J.K. Gilbert and C.J. Boulter (eds.), Developing Models in Science Education, 325–341.

base' and 'to explore the processes of pedagogical reasoning and action within which such teacher knowledge is used' (Shulman, 1987, p.5). This Chapter proposes that a teacher forms a model of pedagogy from this 'knowledge base for teaching'. A model of pedagogy is defined as being composed of at least three interacting models: a model of science (an 'epistemological model', what exists and can be known); a model of learning (how it can be known); and a model of teaching (how it can be made known).

Research in science education has begun to show a range of teacher's models of the nature of science (Brickhouse, 1989, 1991; Dibbs, 1982, Duschl, 1983; Gustafson, and Rowell, 1995; Koulaidis and Ogborn, 1988; Lantz and Kass, 1987; Russell and Munby, 1989). This research has brought about a debate concerning how teachers' models of science influence and shape their teaching of science. For instance, Druva and Anderson (1983) suggest that teaching effectiveness and teaching behaviours are related to teacher knowledge about science, while Lederman (1992) claims that the effects of teacher understanding of the nature of science on science teaching may be overstated.

Koulaidis and Ogborn (1988, 1995) claim that there is not necessarily a direct relationship between teachers' models of science and their practice. These authors, in a review of the body of research on science teachers' views on the nature of science, criticise the conceptual foundations of some such studies. They advise that investigations should assume that teachers might hold 'eclectic or mixed views' (1995, p.280). Further, they suggest that they should include the four main models of science given by the philosophy of science, which are the inductivist, the hypothetico-deductivist, the contextualist, and the relativist. For a more detailed account of these models, see Chalmers (1978), or see the original works by Popper (1979), Kuhn (1970a), Lakatos (1974) and Feyerabend (1983).

The literature proposes that pedagogy is not only influenced by teachers' models of science, but also by their models of learning science (Aguirre and Haggerty, 1995; Gustafson and Rowell, 1995; Huibregtse et al., 1994). Fox (1983) claims that the theory of teaching that a teacher holds determines "the kinds of tasks he sets his students" (p.162). Moreover, students' approach to learning can be influenced by the teacher's model of teaching (Entwistle, 1985) which in turn is influenced by the expectations the teacher has about the students approach to learning (Fox, 1983). Finally, it can be said that the teaching of science not only depends on teachers' understanding of both how children learn, but on teachers' understanding of how knowledge is generated (Aguirre et al., 1990; Duschl and Wright, 1989). Different epistemological models generate different views of the learning process, which in turn yield different approaches to teaching (Swift, 1982). Thus, one can conclude there is a relationship between teachers' models of the nature of science (with an explicit or implicit epistemological model) and their models of learning which generate a model of teaching.

Teachers' models of the nature of science, of learning, of teaching, and the possible interactions among these models to produce their models of pedagogy, have been studied in depth (Zimmermann 1997a, b, Zimmermann and Gilbert, 1998). This work also provided empirical support of Fenstermacher's (1986) theoretical claim that the concept of teaching is inseparable from the concept of learning and that, rather than having a causal dependence, the concept of teaching has an ontological dependence on the concept of learning. Further, the Zimmermann (1997) study suggests that a teacher's models of the nature of science, of learning and of teaching, fit together in different ways resulting in a range of different models of pedagogy, dependent on the 'ecological niche' in which they work.

THE STRUCTURE OF TEACHERS' MODELS OF PEDAGOGY

The five experienced teachers in Zimmermann's (1997a) study, Sergio, Fernando, Mario, Daniel, and Newton[1], are experienced Brazilian secondary physics teachers. Using Koulaidis and Ogborn's (1988) framework for the models of the nature of science, Table 17.1 was produced from the transcripts of interviews with the teachers, which were organised and categorised (Zimmermann, 1997a) with the help of 'NUDIST' qualitative data analysis software (Richards and Richards, 1994).

Table 17.1. Summary of teachers' models of science

	Sergio	*Fernando*	*Mario*	*Daniel*	*Newton*
Models of Science	Mixed: Ranging from inductivist to contextualist.	Contextualist	Inductivist	Mixed: Ranging from inductivist to deductivist.	Eclectic: Ranging from inductivist to relativist.
Scientific Methodology	Three views: 1. One method: from observation to theory; 2. One method: from theory to observation; 3. Scientific community agreement.	One view: Many methods (paradigmatic) methods are a consensus within the scientific community.	One view: One method from observation to theory.	Two views: 1. One method: from observation to theory; 2. One method: from theory to observation.	Three views: 1.Only exact sciences employ the scientific method; 2.Theory to observation and vice-versa (falsification); 3.Anything goes.
Criteria of Demarcation	Two views: 1. Rational: must be proven to be truth; 2. Rational: social acceptance.	One view: Rational: consensus within the scientific community (consensus is made upon a triangular basis: scientist-reality-scientist)	One view; Rational: Scientific knowledge is generated on the basis of the scientific method; it is proved on the basis of experimental results: the truth about the world.	Two views: 1. Rational: the truth about the world; 2. Rational: nearer to truth improvement of knowledge.	Four views: 1. Rational: the truth about the world; 2. Rational: nearer to truth improvement of knowledge; 3. Rational: consensus of the scientific community; 4. Not rational: politicing, a matter of power.
Patterns of Growth	Two views: 1. Compatible growth by accumulation of knowledge; 2. Change is possible, knowledge is tentative.	One view: Discontinuous growth by accumulation intercepted by ruptures with previous knowledge; it is historically and socially determined.	One view: Compatible growth: accumulation of knowledge.	Two views: 1. Compatible growth by accumulation of knowledge; 2. Possibility of change by revision: knowledge is tentative and is improved.	Two views 1. Addition of knowledge Cumulative; 2. Progress by improvement of theories.
Status of Scientific Knowledge	Two views: Special status: 1. Objective account of nature; 2. Imaginative process of discovering knowledge.	One view: Special status: Systematic pattern of thought: Coherent discourse that opposes mere opinion.	One view: Special status: Objective account of nature: Opinion, imagination and speculation play no role.	Two views: Special status 1. Objective account independent from human perceptions; 2. Admits the role of imagination.	Two views: 1. Special; objective account in the natural sciences; 2. Not unique: a knowledge amongst others.

Table 17.2. Summary of teachers' models of teaching and learning

	Sergio	*Fernando*	*Mario*	*Daniel*	*Newton*
Teaching Experience	20 years	5 years	18 years	23 years	15 years
Conceptions of Teaching	Knowledge can be transmitted and might be absorbed.	Knowledge must be constructed: all of us have learnt by constructing.	Knowledge is transmitted: it is only learnt this way.	Knowledge can be transmitted: but students must be involved.	Knowledge can be transmitted, but it is better learnt when it is constructed.
Idealised Approach to Teaching	Finding out students' conceptions and work from these.	Idealised coincides with actual: Constructivist: Conceptual Change.	Idealised coincides with actual: Behaviourist.	Getting students involved in their own learning.	Idealised: changing students' conceptions constructivist.
Role as a Teacher	Deliverer of knowledge and controller of students' behaviour.	Pioneering guide (helps to open spaces).	Coach (instructor who trains students to attain certain skills and abilities).	Actual: transmitter of knowledge; Idealised: coadjutor.	Disseminator of knowledge and of the wonders of science.
Conception of Learning	Building up process (adding knowledge).	A social and individual construction.	Building up process (adding knowledge).	Building up process (adding knowledge).	A process of changing individuals' conceptions, but building up seems easier.
Students' Role	Passive receiver of the right knowledge.	Builders of knowledge (revolutionary and normal scientists).	Passive receiver of the right knowledge.	Idealised: manipulator; Actual: passive absorber.	Passive absorber of knowledge.
Experiments in Science Teaching	To underpin and verify..	Are context bound; Are only meaningful when out of students own questions.	To verify and underpin; to make the abstract concrete; to fix knowledge previously transmitted.	To motivate; To underpin and verify; To learn the process of science.	To encourage students careful and accurate observations; to verify and underpin; to make the abstract concrete.
Actual Approach to Teaching	Behaviourist Transmission reception.	Constructivist Conceptual change.	Behaviourist Transmission reception.	Behaviourist Transmission reception.	Mixture of behaviourist and conceptual change.

There are contrasts and similarities amongst these five teachers. All teach at the same school, thus teaching the same curriculum within the same contextual/cultural demands. All experienced their primary and secondary education under the influence of behaviourism and positivism. They graduated from the same university. Other then Fernando, all have a knowledge of the nature of science that originates from implicit messages they received during their professional lives, for they have never formally studied the nature of science. Defining teaching, learning, and above all, the nature of science, proved to be extremely difficult for nearly all of them. Despite some similarities, these teachers employ different approaches to teaching that range from a reception-transmission model to a conceptual change model (Table 17.2). These variations seem to be due to differences in their reconstruction and interpretation of their own learning experiences *as students*.

Factors such as school philosophy, parents' and students' expectations, do influence their resulting models of pedagogy. However, Zimmermann's (1997a) study suggests that the effective influences on models of pedagogy vary from one teacher to another, depending on the degree of coherence of the teacher's model of science and of science teaching and of science learning. For instance, Daniel and Sergio (see Tables 1 and 2), who hold idealistic models of teaching, which are not coherent with their models of learning, easily abandon their ideals to follow the school's, parents' and students' beliefs as to what and how to teach.

As a way of identifying contrasts among these teachers and the significance of their different views for classroom practice, they were grouped together by the *degree of coherence of their models of pedagogy*. In other words they were grouped together according to the degree in which all the parts of their models of pedagogy fit together to form a united whole (Tables 17.3, 17.4 and 17.5).

Table 17.3. Partially Incoherent Models of Pedagogy: Sergio and Daniel.

	Sergio	***Daniel***
Summary of Models	Mixed models of • the nature of science, • teaching	Mixed models of • the nature of science, • teaching
Scientific knowledge ***First view***	The truth about the world	The truth about the world
Scientific knowledge ***Second view***	Nearer to truth Tentative	Nearer to truth Tentative
Learning ***Only One Model***	Absorbing knowledge Building-up process	Absorbing knowledge Building-up process
Teaching ***First View***	Transmission of knowledge to be absorbed and accumulated	Transmission of knowledge to be absorbed and accumulated
Teaching ***Second View***	Finding out students' conceptions	Getting students involved in their own learning
Actual Teaching	Behaviourist	Behaviourist

Table17.4. Completely Incoherent Model of Pedagogy: Newton.

	Newton
Summary of models	His model of science is eclectic and his model of teaching is eclectic and does not fit in his model of learning.
Scientific Knowledge ***First View***	Sometimes is the truth about the world.
Scientific Knowledge ***First View***	Sometimes is pure theorising (imagination).
Learning ***Unsure***	Not sure whether students conceptions can really be changed Change is too difficult so building up was the way he thinks he has learnt and so his students are also able to build up.
Teaching ***First View***	Transmission of knowledge to be absorbed and accumulated.
Teaching ***Second view***	Conceptual change.
Actual Teaching	Eclectic: amalgamates conceptual change with behaviourism.

Table 17.5. Coherent models of Pedagogy: Fernando and Mario

	Fernando	***Mario***
Summary of models	Consistent model of science: Contextualist. Consistent model of teaching and learning: Constructivist.	Consistent model of science: Inductivist. Consistent model of teaching and learning: Behaviourist.
Scientific knowledge	Tentative.	The truth about the world.
Learning	Changing conceptions.	Building up process.
Teaching	Conceptual change.	Transmitting knowledge and reinforcing behaviour.
Actual Teaching	Teaches accordingly to a conceptual change model and the understanding that knowledge is tentative.	Teaches accordingly to both a behaviourist model and the understanding that knowledge is the truth about the world.

Coherent Model of Pedagogy

Mario and Fernando, who had a consistent model of the nature of science and who also showed that their models of learning are congruent with their models of teaching, were grouped together as having a *coherent model of pedagogy* (Table 17.5). These teachers' ideal approach to teaching coincides with their actual approach to teaching.

Mario holds an inductivist model of science, a *consistent model* of the nature of science, which is congruent with his behaviourist approach to teaching. He believes that scientific knowledge is accumulated. Thus, in harmony with an inductivist model of science, Mario sees his job as transmitting correct knowledge and as planning drill tasks as reinforcement so that his students accumulate more and more knowledge.

Fernando, who is a visionary thinker, holds to a consistent contextualist model of science. He sees science as changing with the passage of time and so he also sees his students' knowledge as changing. Such a belief is in complete congruence with a tentative view of scientific knowledge. He understands that students bring their own conceptions to the classroom and that these can be at odds with established scientific conceptions. They need to change these ideas, so his work is to help them to overcome 'epistemological obstacles' (Bachelard, 1977).

Partially Incoherent Model of Pedagogy

Sergio and Daniel have *mixed models* of science, that is, parts of different models of science combine to form them. They hold mixed models of teaching, which do not coincide with their practices (Table 17.2). They were seen to plan their lessons by making a list of behaviourist objectives with associated reinforcement strategies, a clearly behaviourist approach. Their selection of content is straightforward, their planing consists of a sequence of 'facts' to be transmitted and reinforced. Both teachers see learning as process of building up knowledge (Table 17.2). Sergio and Daniel's explanations of their teaching demonstrate that it is not the concept of teaching itself, but rather the concept of learning, that is pivotal for teaching. Further, because they strongly believe in learning as a 'building up process', they ignore their alternative view of scientific knowledge as being tentative. In this way, they see their work as being to organise students' tasks by emphasising the rules that lead to error-free 'products'. As Sergio and Daniel have *mixed models* of science, as they teach in consonance with an inductivist model of science, as they believe in idealised (non-behaviourist) models of teaching, but actually use a behaviourist approach, they were grouped together as having *partially incoherent models of pedagogy* (Table 17.3).

Completely Incoherent Model of Pedagogy

Newton can be contrasted with none of the other teachers. Holding an *eclectic model* or *hybrid model* (see Chapter 11) of the nature of science, the constituent parts are from different models and do not fit with each other. Thus it is not surprising that the instruction that Newton provides is also highly eclectic. He was observed trying to introduce a conceptual approach to his teaching, but he shifted from the intention of changing students' concepts to the intention of reinforcing their behaviour, which resulted in a muddle. Newton is eclectic in both, his model of science and his approach to teaching. Thus, he was categorised as having a *completely incoherent model of pedagogy* (Table 17.4).

THE DEVELOPMENT OF TEACHERS' MODELS OF PEDAGOGY

Drawing on the ideas of Toulmin's (1972), Chinn and Brewer (1993) and Zimmermann (1997a), this section provides insights into the ways science teachers' models of pedagogy are developed. These insights lead one to ask what knowledge base teachers of science should acquire in order to hold a more consistent and developed model of pedagogy for their teaching. Then, drawing on the work of Shulman and his co-workers, the issue of the knowledge base of teaching is addressed.

According to Toulmin (1972), an individual's concepts and ideas are assembled into a conceptual framework that develops as a result of natural selection, adaptation to the ecological niche in which that person lives. Taking this perspective into account, teachers' models of pedagogy can be seen as an *adaptation* to the *ecological niche* (educational environment that is constituted of social and cultural beliefs, curriculum, goals, accepted practices, etc.) in which they live. This view justifies the claim that a teacher's initial model of pedagogy develops and may change as a consequence of their experience with the curriculum in use (Nott and Wellington, 1996). It also accounts for the tension between idealised perspectives on teaching and pragmatic social pressures. However, it does not account for the nature of individual differences of teachers' models of pedagogy. Thus, it seems worth examining teachers' models of pedagogy at a teacher education level. That is, analysing the model held by an individual teacher before he enters the teaching profession in order to understand how it later changes.

Chinn and Brewer (1993) provide a powerful theoretical framework with which to understand at a macro-level how one's knowledge develops and may change. While the framework is aimed at explaining scientific change by elucidating how scientists or students of science respond to anomalous data, I suggest that the framework is also useful in understanding the development of teachers' models of pedagogy from the time they are students up to the time that they have become experienced teachers.

The role of anomalous data in the context of teacher education can be understood as presenting student-teachers with 'evidence that contradicts their pre-instructional theories' (Chinn and Brewer, 1993, p.2). In Chinn and Brewer's (1993) model, six types of responses to anomalous data enable students to protect their prior beliefs, whilst a seventh type of response allows students' acceptance of evidence with a corresponding change in their prior theory (see Table 17.6).

Table 17.6. Features of each of the seven responses to anomalous data[3]

Type of Response	*Does the individual accept the data?*	*Does the individual explain the data?*	*Does the individual change theories?*
Ignoring	No	No	No
Rejecting	No	Yes	No
Excluding	Yes or Maybe[a]	No	No
Abeyance	Yes	Not yet[b]	No
Reinterpreting	Yes	Yes	No
Peripheral Change	Yes	Yes	Yes, partly[c]
Theory Change	Yes	Yes	Yes[d]

a. The individual may either accept the data as valid or remain agnostic about whether the data are valid.
b. The individual expects that the data will be explainable by the current theory at some future date.
c. Only beliefs in the protective belt are changed.
d. Core beliefs are changed.

Chinn and Brewer's (1993) discussion of the ways an individual's prior knowledge influences how he/she responds to evidence that contradicts his/her beliefs is relevant to the development of teacher's models of pedagogy. They assert that the *entrenchment of the individual's current theory* is especially important 'in influencing how an individual responds to anomalous data' (Chinn and Brewer, 1993, p. 14). It is suggested that 'an entrenched theory is a theory that contains one or more deeply entrenched beliefs' (Ibid, p.15). Zimmermann's (1997a) work shows that the bedrock model that forms one's model of pedagogy is the model of how one learns science, which, in turn, seems to be embedded in one's epistemological belief. One's model of the nature of science, on the other hand, contains ontological and epistemological beliefs that might also be entrenched (Chinn and Brewer, 1993). Zimmermann's (1997a) study shows that some of the beliefs that underpin a teacher's model of pedagogy can be entrenched for reasons derived from the nature of their past experience, from their personal self-image or for social expediency, as suggested by Chinn & Brewer (1993). For instance, Mario's past technical experience prompts him to value the practicality and usefulness of knowledge, causing him to focus on such values during his teaching. Sergio's and Daniel's experiences with parents' and students' expectations lead them to assume that they need to teach directly for the specific goal of getting students greater success in tests and examinations. Fernando's values of critical awareness and democratic citizenship seem to be a reflection of his experiences as an activist and thus he assumes that when students challenge and question knowledge such values are enhanced.

Chinn and Brewer (1993) also point out that one's background knowledge is a crucial factor in influencing one's response to evidence. In their majority, the teachers studied by Zimmermann (1997a) lack *relevant* background knowledge for teaching science (e.g. history of science, sociology of science, relations between science and other realms of knowledge, etc.). Due to this lack, teachers seem to be unable to notice incoherence between the concepts that form their models and so are unable to evaluate evidence and to accept that evidence which shows the limits of their actual models, a prelude to changing them. For instance, Mario and Fernando (Zimmermann, 1997a) present a high degree of coherence between the concepts that form their models of learning and of the nature of science. However, there is a sharp difference between the models of pedagogy of these teachers. By acquiring relevant background knowledge, Fernando has formed a more complex and developed model of pedagogy. Mario seemed to have no relevant background knowledge that could help him to evaluate and change the model he has acquired thorough his behaviourist and positivist past experiences. Finally, looking at Toulmin's model of conceptual ecology and at the case studies presented in Zimmermann's (1997) work, one can discern that the educational niche in which Fernando lived while he was a student-teacher was considerably different from that in which the other four teachers did.

From Chinn and Brewer's (1993) framework, Toulmin's model of conceptual ecology and Zimmermann's (1997) work, it is proposed that meaningful change in a teacher's model of pedagogy may occur when two conditions are met. First, when the teacher is enabled to reflect on his/her concept of learning in the light of relevant background knowledge. This can point to incoherence between his/her conception of learning and the other concepts that form the personal pedagogical framework. Second, the teacher's background knowledge must be relevant to the matter in hand in such a way that it enables the appraisal of evidence to lead to the acceptance of the need to change and development.

THE ECOLOGICAL NICHE EFFECTS

Zimmermann's (1997) research points to three main contextual factors that have a strong effect on teachers' models of pedagogy. These are: (a) the examinations to enter the university in Brazil, called the 'vestibular' (Zimmermann, 1997a); (b) the belief amongst the general population in pseudoscience and the paranormal (Zimmermann and Gilbert, 1998); and (c) teachers' underpayment (Zimmermann 1997b).

Examinations to Enter Tertiary Education

The emphasis given to the *vestibular* (examination students undertake to enter tertiary education) by Brazilian society guides almost all classroom assessment, thus classroom practice (Zimmermann, 1997a). The five teachers agreed that their approaches to teaching are highly directed to help students to get through such examinations. For all teachers but one, this means teaching approved knowledge so that the students can correctly answer the vestibular questions. The difficulties the teachers have in implementing and maintaining changes that would meet their ideals of good teaching are, to some extent, related to these examinations. Vestibular results are publicly announced, and so are the names of the schools that yield the 'best students'. Having a good number of students that get through takes the school to the 'top'. So, there is a hidden, but clearly identifiable, school and parental pressure upon these teachers: they should teach 'correct' knowledge so that the students get good marks. The outcome is a school organisation that favours the kind of learning for which it can display their public accountability. That is, it rejects students' needs for relevant learning in favour of examination passing. As a result, teachers adapt their pedagogy to meet such demands: they teach the students to pass the vestibular.

Pseudoscience and the Paranormal

The Brazilian population is very keen on spiritualist religions, on astrology and other forms of pseudoscience and the paranormal. Such a cultural background has been shown to influence teachers' models of pedagogy in contrasting ways. For instance, whilst Sergio demonstrates a strong wish to dismiss the paranormal and pseudoscience, Newton is very enthusiastic about them (Zimmermann and Gilbert, 1998). To deny any sort of pseudoscientific knowledge, Sergio considers science as rational knowledge that can be trusted, as opposed to irrational mysticism that must be dismissed (Zimmermann and Gilbert, 1998). Newton has an opposite understanding. He believes that there is regularity in the universe that is studied by the sciences. Because he believes in this regularity, he assumes that there is a consonance between the data of a person's birth and the position of the stars.

He says that he is part of this universe and therefore part of this consonance. In short, he sees astrology as a science. If those who believed that eating plenty of eggs was healthy, or that the earth was the centre of the universe, could be so wrong, how can one be sure that the same revolution will not occur in current areas of science? So, why should Newton see this expanded view of science as the road to truth? Such a picture shows the necessity of addressing these issues critically in teacher education programmes, as well as in school science classrooms. For a good review on how to deal with the issues of pseudoscience and the paranormal in science education see Martin (1994).

Teachers' Low Income

In Brazil teaching is not a promising profession in terms of salary (Psacharopoulos, 1987). The teachers studied by Zimmermann (1997a) show that their models of pedagogy are highly affected by their salaries, since they need to take additional employment to supplement their low income. Literature shows that problems such as inadequate salary are important in predicting changes in teachers' models of pedagogy and job stress (Litt and Turk, 1985).

IMPLICATIONS FOR TEACHER EDUCATION

Having described the structure and development of science teachers' models of pedagogy the remaining questions are: (1) What background knowledge is needed so that coherence among the concepts that form one's pedagogical framework can be achieved? (2) What sort of background knowledge is relevant so that a core belief such as how students learn science changes to a more appropriate and developed one? On the whole, what should the *knowledge base* (Shulman, 1987) be so that teachers of science acquire a better developed model of pedagogy? The teacher education courses taken by the teachers presented in this Chapter had little, if any, effect in fostering their knowledge of teaching. These courses seemed to have been poor, superficial and irrelevant with no visible intellectual consequences. The history and philosophy of science had not really been taught during the teacher education courses that these teachers had attended. The majority of the teachers studied here lack knowledge of the psychology of learning. In short, the teacher education courses have *trained*, rather than *educated* them. Although training is part of anyone's education, it does not prepare one to articulate knowledge and to consider aims, methods, and standards for practice. Someone that is only trained to do something learns to do it in a limited way[2]. Even recognising the dangers of generalising from context-specific case studies, this research does raise issues of relevance to those who wish to improve teacher education programmes, specifically:

Sound Subject Matter Knowledge

As pointed out by Fernando, in Zimmermann's (1997a) study, the depth to which he explores any topic in his classroom is dependent upon his subject matter knowledge. He asserts that the content knowledge tackled in teacher education courses should be the sort of knowledge that helps teachers to make the links which students can follow, to discuss the principles and concepts without 'mathematising' them. In other words, it needs to be the sort of knowledge that 'enables teachers with a more accessible language, without, of course, putting the formalism aside'. Fernando points out that teachers who have a lack of subject matter knowledge do not feel comfortable in doing experiments or in undertaking a qualitative discussion of concepts. The proposal made here is to provide sound, perhaps less, subject matter knowledge that would help teachers to acquire a more accessible language for teaching. Moreover, it should be the sort of knowledge that assists them to appreciate the rich intellectual and cultural milieu of the science they teach.

The History and The Philosophy of Science (HPS)

The history and the philosophy of science enables one to understand the nature of science. I suggest that teacher education courses should introduce students to the history and philosophy of science. Although important, possessing informed conceptions of the nature of science does not necessarily lead to improved teaching performance (Lederman, 1992). Thus, such an introduction is meant to help students to analyse HPS critically. This type of critical work might help teachers, in Chinn and Brewer's (1993) words, to accept evidence in order to discard the models of science that are no longer accepted in contemporary philosophy of science. It should not provide teachers with 'frozen' positions, but rather to enable them to be critically aware of the nature of science and so to take a conscious position in order to appreciate science as a tentative endeavour. The claims about the similarities between individual and historical development must be addressed in such a course. By these means, such courses may foster teachers' ability to anticipate students' conceptions. Such a perspective reinforces the claim that HPS should have a stronger role in teacher education. A well-organised knowledge about the nature of science may lead to a more coherent classroom action, that is, a classroom action that is coherent with science and science learning as tentative activities.

Psychology of Learning

A key concept for teaching is learning. Although it can appear pedantic to propose that knowledge of psychology of learning is of paramount importance for any teacher, the suggestion here is, once again, that teacher education programmes should address a sound and relevant psychology of learning. A relevant course in the psychology of learning would make teachers aware that a central task of teaching is to enable the student to perform the tasks of learning, not the tasks of the syllabus content (Fenstermacher, 1986). It would not be to provide teachers with ready-made positions, but rather to help them to construct a collection of elements to take informed and critical actions in their classrooms. So, a key issue in the education of science teachers involves creating the conditions by which teachers can critically and reflexively rethink their own model of learning and, from this, the model of teaching it leads to. In brief, it should be a course that, by presenting theories about the processes of learning, leads to an understanding of what goes on beyond a particular model of learning, namely, its empirical, methodological and epistemological bases.

Pedagogical Content Knowledge

Pedagogical content knowledge seems to be heavily dependent on a clear understanding of both subject matter knowledge (the substantive and syntactic structure of science) and on the psychology of learning science. Teachers' frameworks of pedagogical content knowledge were demonstrated to be low both in complexity (number of concepts included) and coherence (degree to which the concepts were related to one another). Teachers' structure of such knowledge not only needs to be enlarged but also better organised. Moreover, it seems that such knowledge will lead a more developed model of teaching. This can only be attained when subject matter knowledge and pedagogy are treated as associated realms. Thus, drawing on the proposals made by Shulman (1986b) and by Zimmermann (1997a), it is suggested that teacher education programmes should introduce student-teachers to 'the ways of representing and formulating the subject that make it comprehensible to others' Shulman (1986b, p.13).

Integration

The in-source programmes attended by the teachers interviewed display a course organisation made up of discrete, unconnected, disciplines. As the teachers demonstrated, their teacher education programmes involved a large amount of subject matter knowledge disconnected from other realms of knowledge. Their focus is on rapid coverage of physics with little attention, if any, to its origins or applications, and little opportunity to develop an integrated understanding of that knowledge for teaching. Such programmes seem to be of no help for teachers' practice in real classrooms. None of these teachers experienced a teacher education program in physics to aid them in integrating the knowledge learned in twenty-five or more separate discipline-based courses. Disciplines were taught by separate departments. The tendency of students to compartmentalise is well known (Edmondson and Novak, 1993). Thus, this perspective points to the need of addressing 'the perpetual problem of integration' (Grossman and Richert, 1988, p.61).

The suggestion made in the above sections would be useless if one does not think about *integration*, the unity of the parts to form a combined whole in which each part becomes closely linked to each other in the whole system. From Zimmermann's (1997a) work, one can see that it is nearly impossible to separate the content of the discipline from its structure or from the pedagogy needed to teach it, or from the psychology of learning it, and so on. Thus, one could have for instance a course framed by certain topics, themes or problems that require a multidisciplinary approach. Certainly it should be a course that reduces the chances of student teachers failing to grasp any of the inter-relationships that may exist between the component parts of a model of pedagogy. All these parts are tightly connected to each other by interaction and thus cannot be separated into disconnected disciplines; if they are, such a teacher education will not enhance teachers' ability to articulate their knowledge. A proposal of a model for an integrated course would be a valid theme for further research, in which the central questions to be pursued could be: How can our growing knowledge in science education help us develop a professional education programmes for teachers so that student-teachers can integrate what they learn to understand the complexities and realities of classroom life? What is to be selected from the total of all that is known in physics for such a teacher education programme? What should teacher education programmes look like in order to enhance the development of better models of pedagogy by teachers?

Case-study-based Pedagogy

This research points out the significance of the use of case-based pedagogy (Shulman, 1992) in teacher education programmes, that may include case discussions and analyses to assist teachers in their understanding of science, learning and teaching, and thus in their professional development. It is well known that 'teacher education students have negative attitudes towards theory presented to them' (Wubbels, 1992, p.137). An approach of that kind may help to bridge the gulf that exists between theory and practice. Perhaps the analysis of dilemma-based case studies, similar to those found in Sergio's and Daniel's case studies, would make teachers think about the excessive conformity to the norms imposed by school, parents and even by students. Analysis of case studies may provide students with opportunities to recognise problems, and to gain an understanding of how these problems can be faced by considering what the consequences of particular actions are. Moreover, it would give students the opportunity to devise sensible, ethical, moral and educational actions by considering the social, political and cultural consequences of certain behaviours. This approach would help teachers to make more effective decisions.

NOTES

1. The names of the teachers have been changed.
2. Here one can see the reason for my preference to call such courses 'teacher education, rather than teacher training'.
3. In Chinn and Brewer (1993, p.13).

Chapter 18

Challenges and Opportunities of Developing Models in Science Education

Carolyn J. Boulter, John K. Gilbert
The University of Reading, UK

INTRODUCTION

This book is concerned with the nature and role of models and modelling in science and technology and with their significance for education. This has entailed addressing the teaching and learning of models in educational settings and the scope and nature of the representational forms used. In this last Chapter we will draw together and extend these issues to look at some key questions:

Why is a model-based approach to the curriculum and the teaching and learning of science important?

What are the consequences when model-based approaches to teaching and learning are implemented in practice?

What agendas for research and development arise from the answers given to these questions?

THE NEED FOR A MODEL-BASED APPROACH TO TEACHING AND LEARNING

Underlying the contributions to this book are the assumptions that mental modelling is a universal way of thinking, that expressed models are a universal component of communication, and that consensus models are produced by all social groupings which have some degree of permanence.

J.K. Gilbert and C.J. Boulter (eds.), Developing Models in Science Education, 343–362.

The modelling and models associated with the social groups which conduct science and which provide science education are important for three major reasons. First, modelling and models are explicitly recognised in science and science education, even if, as we have argued, their significance has not hitherto been fully explored. Science and science education provide propitious contexts within which to study modelling and models. Second, science is a major cultural achievement of humanity and, as such, its nature and achievements should be widely appreciated. This book has shown that modelling and models play a major role in the nature of science and in its achievements. Third, through its association with technology, science has major economic implications. Chapters 1 and 7 argue that modelling and models play major roles in technology. Thus they have a significant economic status.

For these three reasons, science education (or 'science' as it is usually called in schools) is now a compulsory element in the education of young people in many countries, e.g. UK (DFEE 2000), USA (National Research Council, 1996), Australia (Curriculum Corporation, 1994). The place of modelling as an element of scientific methodology and of models as a major product of and means of communication about science is recognised, to a greater or lesser extent, in these prescriptions. However, they vary a great deal in respect of the level of detail in which they are couched and in the degree of compulsion attached to their implementation.

This recognition of the importance of science education exists side-by-side with an awareness that the split within societies into the 'two cultures' (that of 'the sciences' and of ' the humanities'), as described by C.P. Snow in the 1960s, still makes science unattractive to many. This is typically manifest in the relatively small numbers of students in (for example) Britain wanting to take university degrees in Physics or Engineering. The importance of a proper debate about the nature of appropriate 'scientific literacy' for whole populations and how that should relate to the training of the scientific elite, first formulated by Fensham (1985), is still a matter of urgency. The case made in this book is that a defendable and coherent approach to models and modelling in science education will do much to 'bridge the gap' between the 'two cultures'. The Nuffield 'Beyond 2000' project (Millar and Osborne, 1999), mentioned at the start of Chapter 1, in which models play an evident and significant role, is a recent manifestation of this set of concerns.

The most appropriate basis for the construction of the science curriculum, whether for 'all' or for 'the future scientific elite' is still a matter of debate. There is a growing awareness of a difference between the 'Curriculum-based' approach of Anglo-American research and curriculum construction (which has dominated Anglo-Saxon spheres of influence) and the 'Didaktik-based' approach of continental Europe (which has dominated Germanic, Francophone, and Iberian spheres of influence) (Fensham, 1999). Within the Curriculum tradition there is a separation between teaching for a conceptual understanding of the content of the subject and teaching for an understanding of the processes by which that content is produced. The latter involves students behaving, as far as is possible in school laboratory contexts, like scientists. The Curriculum tradition defines the content knowledge to be taught to all students and takes an evidence-based investigative approach to the process of being scientific. The nature of the science content to be taught and the practical experiences with which school students are to be provided, together with how they are seen to build these into an appropriate body of scientific knowledge, are still major topics for debate and decision. The Didaktik tradition perceives education as the formation of the whole learner and requires science knowledge, a unity of content or process, to be transformed so as to be suitable for this educational process. The nature of this transformation is also still a major topic for debate and decision.

The research on how we come to know through models and modelling can make a significant contribution to the debates within both the Curriculum and the Didaktik traditions. It deals with issues of the effectiveness of practical work by taking a new perspective, asking what the relationship might be between embodied cognition and mental models, and how we come to know through the interaction of the textual, pictorial, mathematical and gestural modes that make up practical experience. An understanding of the way that scientific knowledge can be modelled to meet the needs of the 'whole' learner is also a tantalising prospect. The use of a 'models perspective' as a way of exploring the relationship between the Curriculum and Didaktik traditions can be significant as the possibilities of unified, federated, or confederated, approaches to science education are contemplated within political structures such as the European Union and the United States of America.

The question of scientific literacy for all, connected to the development of general literacy for all, is a matter of crucial importance in those many countries which aspire to have, or to retain, advanced technology-based economies. In English primary schools (for 5 to 11-year-olds) an hour each day is set aside, by government order, for literacy activities, within which literacy in science is playing an increasing part. The emphasis on the acquisition of higher-than-current levels of general literacy shows every sign of spreading throughout the UK school system and even into higher education. The acquisition of science knowledge is, like all knowledge, mediated through oral and written discourse which has rules and conventions. Science discourse has its own literacy, its own vocabulary and forms of representation. Students of science must come to know and be able to use these if they are to form a significant appreciation of science. However, 'scientific literacy' is much more than the 'literacy in science' which is developed in science classrooms and in such 'Literacy Hours'. It is more than the reading of and writing about scientific texts, more than talking about and listening to talk about science, important though these are. It must include an understanding of the nature and processes by which scientific activity is carried out. Here model-based teaching and learning can provide a framework for understanding the nature of the interaction between different modes of representation and the conduct of science itself. Such insights have a direct and practical part to play in developing scientific literacy.

The research, which is relevant to these issues, is, as this book has demonstrated, underway internationally. It has gradually built up over the 1995 to 2000 period, but not from a standing start. In the period 1980 to 1995 a great deal of research into the way that language and practical experience intersect to shape pupils' understandings in science education took place in Britain, e.g. the 'Science Process And Concept Exploration'(SPACE) project, the Children's Learning In Science Project (CLISP), and others elsewhere, e.g. the 'Learning in Science Project'(LISP) in New Zealand. These projects, associated with constructivist approaches to the teaching and learning of science, spawned a number of important curriculum development projects, e.g. Nuffield Primary Science in the UK. Research into children's 'alternative conceptions', which was the major outcome of the research projects, has expanded world-wide to produce a vast body of data. However, the influence of what has come to be known as the Alternative Conceptions Movement (ACM) has recently come under scrutiny (e.g. Solomon, 1994; Osborne, 1996). Some of the issues surrounding the notion of 'constructivism' were rehearsed in Chapter 2: the essentially repetitive nature of current work in the ACM suggests that it has 'run out of steam', lacking that general perception by researchers of theoretical tensions which might drive it forward. Models and modelling provided a perspective with which to evaluate the ACM and its achievements. They are also capable of providing the fresh theoretical and practical insights needed to advance science education in the directions outlined above.

IMPLEMENTING A MODEL-BASED APPROACH TO TEACHING AND LEARNING

The question of how we come to know science and technology through models and modelling is being addressed within CMISTRE (Centre for Models in Science and Technology: Research in Education) through work in a number of different contexts, some of which are described in the following sub-sections. Appropriate methodologies are being developed within each context with which to adopt a 'models and modelling' approach and to try to capture the consequences of that adoption. Schemes of work used in school science are being restructured around a models perspective. Empirical techniques in use include the extended observation of learners as they tackle model-related tasks, the interview of learners during, immediately- and long-after, an address to such tasks. Most importantly of all, action research by teachers and other learning-facilitators is providing successive cycles of research and development. These techniques are being used against a background of scholarship into language acquisition and development, of the history, philosophy, and social psychology of science.

We give an account of primary science classrooms as our first context.

THE CMISTRE PRIMARY TEACHERS' CLASSROOM RESEARCH GROUP

Teachers as researchers have been seen as significant part of developing the professional role of teachers for some time (Elliot et al., 1996). In England the official Teacher Training Agency sees encouraging what they call the 'research and evidence based' professionalism of teachers as a significant part of their work in forming training policy (see www.teach-tta.gov.uk/research/evidence). This professionalism includes teachers generating findings through doing research in their own classrooms as well as in knowing about research that others have done. In 1995, shortly after the main CMISTRE research group was formed, a small group of primary school teachers keen on researching their own practice in school was convened by Carolyn Boulter. The aims of setting up this group were to empower these teachers to do their own research around a model-based theme, to build their confidence in their ability to do it and to enable them to have enough support and time to carry it through. These aims were met, in the tradition of 'teachers-as-researchers', through collaborative leadership of the group so that teachers raised their own questions and took ownership of the methods they used to find answers. As they began to find out about aspects of their practice through research their confidence grew. Enabling teachers to have the time and emotional space in which to think about and carry out research has proved to be the most difficult of the aims to fulfil.

In order to further these aims it was considered very important that any work should be made publicly available both to enhance the confidence of the teachers and to enable other teachers to see what had been achieved. The first phase of the work culminated in a publication of the various pieces of their research in an in-house form (Marsh and Boulter, 1997). In this first phase the group concentrated on aspects of 'scale' in the representations that they used in their classrooms. This topic was raised by the teachers as one which was important because it had many cross-curricular connections (Boulter, 1998). Individual members of the group then investigated particular aspects of their own practice. This included: looking at children's understandings of the word scale, the place of scale in a National Assessment Task for 7-year-olds which required them to group pictures of animals; the understandings of weight and length scales shown in model bridge building by 10-year-olds; the use of orreries with different scales and representational features in teaching; the implications of scale in drawings of the solar system; pupils' drawing of small animals to scale. The group was introduced to the developing understandings of modelling and came together to discuss plans and to write up results during 'in-service education days'. Throughout the teachers themselves were and are responsible for initiating, conducting and managing their own research with the support of the co-ordinator.

The group realised that publishing in this way was not as accessible to other teachers as they wished. Two articles covering the work of this phase were therefore prepared through 'writing days' and appeared in the Primary Science Review (Marsh and Boulter, 1997,1998). Presentation of the work at conferences in form of talks and workshops has also been important both in disseminating the work of the group and in emphasising the importance of teacher/researchers prepared to investigate their own practice. The teachers have been empowered to conduct research and, though the turnover of involved teachers is considerable because teaching has become such a time-consuming task, some members have continued to participate for several years. This has provided continuity for and a developmental aspect to the group, reflected in a growing ability to take on the support of other members.

In the second phase of the work (since 1998) the emphasis has moved towards: investigating representations of the very large, very small and of the human scale; the drawings that 7-year-olds make of objects down the microscope; the sizes and features of human scale objects drawn by children throughout the primary range; and children's and adult's understandings of the solar eclipse and its scale. In the course of this work the teachers have come to understand more of the nature of models in science and the importance that modelling has in learning science. On several occasions insights from this group of practitioners has fed into the work of the CMISTRE group as a whole. For example, the analysis of the discussion of a group of 9-year-olds about which orrerry best showed the structure and behaviour of the solar system illustrated how relatively young pupils are critical of the various models for showing specific aspects of a phenomenon and can articulate these very clearly if given several models of the same phenomenon to critique. Members working on astronomy exhibits in museums took up these insights.

As the millennium approached the members of the group realised that they were also becoming more interested in how to encourage other teachers, both within the UK. and abroad, to become teacher researchers. Interest shown by teacher trainers and others in other parts of the world has also developed. It was also realised that the group's research is concerned with two aspects of modelling. The first is in finding ways of capturing and analysing the models that children use in classroom interactions. The second is in finding ways of changing classroom practice in line with the emerging understanding of model-based teaching and learning.

What has been learnt by the group can be presented under the following headings:

About the collaborative process within the group:

- The teacher's confidence both in doing research and in writing about it have increased dramatically so that they feel empowered in these respects. What continues to be most problematic is the provision of sufficient high-quality time for the work to be done.

- Co-ordination by a person who understands the role as one of enabling is essential. It is not clear, given this requirement, how we can encourage other groups to be set up and sustained on a large scale.

- A visit by the co-ordinator to group members in their schools greatly aids the process of research and enhances the status of the research there.

About introducing the teachers to ideas of modelling:

- Teachers don't have time to read much but they do read each other's research reports, particularly if they have been engaged in their production. They also read research related to their own questions but don't have time to visit libraries and search for materials. The co-ordinator therefore needs to have access and knowledge of suitable research to introduce at appropriate moments and to ensure that members get each other's draft reports.

- Concepts of modelling are best caught and not taught. Teachers have gradually fallen into the terminology supported by very occasional seminars.

About model-based teaching and learning:

- A wide range of different modes of representation for models of the same phenomenon presented in classrooms aids the development of enriched explanations.

- The ways that teachers move between different modes of representation in the classroom (e.g. pass from a concrete model of the solar system to a 2D drawing of it on the blackboard) is vital in supporting pupils' interpretations of these representations.

About pupils' understandings of scale and how teaching about it might be changed:

- Choosing stories for 5 to 7-year-olds with a view to introducing the qualitative ideas of 'scaling up' and 'scaling down' and quantitative terms like 'doubling' and 'halving' is likely to increase their understanding of scale in science.

- For 7 to 11-year-olds, the introduction into classroom talk of integer and fractional scale factors as part of modelling of quantitative scaling, and encouraging them to estimate scaling up and scaling down in all sorts of contexts, is likely to improve their understanding in preparation for the assumptions made by science teachers at later stages.

- When children experience difficulty in assessments, which involve classification, attention should be paid to the relative sizes of the illustrated objects and to the possibility of them being physically manipulated.

- Using several different concrete models, for example of the solar system, and the examination of how good the model is, enables children to explain their mental models more clearly. It also helps them to realise the ways that models work and that each model has strengths and limitations. It assists in the representation of specific structures or behaviours.

- The physical manipulation, for example of orreries, and talking in a small group discussion situation about how effective they are, allows children to challenge each other's ideas and to develop their own mental models, in this case of the solar system.

- Pupils of 7 to 11 years experience difficulty in understanding 'scale' when they use microscopes and draw what they see. Using quantitative measures to compare sizes of objects observed with everyday objects seems to improve pupils' ability to judge the sizes of objects they see through a microscope.

- In observational drawing sessions in class, pupils aged from 5 to 11 added imaginary parts to their drawings which were not visible in the object. The drawings also tended to be the same size or smaller than life size. When asked later to recall the object and to draw it using only their mental model, the drawings tended to be larger than life and some features of the observation were present. Improving children's observational drawing may be an important part of model-based teaching and learning and in challenging pupil's mental models of objects.

The second context is that of the secondary school science classroom.

THE HAMPSHIRE/SOUTHAMPTON/READING PROJECT: MODELS AND MODELLING FOR KEY STAGE 3 STUDENTS OF SCIENCE

The second activity reported here is one in which a group of secondary school (11 to 16 years) teachers, selected by the Science Inspector for their local government area, addressed model-related issues in an attempt to raise standards in their own schools, providing mutual support through a programme of meetings at which experience was shared.

The project must be seen against a background of the National Curriculum for England and Wales, which is compulsory for all students of compulsory schooling age (5 to 16 years) in state-funded schools. What is to be taught in all subjects is prescribed in great detail. For every subject, students are ideally expected to progress though eight 'Levels of Attainment' between the ages of 5 and 16. The curriculum is divided into four parts: Key Stage 1 (KS1), covered between 4/5 to 7 years; Key Stage 2 (KS2), covered between 7 to 11 years; Key Stage 3 (KS3), covered between 11 to 14 years; Key Stage 4 (KS4), covered between 14 to 16 years. National 'Standard Assessment Tests' are taken at the end of KS1, 2, and 3. The results for each school in respect of Science, English, Mathematics, are published for KS2 and KS3, with schools being rank-ordered by the percentage of students gaining the target 'levels of attainment' for the Key Stage. The outcome of this regime, augmented by a system of school inspections, is that, in any locality, any school which is 'low ranked' is not chosen by parents and will consequently loose income. If perceived to be 'failing' it may be closed altogether. All schools are therefore anxious to improve, or at least to retain, their ranking.

The project, which started in 1997 and which was ongoing at the time that this book went to press, is taking place in two adjacent local government areas in the south of England. It was jointly initiated by the respective Inspectors of Science of the Hampshire Inspection and Advisory Service (Jill Moore) and the Southampton Education Service (Mike Evans) and by the co-ordinator of CMISTRE (John Gilbert) on the basis of long-standing professional relationships. The focus was on the teaching and learning of science at KS3 because of evidence from the National Assessment Scheme and from school inspections that pupils were making less-than-expected progress in those years. All 54 state schools in those areas with pupils of the relevant age group were invited to participate. All of them in fact did apply: the 14 chosen were selected on the basis of their Heads of Science having previously taken part in successful research and development work.

The overall aim of the project has been that the participating teachers engage in action research, designed and conducted by themselves and supported by the joint co-ordinators, on the theme of models and modelling. The intention has been that all the science teachers in the 14 schools, under the leadership of the person who was directly involved in the project, would benefit by association with it. The project has three aims. First, for the teachers to improve their capability to teach the curriculum, which increasingly consists of abstract ideas as KS3 progresses, in such a way as to stimulate students' interest in science. Second, for the teachers to broaden their knowledge of teaching strategies which make use of models and modelling. Third, to develop methods with which to identify students' expressed models of key scientific ideas. These measures would, it was anticipated, lead to pupils providing improved scientific/historical explanations in class of a wide range of biological, chemical, and physical, themes. It was also intended that students would show improved capacities to make predictions and to evaluate the outcomes of their experimental investigations. If these two aims were attained, it was anticipated that students would achieve better scores in the end-of-Key Stage National Test, although it was realised that such improvement would rest on a broad range of factors.

At the beginning of each school year, a whole-day project meeting is held to establish the agenda of work for the year. This is supported by 4 subsequent half-day whole-project meetings and with visits to the teachers and their science colleagues in their schools. The teachers were free to adopt approaches and to develop materials and schemes of work which, they believed, would lead to 'the raising of standards' (getting better Test scores) in their schools. They did so within a framework of 'good practice' in respect of models and modelling which was built up from the ideas given earlier in this book, namely: to structure the curriculum, wherever appropriate, around historical/consensus models; to match the model used to the particular learning objective that the National Curriculum required; to use suitable teaching models to develop an understanding of these historical/consensus models; to teach students about the role of models and modelling in science; to teach students how to evaluate the scope and limitations of all models, particularly of teaching models.

Each school produced an annual written report on what had been done and what achieved. Taking an overview of these reports enabled categories of achievement for the project to be identified. The project has enabled teachers to:

Take a sharper focus on models in curriculum planning

For example:

(a) The four major themes which underpin the detailed prescription of KS3 have been identified as 'the particulate nature of matter', 'energy', 'biological cells', 'force'.

(b) A models-oriented algorithm has been developed with which to identify the learning and teaching requirements for these themes. Its sections are: a de-structuring analysis of the Programme of Study; the identification of learning objectives which will lead to the various Levels of Attainment which are possible; a statement of the explanatory purposes which laboratory (and other) activities should serve; a description of these activities; the identification of a suitable scientific/curriculum model to be understood; the identification of teaching models which may be useful.

(c) This algorithm has been applied to each of the core themes and the production of 'schemes of work', based on the notion of a 'spiral curriculum' to cover a school year (one of: 7,8,9) which fall within KS3.

(d) A 'programme of revision teaching' has been produced, to help students prepare for the National Assessment Test, based around core scientific/ curriculum models.

Use a range of teaching models within any given topic

Work included:

(e) Looking at teaching models which had been published in the professional journal *School Science Review* to analyse their strengths and weaknesses (scope and limitations). This enabled teaching models to be identified which were particularly valuable for specific purposes.

(f) The conduct of a teaching experiment to see whether 2D or 3D representations of a 'biological cell' were more useful in promoting understanding of the theme. The construction of 3D models proved more helpful to students.

(g) The identification of students' 'preferred learning style' and an attempt to match this with specific model-based activities. It was found that teachers who were successful in explaining complex models to a particular student had the same preferred learning style as the student.

Develop methods to elicit students' pre-instruction expressed models

(h) Tests designed to identify students' understanding of the scientific/curricular models, which underpin each of the four core themes of KS3, were developed and applied. Whilst the results replicated what was known from the literature, they provided a basis for teachers to evaluate the progress that individuals had made during the lessons that followed.

Use models as the basis for intra-departmental professional development

i) The teachers who directly participate in the project report that the work done in school provides a valuable focus for professional development activities. As is perhaps common in all school situations, there is a tendency for staff meetings to concentrate on items of specific administration, to the neglect of issues of teaching and learning. The pooling of ideas of which teaching models were valuable in the teaching of a specific theme did show that staff varied in their own understanding of the underlying scientific model. The non-threatening context of discussing how to teach an idea provides a forum in which actual understandings of that idea can be agreed upon. This is particularly valuable, given that, whilst individual secondary teachers in the UK tend to study one science in great depth at university, they are required to teach all the three major sciences (physics, chemistry, biology) in school for the 11 to 16 age band. Moreover, as the majority of science teachers in most of the schools involved in the project studied biology at university, themes associated with chemistry and physics are often found challenging by them. In short, addressing Shulman's (1986) 'pedagogic subject knowledge' provides a route to improving the 'content knowledge' of many teachers.

In addition to work in these formal contexts of science education, work has also taken place in two non-formal contexts, where participation is voluntary. The next account is of work in an interactive 'science and technology centre'.

THE QUESTACON PROJECT: MENTAL MODELLING IN SCIENCE AND TECHNOLOGY CENTERS

The most significant trend in the science and technology museums movement in recent years has been the introduction of 'science and technology centres' (STCs). These consist of 'interactive exhibits' (IEs), in which a visitor, often in collaboration with others, takes some action in response to written instructions. The response produced by the IE often leads to a train of additional actions and responses. Such exhibits have proved very popular with the public and offer a way of bringing the entertainment and educational functions of STCs together. They seem to lead to a further development of the 'public awareness of science and technology' which may be defined as 'a set of attitudes, dispositions towards science and technology, which are manifest in behavioural intentions and in the skills with which these are discharged'.

The Questacon Project, which began in 1997, is a collaboration between the National Centre for the Public Awareness of Science at Australian National University, Questacon (The Australian National Science and Technology Centre, Canberra) and John Gilbert. Its aims are to explore the short- and long-term consequences of visitor use of IEs and to develop a unified model of entertainment and education in STCs. Data are being collected by a research assistant and by five of the 'Explainers' employed by Questacon to support visitors' experiences. They are interviewing adult visitors, some with accompanying children, both immediately after they have used an exhibit and, by telephone, some three months afterwards. The project is based on the notion that visitors develop mental and expressed models of the teaching models, which represent the scientific/historical models on which exhibits are based. The research questions being addressed are: Why do visitors choose exhibits? What are the short- and long-term outcomes of using exhibits? A unified model for education and entertainment has emerged which has the following categories: the opportunities presented to visitors; the basis for choice of exhibits; visitor's manner of use of exhibits; the short-term consequences of exhibit use; the long-term consequences of exhibit use. It is hoped that such insights will be able to directly influence the design and deployment of IEs (Gilbert, in press).

To illustrate the kinds of results which are emerging from the project, the patterns for 'short-term and long-term responses' will be summarised. These were conceived of as lying within four categories: conative (the consequences for personal initiative and planning), behavioural (what the visitors actually did at an exhibit), cognitive (the kinds of explanations that visitors gave, see Chapter 10 for the typology used), and affective (feelings of greater access to science, of aesthetic appreciation of the exhibit). The only conative response found was where parents volunteered that their child had enjoyed the self-directional aspect of the experience, contrasting it with the omnipresent teacher directedness of school activities. The parents of children from rural areas volunteered that their children had enjoyed actually doing something, contrasting Questacon with the lack of resources for practical work in country schools. As would be expected, cognitive responses were much in evidence. The fact that the explanations for exhibits offered by interviewed visitors were very largely of the 'descriptive' variety may, in part at least, be attributed to the fact that the interviews took place immediately after the exhibit had been used. There was an interestingly high incidence of affective responses, usually of a kinaesthetic or aesthetic variety. However, of perhaps greater interest has been the long-term responses: a random 15% sample of visitors was interviewed by telephone some three months after their visit. Almost all not only remembered their visit in some detail, but had also further developed the associated mental models since the visit. The affective responses had broadened to link the visit to other pleasant experiences before and since the visit. Moreover, the range of explanatory types used in discussing exhibits had expanded as a result of discussing the visit with friends who were thinking of going to Questacon.

The second non-formal context in which work took place was a classical museum into which some interactive exhibits had been introduced.

The 'Museum of Astronomy and Allied Sciences' (MAST) project

The 'Mental Models in Science and Technology Education' project is a collaboration between four Brazilian institutions (The Catholic University of Rio de Janeiro, The Federal University of Fluminense, The Museum of Astronomy and Allied Sciences (MAST) and The Federal University of Santa Catarina) and the University of Reading. Funded by the British Council and by Brazilian Sources, the project, initially over the period 1998 to 2001, has four aims: to promote further the development of a science and technology research culture in the five participating institutions; to develop a 'model of mental model' for use by the CMISTRE group as a whole (see Chapter 5); to develop approaches for the in-service education of teachers in respect to models and modelling; to analyse features of the models used in science exhibitions; and to analyse the interaction of visitors with these models. This section presents some results on this last topic.

Gouvea et al. (1998) investigated what students retained after a visit to an interactive science exhibition. The students who participated in the research came to the museum with their teacher as part of a schools' visiting program. Visits were tape-recorded and a research assistant took notes on students' interactions with the exhibition. One month after the visits, 12 teachers from different schools and 24 students (2 from each school) were interviewed about their visit to the Museum. The interviews, which were audio-recorded, were conducted in the students' schools. The criteria used in the choice of the schools included coverage of different age groups, both state-funded and private schools, and evidence of a genuine engagement in the visit as detected by observation at the time. Once a school had been selected, the choice of the two students to be interviewed was made at random. Interviews with students started by asking them to describe their visits. When a teaching model used in the exhibition was spontaneously mentioned by a student, the researcher asked for information about what student understood by the model. When necessary, this conversation was supported by a photograph of the model in question which was provided by the researcher. Next, the interviewer engaged students in conversation to elicit their conceptions and models related to the subject matter represented by the teaching model. When a student finished recalling all the teaching models that she/he was able to remember, the researcher showed him/her photographs of the other teaching models present in the rooms visited by that particular school but not remembered by the students.

The results showed that, even when interviewed one month after their visit, students remembered a lot about their visit. But they did not remember everything. There was a tendency for some teaching models to be frequently mentioned while others were forgotten. Remembering them seemed to depend very much on specific features. These features are related to what Gouvea et al. (1998) call 'the power of attraction' of the model. However, there were teaching models which attracted them, that were often remembered, but about the use of which students were unable to enter into detailed conversation. This result emphasised that, in the evaluation of science exhibitions, it is not only the power of attraction of an exhibit should be considered but also its 'power of retention'.

Were the students able to make sense of the science topic underpinning the teaching model? Could they identify the science topic? Could they talk about it in a manner that shows a coherent appreciation of relationships between the science topic and the phenomenon exemplified by the teaching model? Could they express their mental models? To what extent could they test these mental models whilst they were at the museum? Could they test them during the interview? All these questions are related to what can be called 'the level of engagement' of the student with the exhibition. In respect to these questions the results obtained were that:

- As the most mentioned and the better-described teaching models were almost identical for all the students, we may conclude that the power of attraction and the power of retention depend almost exclusively on specific features of the teaching models. At least this is so for the sample enquired into, containing as it did students from schools in the same town, attending schools which are keen to take pupils to museums.

- The level of engagement depends both on the teaching model and the visitor. This should be not a great surprise, since level of engagement is defined as a pattern of relationship between parts of the exhibition and visitors. However, it is important to stress that there were some teaching models that are much more likely to provoke engagement than others. On the other hand, there were some teaching models, which provoke no engagement at all. The definition of the features, which can cause a good level of engagement, is still an issue for the MAST group. A high level of physical interaction does frequently play a role in this process, but this physical interaction does not always lead to the engagement of mental models with the exhibit.

- Both when describing the teaching models and when engaged in conversation about the science topic related to them, students used wide range of expressed models. Sometimes the students' expressed models were congruent with scientific models, but several times they expressed 'alternative models'. In most cases the research was not capable of clarifying to what extent the development of science-compatible models were related to the visit to the museum. Despite this, the research findings showed several situations in which students revised their models, both during the visit and at the later interview. Most importantly, the research findings stressed that the visit offered an opportunity for students to engage in activities which stimulates them not only to use their expressed models but also sometimes to test and revise them.

These research findings have been used as a guideline for the design of a subsequent exhibition at MAST. The theme of this exhibition is 'The Moving Earth'. Teaching models were designed to maximise the power of attraction, the power of retention, and the level of engagement. A discussion of some features of this exhibition has been presented elsewhere (Falcao et al. 1999).

AGENDAS

A review of the earlier Chapters in this book and particularly of the case studies of applications and implications which have been presented above enable us identify areas for future development and research.

Agendas for the Practice of Model-based Teaching and Learning

The models' perspective can, as we have shown above, provide a structure for the design of schemes of teaching in schools. It can also affect the ways in which teaching takes place when teachers make explicit the nature of the models that they are using. The different forms of representation used can positively affect pupils' learning. A collaborative learning scenario can enhance pupils' building of their own models to explain phenomena. This work suggests that following agenda for action and enquiry:

- Evaluation of the effectiveness of the model-based teaching and learning approach in all the major formal phases of education: primary (or elementary), secondary (high school), higher education (university).

- Finding effective ways of introducing teachers to and supporting them in the use of models and modelling in science teaching and learning.

- Engaging with the writers and designers of textbooks (and other resource materials) so as to ensure that the potential of models and modelling is developed to best effect.

- Analysing existing national curricula so as to highlight the implications for models and modelling contained within them.

- Engaging with national curriculum developers to ensure that models and modelling are appropriately represented in national prescriptions.

The work in non-formal contexts described in this Chapter is helping in exhibit design as well as in the analysis of the interaction between visitors and exhibits. This also raises a large practical agenda for the future. Such research would seek to answer basic questions such as:

- What sequences of representations assist in model-based understanding of a phenomenon?

- What differences are there between the mental models of the 'writers' and 'readers' of exhibits?

- In what ways does an exhibit based on the models' perspective enhance scientific literacy in visitors?

- Does a models-based perspective help visitors to understand how to select and engage with exhibits?

- Does a model-based approach help in designing support materials for visitor use?

These questions apply with equal force where the museum is virtual, existing in hyperspace, and accessed from the computer screen. In this latter learning situation the additional comparative question is:

- What are the differences between the natures of 'appropriate sequences' for learning, in the writing and reading of experiences, in the enhancement of scientific literacy and the engagement of learners, when presented literally and virtually?

Agendas for the Development of the Theory

Questions concerning the practice of model-based teaching and learning raise agendas for theoretical research which fall into the following categories:

- The insights of an expanded range of philosophers and paradigms of enquiry when brought to bear on the theme of models and modelling in science and technology education.

- The relationship between phenomena, models and representations.

- The characterisation of the historical/consensus models which have been/are being used in scientific enquiry into a wide range of phenomena.

- The notion of progression in the learning of historical/consensus models.
- The development of reflective practice in science and technology teachers in respect of the provision of explanations, which are based on the use of models.

CONCLUDING REMARKS

This book has attempted to show how a 'models and modelling' perspective on science and technology education has developed so far through research within the CMISTRE research group. It has also shown how this theoretical work is being applied in both out-of-school and in-school contexts. It springs from a need to address the development of a curriculum for school pupils and appropriate out-of-school experiences that will enhance scientific literacy. In this way it is looking beyond 2000 to a world where the boundaries between the visual and the verbal, the formal and informal, even the phenomenon and its representation, are broken down. It is in within this future that children and teachers will come to understand science and where they will develop and use that knowledge within their life-long learning. There is much still to do which is relevant to the advancement of science and technology education, wherever it takes place.

REFERENCES

Abel, S., Roth, M. "Reflections on a fifth-grade life science lesson: making sense of children's understanding of scientific models." *International Journal of Science Education* 17, no. 1 (1995): 59-74.

Abrantes, P. "Kuhn and his idea of 'exemplar'." *Principia* 1, no. 2 (1998): 61-102.

Adey, P., Shayer, M. *Really Raising Standards*. London: Routledge, 1994.

Adler, A. "Mathematics and Creativity." In *The World Treasury of Physics, Astronomy and Mathematics*, edited by T. Ferris, 435-446. Boston: Little, Brown and Co., 1972.

Aguirre, J.M., Haggerty, S.M. "Preservice teachers' meanings of learning." *International Journal of Science Education* 17, no. 1 (1995): 119-131.

Aguirre, J.M., Haggerty, S.M., Linder, C.J. "Student-Teachers' Conceptions of Science, Teaching and Learning: A Case Study in Pre-service Science Education." *International Journal of Science Education* 12, no. 4 (1999): 381-390.

Aikenhead, G. "Science Education: border crossing into the subculture of science." *Studies in Science Education* 27 (1996): 1-52.

Alexander, P., Judy, J. "The Interaction of Domain-Specific and Strategic Knowledge in Academic Performance." *Review of Educational Research* 58, no. 4 (1988): 375-404.

Alexander, R. Rose, J., Woodhead, C. "Curriculum organisation and classroom practice - A discussion paper.". London: DES, 1992.

Allard, F. "Cognition, Expertise and Motor Performance." In *Cognitive Issues in Motor Expertise*, edited by J.L. Starkes and F. Allard, 17-33. North Holland: Elsevier Science Publishers B.V., 1993.

Allchin, D. "How NOT to Teach History in Science." Paper presented at the Proceedings of the Third International History, Philosophy and Science Teaching Conference, Minneapolis 1995.

Anning, A., Driver, R., Jenkins, E., Kent, Layton, D., Medway, P. "Towards an agenda for research in technology education: Occasional paper No.3 Education for Capability Research Group.". Leeds: University of Leeds, School of Education, 1992.

Archer, B. *A definition of cognitive modelling in relation to design and technology, Design: Occasional Paper No.1, Occasional Paper No.1*. Loughborough: Loughborough University of Technology, 1992.

Archer, B. *The Designerly Approach to Knowledge. Lecture Course*. London: Royal College of Art, 1979.

Archer, B. "The mind's eye: not so much seeing as thinking." *Designer* (1980): 8-9.

Archer, B. "The nature of research into design and design Education." Paper presented at the Proceedings of IDATER 91 Conference, Loughborough 1991.

Arons, A.B. "Historical and philosophical perspectives attainable in introductory physics courses." *Educational Philosophy and Theory* 20, no. 2 (1988): 13-23.

Atkins, J., Helms, J. "Getting serious about priorities in science education." *Studies in Science Education* 21 (1993): 1-20.

Atkinson, D.E. "Spreadsheets for doing and teaching chemistry." In *Using Computers in Chemistry and Chemical education American Chemical Society*, edited by T.J. Zielinski and M.L. Swift, 143-162. Washington DC: American Chemical Society, 1997.

Bachelard, G. *O Racionalismo Aplicado*. Translated by Nathanael C. Caixeiro. 5th ed. Rio de Janeiro: Zahar Editores, 1977.

Baeninger, M., Newcombe, N. "The role of experience in spatial test performance: A meta-analysis." *Sex Roles* 20, no. 5, 6 (1989): 327-344.

Bailey, K.D. *Typologies and Taxonomies: An Introduction to classification techniques*. Thousand Oaks, California: Sage Publications, 1994.

Baker, R.S., Talley, H.L. "Visualization skills as a component of aptitude for chemistry - a construct validation study." *Journal of Research in Science Teaching,* 11, no. 12 (1974): 95-97.

Banks, F, ed. *Teaching Technology*. London: Routledge/Open University Press, 1994.

Barnea, N., Dori, Y.J. "Computerised molecular modelling as a tool to improve chemistry teaching." *Journal of Chemical Information and Computer Sciences* 36, no. 4 (1996): 629-636.

Barnea, N. "The use of computer based analog models to improve visualisation and chemical understanding." In *Exploring Models and Modelling in Science and Technology Education*, edited by J.K. Gilbert, 145-161. Reading: University of Reading New Bulmershe Papers Series, 1997.

Barnes, B. "The Science-technology relationship: a model and a query." *Social Studies in Science*, no. 12 (1982): 166-172.

Barnes, D., Todd, T. *Communication and Learning in Small Groups*. London: Routledge and Kegan Paul, 1977.

Barnes, D., Britton, J., Rosen, H. *Language, the Learner and the School*. London: Penguin, 1969.

Baynes, K., ed. *Attitudes in Design Education*. London: Lund Humphries, 1969.

Baynes, K. *Design Education*. London: Design Council and Heineman, 1976.

Baynes, K. "The ethics of representation." In *Modelling: the language of designing, Design: Occasional Paper No.1*, edited by B. Archer P. Roberts, K. Baynes. Loughborough: Department of Design and Technology, University of Loughborough, 1992.

Beetlestone, F. *Creative Children, Imaginative Teaching*. Buckingham: Buckingham Open University, 1998.

Bent, H.A. "Uses of the Past." In *Teaching the History of Chemistry - A Symposium*, edited by G.B. Kauffman, 127-147. Budapest: Akademiai Kiado, 1971.

Bentley, R., Horstmann, T., Trevor, J. "Basic Support for Cooperative Work on the World Wide Web." *International Journal of Human-Computer Studies* 46, no. 6 (1997): 827-846.

Bernstein, B. *On the Classification and Framing of educational Knowledge*. Beverley Hills, CA: Sage, 1971.

Betts, D.M. "Developing an appreciation of the nature of science using the history of the periodic law in a secondary school chemistry course." In *The History and Philosophy of Science in Science Education*, edited by S. Hills, 105-113. Kingston, Ontario: Queen's University, 1992.

Bezzi, A. "A Macintosh program for improving three dimensional thinking." *Journal of Geological Education* 39 (1991): 284-288.

Bhaskar, R. *A Realist Theory of Science*. London: Harvester Wheatsheaf, 1978.

Billig, M. *Arguing and Thinking: A rhetorical approach to social psychology*. Cambridge: Cambridge University Press, 1987.

Bird, A. *Philosophy of Science*. London: University College Press, 1998.

Bishop, E.J. "Developing students' spatial ability." *The Science Teacher* 45, no. 8 (1978): 20-23.

Bleaney, B.J., Bleaney, B. *Electricity and Magnetism*. 2nd Edition ed. Oxford: Oxford University Press, 1965.

Bliss, J. "From Mental Models to Modelling." In *Learning with Artificial Worlds*, edited by H et al Mellar, 27-32. London: Falmer, 1994.

Bolcer, J.D., Hermann, R.B. "The development of computational chemistry in the United States." In *Reviews in Computational Chemistry*, edited by K.B. Lipkowitz and D.B Boyd, 1-63, 1994.

Boohan, R. "Computer modelling and dynamic processes in science education." In *Exploring Models and Modelling in Science and Technology Education*, edited by J.K. Gilbert, 126-144. Reading: The New Bulmershe Papers, University of Reading, 1997.

Boohan, R. "Creating Worlds from Objects Events." In *Learning with Artificial Worlds*, edited by H. Mellar et al, 171-180. London: Falmer, 1994.

Boulter, C., Gilbert, J. "Argument and Science Education." In *Competing and Consensual Voices: The theory and practice of argument*, edited by P. Costello and S. Mitchell. London: Multilingual Matters, 1995.

Boulter, C., ed. *Aspects of Primary Children's understandings of scale: Work by the teacher/researchers within the MISTRE Primary Classroom Research Group*. Reading: University of Reading, New Bulmershe Papers, 1997.

Boulter, C. "Collaborating to Investigate Questions: A Model for Primary Science." unpublished doctoral dissertation, University of Reading, 1992.

Boulter, C., Gilbert, J. "Texts and Contexts: Framing Modelling in the Primary Science Classroom." In *Research in Science Education in Europe: Current Issues and Themes*, edited by G. Welford, J. Osborne and P. Scott, 177-188. London: Falmer, 1996.

Boulter, C., Marsh, G. "What do you mean 'what scale is it'? P*rimary Science Review* 50, no. Nov/Dec (1997): 12-14.

Boulter, C., Prain, V., Armitage, M. "What's going to happen in the eclipse tonight?: Rethinking perspectives on primary school science." *International Journal of Science Education* 20, no. 4 (1998): 487-500.

Boulter, C.J. "Language, Models and Modelling in Primary Science Classroom." In *Exploring Models and Modelling in Science and Technology Education*, edited by J.K. Gilbert, 180-200. Reading: Faculty of Education and Community Studies, University of Reading, 1997.

Brand, C., Radar, C., Carlone, H., Lewis, C. "Prospects and Challenges for Children Creating Science Models." Paper presented at the 1998 Meeting of the National Association for Research in Science Teaching, San Diego, California, April 19-22 1998.

Brickhouse, N.W. "Teachers' Beliefs about the Nature of Science and their Relationship to Classroom Practice." *Journal of Teacher Education* 41, no. 3 (1991): 53-62.

Brickhouse, N.W. "The teaching of the philosophy of science in secondary classrooms: case studies of teachers' personal theories." *International Journal of Science Education* 11, no. 4 (1989): 437-449.

Brock, W.H. *The Fontana History of Chemistry*. London: Fontana Press, 1992.

Brosnan, T. "Using Spreadsheets to Develop Understanding in Science." In *Learning with Artificial Worlds*, edited by H. et al Mellar, 76-84. London: Falmer, 1994.

Brown, E.J. "Highly Intelligent but not Necessarily Highly Creative." *School and Community* 49 (1962).

Brown, G., Edmondson, R. "Asking questions." In *Classroom Teaching Skills*, edited by E. Wragg, 97-120. London: Croom Helm, 1984.

Brown, J.R. *The Laboratory of the Mind*: Routledge, 1991.

Brown, J.R. "The Structure of Thought Experiments." *International Studies in The Philosophy of Science: The Dubrovnik papers* (1986): 1-15.

Bruckman, A. *Community Support for Constructionist Learning, Computer Supported Collaborative Work 7*, 1998.

Bruckman, A., de Bonte, A. "Moose goes to School: A Comparison of Three Classrooms using a CSCL Environment." Paper presented at the CSCL 97, Toronto 1997.

Bruner, J. *Child's Talk*. London: Open University Press, 1983.

Bruner, J. *Towards a Theory of Instruction*. Harvard: Harvard University Press, 1966.

Brush, S.G. "History of Science and Science Education." *Interchange* 20, no. 2 (1989): 60-70.

Brush, S.G. "Why Chemistry needs History - and how it can get some." *Journal of College Science Teaching* 7 (1978): 288-291.

Buckley, B., Boulter, C. "Analysis of representations in model-based teaching and learning." In *Visual Representations and Interpretations*, edited by R. Paton and I. Neilson, 289-294. Liverpool: Springer, 1999.

Buckley, B.C. "A case of intentional model-based learning." Paper presented at the Annual meeting of the American Educational Research Association, San Francisco 1995.

Buckley, B.C. "Multimedia, misconceptions and working models of biological phenomena: Learning about the circulatory system." unpublished doctoral dissertation, Stanford University, 1992.

Buckley, B.C. "The role of representations in model-based learning: The circulatory system." Paper presented at the Annual meeting of the American Association of Educational Research, San Diego 1998.

Buckley, B.C., Boulter, C.J., Gilbert, J.K. "Towards a typology of models for science education." In *Exploring Models and Modelling in Science and Technology in Science Education*, edited by J.K. Gilbert, 90-105. Reading: Faculty of Education and Community Studies, 1997.

Bunge, M. *Philosophy of Physics*. Dordrecht: Reidel, 1973.

Bunge, M. *Theory and Reality*. Sao Paolo, Brazil: Editora Perspectiva, 1974.

Bunge, M. *Treatise on Basic Philosophy, Volume 3: The furniture of the World*. Dordrecht: Reidel, 1977.

Bynam, W.F., Browne, E.J., Porter, R., ed. *Dictionary of the History of Science*. London: Macmillan, 1981.

Campbell, N.R. *The Elements of Physics*. Cambridge: Cambridge University Press, 1920.

Capani, A. "The Philosophy of Mario Bunge and the subject of positivism." *Manuscript* XIV, no. 2 (1991): 113-142.

Carlsen, W. "Questioning in classrooms: a sociolinguistic perspective." *Review of Educational Research* 61, no. 2 (1991): 157-178.

Carr, M. "Model confusion in chemistry." *Research in Science Education* 14 (1984): 97-103.

Carson, R.N., ed. *If science is presented as historical narrative: A sample chapter*. Edited by S. Hills. Vol. 1, *The history and philosophy of science in science education*. Kingston, Ontario: Queen's University, 1992.

Chalmers, A.F. "The Heuristic Role of Maxwell's Mechanical Model of Electromagnetic Phenomena." *Studies in the History and Philosophy of Science* 17 (1986): 415-427.

Chalmers, A.F. *What is This Thing Called Science*. 2nd ed. Milton Keynes: Open University Press, 1982.

Chan, C.K.K., Burtis, P.J., Scardamalia, M., Bereiter, C. "Constructive Activity in learning from Text." *American Educational research Journal* 29, no. 1 (1992): 97-118.

Chi, M.T.H., Chiu, M.-H., deLeeuw, N. *Learning in a non-physical science domain: The human circulatory system*. Pittsburgh, PA: Learning Research and Development Center, 1991.

Chinn, C.A., Brewer, W.F. "The role of anomalous data in knowledge acquisition: A theoretical framework and implications for science instruction." *Review of Educational Research* 63, no. 1 (1993): 1-49.

Chisti, Y., Moo-Young, M. "Fermentation technology, bioprocessing, scale-up and manufacture." In *Biotechnology. The Science and the Business*, edited by R. Cape V. Moses. New York: Harwood Academic Publishers, 1991.

Clement, J. "Expert Novice Similarities and Instruction using Analogies." *International Journal of Science Education* 20, no. 10 (1998): 1271-1286.

Clement, J. "Learning via model construction and criticism: Protocol evidence on sources of creativity in science." In *Handbook of Creativity: Assessment, Theory and Research*, edited by R.R. Ronning J.A.Glover, C.R. Reynolds, 341-381. New York: Plenum Press, 1989.

Clement, J., Brown, D.E., Zietsman, J.A. "Not all preconceptions are misconceptions: Finding "anchoring conceptions" for ground instruction on students' intuitions." Paper presented at the Paper presented at the Annual meeting of the American Educational Research Association, San Francisco 1989b.

Clement, J. "Use of physical intuition in expert problem solving." Paper presented at the the Workshop on Implicit and Explicit Knowledge, University of Tel Aviv 1988.

Cobern, W. "Science and a social constructivist view of science education." In *Socio-Cultural Perspectives on Science Education*, edited by W. Cobern, 7-23. Dordrecht: Kluwer, 1998.

Cohen, M.R., Drabkin, I.E. *A Source Book in Greek Science*: Harvard University Press, 1948.

Cohen, R.S. "Individuality and common purpose: the philosophy of science." *Science & Education* 3, no. 4 (1994): 393-407.

Coren, C. "How one man inspires other to 'Do a Dyson'." *The Times*, October 28th 1996 1996.

Cosgrove, M. "A Study of Science-in-the-Making as Students Generate an Anology for Electricity." *International Journal of Science Education* 17 (1995): 295-310.

Cox, M., Webb, M. "Developing Software and Curriculum Materials: The Modus Project." In *Learning with Artificial Worlds*, edited by H. et al Mellar, 188-198. London: Falmer, 1994.

Crabbe, M., Rées Lay, C., Appleyard, J. Desktop Molecular Modeller Oxford Version 3.0 Version 3.0. Oxford University Press, Oxford, 1994

Craik, K. "The Nature of Explanation." In *Mental Models*, edited by P.N. Johnson-Laird. Cambridge: Cambridge University Press, 1983/1943.

Cros, D., Maurin, M., Amouroux, R., Chastrette, M., Leber, J., Fayol, M. "Conceptions of first-year university students of the constituents of matter and the notions of acids and bases." *European Journal of Science Education* 8, no. 3 (1986): 305-13.

Crowder, E.M. "Gestures at work in sense-making science talk." *Journal of the Learning Sciences* 5, no. 53 (1996): 173-208.

Cupani, A. "The Philosophy of Mario Bunge and the Subject of Positivism (in Portuguese)." *Manuscrito* XIV, no. 2 (1991): 113-142.

Curriculum, Corporation. *Science: A Curriculum Profile for Australian Schools*. Carlton, Victoria: Curriculum Corporation, 1994.

Dagher, Z. "Analysis of Analogies used by Science Teachers." *Journal of Research in Science Teaching* 32 (1985b): 259-270.

Dagher, Z. "Does the use of analogies contribute to conceptual change?" *Science Education* 78, no. 6 (1994): 601-614.

Dagher, Z. "Review of Studies of the Effectiveness of Instructional Analogies in Science Teaching." *Science Education* 79 (1985a): 295-312.

Dagher, Z., Cossman, G. "Verbal explanations given by science teachers: their nature and implications." *Journal of Research in Science Teaching* 29, no. 4 (1992): 361-374.

Dampier, W.C. *A History of Science*. 4th ed. Cambridge: Cambridge University Press, 1966.

Davies, T. "Creativity and Modelling in Design and Technology." In *Exploring Models and Modelling in Science and Technology Education: Contributions from the MISTRE Group*, edited by J.K. Gilbert, 208-227. Reading: University of Reading, 1997.

Davies, T. "Modelling and Creativity in Design and Technology." Paper presented at the IDATER 96 Conference, Loughborough 1996.

de Bonte.. A. Personal Communication. 1998b.

deKleer, J., Brown, J.S. "Assumptions and ambiguities in mechanistic mental models." In *Mental Models*, edited by D. Gentner A.L. Stevens, 155-190. Hillsdale, NJ: Lawrence Erlbaum Associates, 1983.

deKleer, J., Brown, J.S. "Mental models of physical mechanisms and their acquisition." In *Cognitive Skills and their Acquisition*, edited by J.R. Anderson, 285-309. Hillsdale, NJ: Lawrence Erlbaum, 1981.

Demeo, S. "An Immaculate Conception: Le Chatelier and Equilibrium." In *The History and Philosophy of Science in Science Teaching*, edited by S. Hills, 227-237. Kingston, Ontario: Queen's University, 1992.

Department of National Education. "A Curriculum model for education in South Africa.", 1991.

DES, (Department of Education and Science). *General Certificate of Secondary Education: a general introduction*. London: HMSO, 1985.
DES, (Department of Education and Science). *Technology for Ages 5 to 16, Final Report*. London: HMSO, 1988.
Design Council. "Design in Secondary Education: A Report." . London: Design Council, 1980
Deutsch, D. *The Fabric of Reality*. London: Allen Lane, 1997.
DFEE. *Science in the National Curriculum*. London: HMSO, 1995a.
DFEE. *Design and Technology in the National Curriculum*. London: Department for Education, 1995b.
DFEE. *Information Technology in the National Curriculum*. London: HMSO, 1995c.
DFEE. *Excellence in Schools*. London: HMSO, 1996.
DFEE. *Science in the National Curriculum*. London: HMSO, 2000.
Dibbs, D.R. "An Investigation into the Nature and Consequences of Teachers Implicit Philosophies of Science." Unpublished Doctoral Dissertation, University of Aston, 1982.
Dillon, P. "Teaching and Learning with Telematics: an overview of the literature." *Journal of Information Technology for Teacher Education* 7, no. 1 (1998): 33-49.
Dingalo, R.D. "The selection of fabrication materials for design and technology in Botswana." Paper presented at the IDATER 95 Conference, Loughborough 1995.
Donaldson, M. *Children's Minds*. London: Fontana, 1978.
Dori, Y.J., Barnea, N. "In-Service Chemistry teachers Training: The impact of introducing computer technology on teachers' attitudes and classroom implementation." *International Journal of Science Education* 19, no. 5 (1997): 577-592.
Downey, M.E., Kelly, A.V. *Theory and Practice of Education: An Introduction*. 3rd ed. London: Harper and Row, 1986.
Druva, C.A., Anderson, R.D. "Science teacher characteristics by teacher behaviour and by student outcome: A meta-analysis of research." *Journal of Research in Science Teaching* 20 (1983): 467-479.
Duhem, P. *La théorie physique, son objet, sa structure*. Paris: Rivière, 1914.
Duit, R. "On the Role of Analogies and Metaphors in learning Science." *Science Education 75* 6 (1991): 649-672.
Duschl, R.A., Wright, E. "A case study of high school teachers' decision making models for planning and teaching science." *Journal of Research in Science Teaching* 26, no. 6 (1989): 467-501.
Duschl, R.A. "Research on the History and Philosophy of Science." In *Handbook of Research on Science Teaching and Learning*, edited by D.L. Gabel, 443-465. New York: MacMillan, 1994.
Duschl, R.A. "Science Teachers' Beliefs about the Nature of Science and the Selection, Implementation and Development of Instructional Tasks: A Case Study." In *Dissertation Abstracts International*, 442-A. Maryland: University of Maryland, 1983.
Dwyer, F.M. *Strategies for Improving Visual Learning*: State College Learning Services, 1978.
Edmondson, K., Novak, J. "The Interplay of Scientific Epistemological Views, Learning Strategies and Attitudes of College Students." *Journal of Research in Science Teaching* 30, no. 6 (1993): 547-559.
Edwards, A., Furlong, V. *The Language of Teaching*. London: Heinemann, 1978.
Edwards, D., Mercer. *Common Knowledge: The Development of Understanding in the Classroom*. London: Routledge Kegan Paul, 1987.
Eggleston, J. *Developments in Design Education*. London: Open Books, 1976.
Einstein, A. "Autobiographic notes." In *Einstein, philosopher and scientist*, edited by P.A. Schilpp. La Salle: Open Court, 1970.
Einstein, A., Infield, L. *The Evolution of Physics*. Cambridge: Cambridge University Press, 1938.
Einstein, A. "Notes on the origin of the general theory of relativity." In *Einstein, A Centenary Volume*, edited by A.P. French. Cambridge, MA: Harvard University Press, 1979.
Einstein, A. *Relativity, the Special and the General Theory*. New York: Crown, 1961.
Einstein, A. *Relativity: The Special and General Theory*. London: Methuen, 1960.
Eliot, J.K., Hauptman, A. "Different dimensions of spatial ability." *Studies in Science Education* 8 (1981): 45-66.
Ellis, P. "Practical Chemistry in a Historical Context." In *Teaching the History of Science*, edited by M. Shortland and A. Warwick, 156-167. Oxford: Blackwell, 1989.
Elmer, R. "Models in design and technology education." In *Contributions from the MISTRE Group Exploring Models and Modelling in Science and Technology Education*, edited by J.K. Gilbert, 58-75. Reading: University of Reading, 1997.
Elmer, R. "Whose need?: a study of reflective practice of eight students on a BA (Hons) Design and Technology course." Unpublished PhD Thesis, University of Reading, 1996.

Entwistle, N. "Contribution of Psychology to Learning and Teaching." In *New Directions in Educational Psychology: Learning and Teaching*, edited by N. Entwistle, 3-8. London: The Falmer Press, 1985.

Eriksson, A. *Pedagogisk-psykologiska problem-lararenkater om kreativitet I skolan*. Translated by Inga Betts. Malmo, Sweden: Malmo University Press, 1970.

Espinet, M. "Newton's and Goethe's Processes of Inquiry: Two alternative ways of knowing in science, more history and philosophy of science in science teaching." Paper presented at the First History and Philosophy of Science in Science Teaching Conference, Tallahassee 1991.

Falcao, D., Gouvea, G., Cazelli, S., Valente, M.E., Cazelli, S., Queros, G. "Models and learning in science and technology museums." Paper presented at the 3rd European Science Education Research Association 1999.

Farmer, B. "From science teacher to technology facilitator: a case study of Katherine." *Research in Science Education* 24 (1994): 68-75.

Fauvel, J., Flood, R., Shortland, M., Wilson, R., ed. *Let Newton Be!* Oxford: Oxford University Press, 1988.

Feher, E., Rice Meyer, K. "Children's conceptions of colour." *Journal of Research in Science Teaching* 29, no. 5 (1992): 389-406.

Feldman, A. "Reconstructing the Image of Science Teachers in the Teaching and Learning of the History and Philosophy of Science." Paper presented at the Proceedings of the third International History, Philosophy and Science Teaching Conferences, Minneapolis 1995.

Feldman, D., Csikzenmihayli, M., Gardner, H. *Changing the World: A Framework for the Study of Creativity*. USA: Praeger, 1994.

Fensham, P. "The Didaktik meets Curriculum Conversation: Some Implications for Research in Science Education." Paper presented at the NARST Annual Meeting, Boston, MA 1999.

Fensham, P. "Science for all: A Reflective Essay." *Journal of Curriculum Studies* 17, no. 4 (1985): 415-435.

Fenstermacher, G.D. "Philosophy of Research on Teaching: Three Aspects." In *Handbook of Research on Teaching*, edited by M.C. Wittrock. New York: Macmillan, 1986.

Feyerabend, P. *Against Method*. (3rd) ed. London: Verso, 1983.

Feynman, R.P. *The Character of Physical Law*. New edition with original introduction by Davies, P ed. London: Penguin Books, 1965.

Feynman, R.P., Leighton, R., Sands, M. *The Feynman Lectures on Physics*. Vol. I. Reading, Mass: Addison-Wesley, 1970.

Feynman, R.P. *QED: The Strange Theory of Matter and Light*. Princeton: Princeton University Press, 1985.

Fischbein, E. *Intuition in Science and Mathematics*. Dordrect: Reidel, 1987.

Flanders, N. *Analysing Teacher Behaviour*. Reading, Mass.: Addison Wesley, 1970.

Fox, D. "Personal theories of Learning." *Studies in Higher Education* 8, no. 2 (1983): 151-163.

France, B. "Realising the technology curriculum: professional development in biotechnology education." Unpublished Doctoral Thesis, University of Waikato, 1997.

France-Farmer, B. "Biotechnology: applied science or technology?" In *Developing the Science Curriculum in Aotearoa New Zealand*, edited by R. Baker B. Bell. Auckland: Addison Wesley Longman, 1998.

Franco, C., Barros, H.L., Krapas, S., Queiroz, G., Colinvaux, D., Alves, F. "From scientists' and inventors' minds to some scientific and technological produces: relationships between theories, models, mental models and conceptions." *International Journal of Science Education* 21, no. 3 (1999): 277-291.

Franco, C. "Individual and historical development in science." Unpublished Doctoral Dissertation, University of Reading, UK, 1993.

French, P., Maclure, M. "Getting the Right Answer and Getting the Answer Right." *Research in Education* 22, no. 1-23 (1980).

Fryer, M. *Creative Teaching and Learning*. London: Paul Chapman, 1996.

Gabel, D.L. "Modelling with magnets - A unified approach to chemistry problem solving." *The Science Teacher* March (1992): 58-63.

Galilei, G. *Dialogue Concerning the Two New Sciences*. Translated by Stillman Drake Translation. Madison, Wisconsin: University of Wisconsin Press, 1974.

Galilei, G. "Siderius Nuncius." In *Le opere di Galileo Galilei*, edited by A. Favero. Firenze: Barbera, 1968.

Gallagher, S.A., Johnson, E.S. "The effect of time limits on performance of mental rotation by gifted adolescents." *Gifted Child Quarterly* 36, no. 1 (1992): 19-22.

Galton, M., Simon, B., Croll, P. *Inside the Primary School*. London: Routledge Kegan Paul, 1980.

Gardner, H. "Creativity, New Views form Psychology and Education." *RSA Journal*, no. May (whole issue) (1995).

Gardner, P. "The relationship between technology and science: some historical and philosophical reflections, Part II." *International Journal of Technology and Design Education* 5, no. 1 (1995): 1-33.

Garnet, P.J., Hackling, M.W. "Students alternative conceptions in chemistry: A review of research and implications for teaching and learning." *Studies in Science Education* 25 (1995): 69-95.

Gasking, D. "Mathematics and the World." In *The World of Mathematics, Volume III*, edited by J.R. Newman, 1708-1722. London: George Allen and Unwin, 1960.

Gentner, D., Gentner, D.R. "Flowing waters or teeming crows: Mental models of electricity." In *Mental Models*, edited by D. Gentner and A.L. Stevens, 99-130. Hillsdale: Lawrence Erlbaum, 1983.

Gentner, D., Stevens, A.L., ed. *Mental Models*. Hillsdale: Lawrence Erlbaum, 1983.

Gentner, D. "A theoretical framework for analogy." *Cognitive Science* 7 (1983): 155-170.

Getzels, J., Jackson, P. *Creativity and Intelligence: Explorations with Gifted Children*. London and New York: John Wiley & Sons, 1962.

Giere, R. *Explaining Science*. Chicago: The Univeristy of Chicago Press, 1988.

Gilbert, J., Osborne, R. "The use of models in science and science teaching." *European Journal of Science Education* 2, no. 1 (1980): 3-13.

Gilbert, J.K. "The interface between science education and technology education." *International Journal of Science Education* 14, no. 5 (1992): 563-578.

Gilbert, J. "Models in science and science education." In *Contributions from the MISTRE group Exploring Models and Modelling in Science and Technology Education*, edited by J.K. Gilbert, 5-19. Reading: The University of Reading, 1997.

Gilbert, J.K., ed. *Exploring models and modelling in science and technology education; contributions from the MISTRE group*. Reading: Faculty of Education and Community Studies, University of Reading, 1997.

Gilbert, J.K., Boulter, C.J. "Learning science through models and modelling." In *The International Handbook of Science Education*, edited by K. Tobin B. Frazer, 53-66. Dordrecht: Kluwer, 1998.

Gilbert, J., Boulter, C., Rutherford, M. "Models in Explanations, Part 1: Horses for Courses." *international Journal of Science Education* 20, no. 1 (1998a): 83-97.

Gilbert, J., Boulter, C., Rutherford, M. "Models in Explanations, Part 2: Whose voice? Whose ears?" *International Journal of Science Education* 20, no. 2 (1998b): 187-203.

Gilbert, J.K. "On the explanation of change in science education." *Science and Education* 8, no. 5 (1999): 543-557.

Gilbert, J. "Towards a unified model of education and entertainment in science and technology centres." In *Science Communication in Theory and Practice*, edited by C. Bryant, M. Gore and S. Stocklmayer. Dordrecht: Kluwer, in press.

Gilbert, S.W. "Model building and a definition of science." *Journal of Research in Science Teaching* 28, no. 1 (1991): 73-79.

Gilmore, D., Pheasey, K., Underwood, J., Underwood, G. *Learning graphical programming: An evaluation of KidSim*: http://www.psychology.nottingham.ac.uk/staff/David.Gilmore/KidSim/Interact/Kidsimdg.html, 1997.

Glaser, B., Strauss, A. *The Discovery of Grounded Theory: Strategies for Qualitative Research*. Chicago: Aldine, 1967.

Glynn, S., Takahashi, T. "Learning from Analogy-Enhanced Science Text." *Journal of Reserach in Science Teaching 35*, no. 10 (1998): 1129-1149.

Gobert, J., Clement, J. "Effects of student-generated diagrams versus student-generated summaries on conceptual understanding of causal and dynamic knowledge in plate tectonics." *Journal of Research in Science Teaching* 376, no. 1 (1999): 39-53.

Goldsmith, E. *Research into Illustration*. Cambridge: Cambridge University Press, 1984.

Golin, G. "Some experiences in teaching the history of science to future high school teachers in Universities of the Former Soviet Union (1970-1990)." Paper presented at the Proceedings of the Third International History, Philosophy and Science Teaching Conference, Minneapolis 1995.

Gooding, D. "The procedural turn: or, why do thought experiments work?" In *Cognitive models of science (Minnesota Studies in the Philosophy of Science Vol. XV)*, edited by R.N. Giere, 45-76. Minneapolis: University of Minnesota Press, 1992.

Goodman, D., Russell, C.A. *The Rise of Scientific Europe 1500-1800*. London: Hodder and Stoughton, 1991.

Gorson, C.K. "Software for 3D Molecular Graphics and Modelling." In *Chemical Structure Software for Personal Computers*, edited by D.E. Meyer, Worre, W.A., Love, R.A., 53-58. Washington, DC: ACS Professional Reference Book, 1988.

Gouk, P. "The Harmonic Roots of Newtonian Science." In *Let Newton Be!*, edited by R. Flood J. Fauvel, M. Shortland, R. Wilson, 101-126. Oxford: Oxford University Press, 1988.

Gouvea, G., Franco, C., Valente, M.E., Cazelli, S., Marandino, M., Falcao, D. "On evaluation research in science museums." In *Evaluation and museum education*, edited by C. Dufresne-Tasse, 169-178. Montreal: ICOM, 1998.

Graesser, A., Person, N. "Question asking tutoring." *American Educational Research Journal* 3, no. 1 (1994): 104-137.

Green, P. *Design Education: Problem Solving and Visual Experience*. London: Batsford, 1974.

Gribbin, J. *In Search of Schrodinger's Cat*. New York: Bantam Books, 1988.

Grosslight, L., Unger, C., Jay, E., Smith, C.L. "Understanding models and their use in science; Conceptions of middle and high school students and experts." *Journal of Research in Science Teaching* 28, no. 9 (1991): 799-822.

Grossman, P., Richert, A. "Unacknowledged Knowledge Growth: A Re-examination of the effects of teacher educaiton." *Teaching and Teacher Education* 4, no. 1 (1988): 53-62.

Guilford, J.P. "Creativity." *American Psychologist* 5 (1950): 444-54.

Guilford, J.P. "Creative Abilities in the Arts." *Psychological Review* 64 (1957): 110-118.

Gulinska, H., Lewicki, R., Burewicz, A. "Interactive computer video programs used in the process of chemistry teaching." *Journal of Chemical Education* 68, no. 6 (1991): 490-492.

Gustafson, B.F., Rowell, P.M. "Elementary preservice teachers; constructing conceptions about learning science, teaching science and the nature of science." *International Journal of Science Education* 17, no. 5 (1995): 589-605.

Hackfort, C. "Newton's Optics: the changing spectrum of science." In *Let New Be!*, edited by R. Flood J. Fauvel, M. Shortland, R. Wilson, 81-100. Oxford: Oxford University Press, 1988.

Hagarty, M. "Constructing Mental Models of Machines from texts and diagrams." *Journal of Memory and Language* 32 (1993): 717-742.

Hand, B. "Student understanding of acids and bases: a two year study." *Research in Science Education* 19 (1989): 133-44.

Hand, B.M., Treagust, D.F. "Application of a conceptual conflict teaching strategy to enhance student learning of acids and bases." *Research in Science Education* 18 (1988): 53-63.

Hardwicke, A. "Using molecular models to teach chemistry, Part 1 Modelling Molecules." *School Science Review* 77 (1995): 59-64.

Hardy, M., Taylor, P. "Von Glaserfeld's Radical Constructivism: A critical review." *Science and Education* 6 (1997): 135-150.

Harré, R. *An Introduction to the Logic of the Sciences*. London: Macmillan, 1967.

Harré, R. *Varieties of Realism*. Oxford: Blackwell, 1986.

Harrison. "Modelling in Key Stages 1 and 2." Paper presented at the IDATER 92 Conference, Loughborough 1992.

Heimann, P.M. "Maxwell and the modes of consistent representation." *Archives for the History of Exact Sciences* 6 (1970): 171-213.

Hempel, C. *Aspects of Scientific Explanation*. New York: Free Press, 1965.

Herivel, J. *The Background to Newton's Principia. A study of Newton's dynamical researches in the years 1664-1684*. Oxford: Clarendon Press, 1965.

Hesse, M. *Models and Analogies in Science*. London: Sheen and Ward, 1963.

Hesse, M. *Revolutions and Reconstructions in the Philosophy of Science*. Brighton: Harvester, 1989.

Hodson, D. "Toward a philosophically more valid science curriculum." *Science Education* 72, no. 1 (1988): 19-40.

Hodson, D. "Re-thinking old ways: towards a more critical approach to practical work in school science." *Studies in Science Education*, no. 22 (1993): 85-142.

Hodson, D. "In search of a meaningful relationships: an exploration of some issues relating to integration in science and science education." *International Journal of Science Education* 14, no. 5 (1992): 541-562.

Hoeve-Brouwer, G. van, de Vos, W. "Chemical bonding or chemical structure?" In *Problem Solving and Misconceptions in Chemistry and Physics*, edited by H.-J. Schmidt, 238-245. Hong Kong: International Council of Associations for Science Education, 1994.

Hohmann, M., Banet, B., Weikart, D. *Young children in action*. Highscope: Ypsilanti, Michigan, 1997.

Holton, G. "On trying to understand Scientific Genius." In *Thematic Origins of Scientific Thought*, edited by G Holton. Cambridge: Harvard University Press, 1988.

Horowitz, P., Schwartz, J., Kindfield, A.C.H., Yessis, L.M., Hickey, D.T., Heidenberg, A., Wolfe, E.W. "Implementation and evaluation of the GenScope^{TM} learning environment: Issues, solutions, and results." Paper presented at the Third International Conference of the Learning Sciences, Charlottesville, VA in press.

Hoyningen-Huene, P. *Reconstructing scientific revolutions*. Chicago: The University of Chicago Press, 1993.

Hudson, L. *Frames of Mind*. Harmondsworth: Penguin, 1968.

Huibregtse, I., Korthagen, R., Wubbels, T. "Physics teachers' conceptions of learning, teaching and professional development." *International Journal of Science Education* 26, no. 5 (1994): 539-561.

Ingham, A.M., Gilbert, J.K. "The use of models by students of chemistry at higher education level." *International Journal of Science Education* 13, no. 2 (1991): 193-202.

Isaacson, W. "The Biotech Century, Time, Special Issue." *The Future of Medicine* January 11 (1999): 22-23.

Iverson, J., Goldin-Meadow, S. *Nature, 396* (1998): 228

Jamieson, I., Miller, A., Watts, A. "Production simulations." In *Teaching and Learning Technology*, edited by P. Murphy R. McCormick, M. Harrison, 251-267. Wokingham: Addison-Wesley, 1992.

Janda, M. "Collaboration in a Traditional Classroom Environment." *Written Communication* 7, no. 3 (1990): 291-315.

Johnson, M. *The Body in the Mind*. Chicago: The University of Chicago Press, 1987.

Johnson-Laird, P.N. *Mental Models*. Cambridge: Cambridge University Press, 1983.

Johnstone, A.H. "Why is science difficult to learn? Things are seldom what they seem." *Journal of Computerised Assisted Learning* 7 (1991): 75-83.

Jones, A., Carr, M. *Analysis of student technological capability Volume 2: Working papers from the second phase of the Learning in Technology Education Project*. Hamilton, New Zealand: Centre for Science and Mathematics Education Research, University of Waikato, 1993.

Jones, J.C. *Design Methods*. Chichester: Wiley, 1970.

Joseph, J.H., Dwyer, F.M. "The instructional effectiveness of integrating abstract and realistic visualization." Paper presented at the Paper presented at the Annual Meeting of the Association for Educational Communications and Technology, Research and Theory Division, Dallas, Texas 1982.

Justi, R., Gilbert, J.K. "History and philosophy of science through models: the case of chemical kinetics." *Science and Education* 8, no. 2 (1999a): 287-307.

Justi, R., Gilbert, J.K. "A cause of ahistorical science teaching: the use of hybrid models." *Science Education* 83, no. 2 (1999b): 163-178.

Justi, R.S. "Models in the teaching of chemical kinetics." Unpublished Doctoral Dissertation, University of Reading, 1997.

Justi, R., Gilbert, J.K. "History and philosophy of science through models: the challenges in the case of 'the atom'." *International Journal of Science Education* (in press).

Kail, R., Stevenson, M., Black, K. "Absence of a sex difference in algorithms for spatial problem solving." *Intelligence* 8 (1984): 37-46.

Kauffman, G.B. "History in the Chemistry Curriculum." *Interchange* 20, no. 2 (1989): 81-94.

Kelly, A.V., Kimbell, R.A., Patterson, V.J., Stables, K. *Design and Technological Activity: A Framework for Assessment*. London: Assessment of Performance Unit/HMSO, 1987.

Kelly, G. *The Psychology of Personal Constructs*. New York: Norton, 1955.

Kemp, K., Brown, T., Allen, L., Allen, C. *Feisty Females: Inspiring Girls to Learn Mathematics*. London: Heinemann, 1998.

Kennedy, M., Davies, R. *The Impact of Biotechnology on New Zealand Industry 1984-1994, a decade of rapid change*. Wellington: New Zealand Biotechnology Association, SIR Publishing, 1994.

Kimbell, R.A., Stables, K., Wheeler, T., Wosniak, A., Kelly, V. *The Assessment of Performance in Design and Technology*. London: Assessment of Performance Unit/HMSO, 1991.

Kimbell, R.A., Stables, K., Green, R. *Understanding Practice in Design and Technology*. Milton Keynes: Open University, 1996.

Kimbell, R. *Assessing Technology: International trends in curriculum and assessment*. Buckingham: Open University, 1997.

Kindfield, A.C.H. "Biology diagrams: Tools to think with." *Journal of the Learning Sciences* 3, no. 1 (1993/1994): 1-36.

Kindfield, A.C.H. "Understanding a basic biological process: Expert and novice models of meiosis." *Science Education* 78 (1994): 255-283.

Kipnis, N. "The 'Historical-Investigative' Approach to Teaching Science." *Science & Education* 5, no. 3 (1996): 277-292.

Kiser, L. "Interaction of spatial visualisation with computer enhanced and traditional presentations of linear absolute value inequalities." *Journal of Computers in Mathematics and Science Teaching* 10, no. 1 (1990): 85-96.

Koestler, A. *The Act of Creation*. London: Pan Books, 1964.

Koslow, R.E. "Sex-related differences and visual spatial mental imagery as factors affecting symbolic motor skill acquisition." *Sex Roles* 17 (1987): 521-527.

Kosslyn, S.M. "Seeing and imagining in the cerebral hemisphere: a computational approach." *Psychological Review* 94 (1987): 148-175.

Koulaidis, V., Ogborn, J. "Philosophy of Science an Empirical Study of Teachers' Views." *International Journal of Science Education* 11, no. 2 (1989): 173-184.

Koulaidis, V., Ogborn, J. "Science Teachers' Philosophical Assumptions: How well do we understand them?" *International Journal of Science Education* 17, no. 3 (1995): 273-283.

Kuhn, T.S. *The Structure of Scientific Revolutions (first edition)*. Chicago: The University of Chicago Press, 1962.

Kuhn, T.S. *The Structure of Scientific Revolutions (second edition)*. Chicago: The University of Chicago Press, 1970a.

Kuhn, T.S. "Logic of discovery or psychology of research." In *Criticism and the Growth of Knowledge*, edited by A. Musgrave I. Lakatos. Cambridge: Cambridge University Press, 1970b.

Kuhn, T.S. "Reflections on my critics." In *Criticism and the growth of knowledge*, edited by A. Musgrave I. Lakatos. Cambridge: Cambridge University Press, 1970c.

Kuhn, T.S. "Discussion." In *The Structure of Scientific Theories*, edited by F. Suppe. Urbana: The University of Illinois Press, 1977a.

Kuhn, T.S. "Second thoughts on paradigms." In *The Structure of Scientific Theories*, edited by F. Suppe. Urbana: The University of Illinois, 1977b.

Kuhn, T.S. "A function for thought experiments." In *The Essential Tension*, edited by T.S Kuhn. Chicago: University of Chicago Press, 1977d.

Kuhn, D. "Thinking as Argument." *Harvard Educational Review* 62, no. 2 (1992): 155-178.

Lakatos, I. "Falsification and the methodology of scientific research programmes." In *Criticism and the Growth of Knowledge*, edited by A. Musgrave I. Lakatos. Cambridge: Cambridge University Press, 1974.

Lakatos, I. "History of science and its rational reconstruction." In *The Methodology of Scientific Research Programmes*, edited by J. Worrall and G. Currie, 102-138. Cambridge: Cambridge University Press, 1978.

Lakin, S., Wellington, J. *Teaching the Nature of Science: A Study of Teachers' Views and their Implications for Science Education*. Sheffield: University of Sheffield, 1991.

Lakoff, G. *Women, Fire and Dangerous Things: What Categories Reveal about the Mind*. Chicago: University of Chicago Press, 1987.

Lantz, O., Kass, H. "Chemistry teachers' functional paradigms." *Science Education* 71 (1987): 117-137.

Larkin, J.H., Simon, H.A. "Why a diagram is (sometimes) worth ten thousand words." *Cognitive Science* 11 (1987): 65-99.

Larkin, J.H. "Display-based problem solving." In *Complex Information Processing: The impact of Herbert A. Simon*, edited by K. Kotovsky K. Klarh, 319-341. Hillsdale, NJ: Lawrence Erlbaum, 1989.

Lather, P. "Critical frames in educational research: feminist and post-structural perspectives." *Theory into Practice* 31, no. 2 (1992): 87-99.

Laudan, L. *Progress and its Problems*. London: Routledge and Kegan Paul, 1977.

Lave, J. *Situated Learning in Communities of Practice*. Edited by J. Levine L. Resnick, S. Teasley, *Shared Cognition: Thinking as Social Practice, Perspectives on Socially shared Cognition*. Washington DC: American Psychological Association, 1991.

Layton, D. "The Secondary School Curriculum and Science Education." *Physics Education* 18, no. 1 (1973): 19-23.

Layton, D. *Technology's Challenge to Science Education*. Buckingham: Open University Press, 1993.

Layton, D. *A School Subject in the Making? The Search for Fundamentals*. Edited by D. Layton. Vol. V, *Innovations in Science and Technology Education*. Paris: United Nations Educational, Scientific and Cultural Organisation, 1994.

Lazarowitz, R., Naim, R. "The use of three dimensional models in teaching "The Cell" to ninth grade biology students in an individualised approach." Paper presented at the The International Association for the Study of Cooperation in Education (IASCE), 3rd International Conference, Regina, Saskatchewan, Canada, July 1985.

Leatherdale, W. *The Role of Analogy, Model and Metaphor in Science*. Amsterdam: North Holland, 1974.

Lederman, N.G. "Students' and Teachers' Conceptions of the Nature of Science: A Review of Research." *Journal of Research in Science Teaching* 29, no. 4 (1992): 331-359.

Lemke, J. *Talking Science: Language, Learning and Values*. Norwood, NJ: Ablex, 1993.

Lexa, M. *Virtual Cell*: http://ampere.scale.uiuc.edu/~m-lexa/cell/, 1997.

Liddament. "Using models in design and technology education; some conceptual and pedagogic issues." Paper presented at the IDATER 93 Conference, Loughborough 1993.

Linn, M., Hyde, J.S. "Gender, mathematics and science." *Educational Researcher* 18, no. 8 (1989): 17-27.

Litt, M.D., Turk, D.C. "Sources of Stress and Dissatisfaction in experienced High School Teachers." *Journal of Educational Research* 78, no. 3 (1985): 178-185.

Liu, Y.T. "'What' and 'Where' is design creativity: a cognitive model for the emergence of creative design." Paper presented at the IDATER 96 Conference, Loughborough 1996.

Lohman, D.F. "The effect of speed accuracy trade-off on sex differences in mental rotation." *Perception and Psychophysics* 39 (1986): 427-436.

Lord, T.R. "Enhancing the visuo-spatial aptitude of students." *Journal of Research in Science Teaching* 22, no. 5 (1985): 395-405.

Lord, T.R. "A look at spatial abilities in undergraduate women science majors." *Journal of Research in Science Teaching* 24, no. 8 (1987): 757-767.

Lowe, R. "Selectivity in diagrams: reading beyond the lines." *Educational Psychology* 14, no. 4 (1994): 467-491.

Lowe, R. *Successful Instructional Diagrams*. London: Kogan Page, 1993.

Lubben, F., Millar, R. *A survey of the understanding of children aged 11-16 of key ideas about evidence in science (PACKS Research paper 3)*: University of York/University of Durham, 1994.

Lythcott, J. "Aristotelian was given as the answer but what was the question." *American Journal of Physics* 53 (1985): 279-281.

MacCoby, E.E., Jacklin, C.N. *The Psychology of Sex Differences*. Stanford: Stanford University Press, 1974.

Macer, D. "Genetic engineering: A perspective on Current Issues." Christchurch: DSIR Crop Research, 1990.

Mach, E. *Knowledge and Error*. Translated by C.M. Williams. Dordrecht: D.R. Reidel, 1976.

Mach, E. *Popular Scientific Lectures*. Translated by McCormack, T. 5th ed: La Salle: Open Court, 1943.

Manuel, F.E. *A Portrait of Isaac Newton*. London: Frederick Muller, 1968.

Marsh, G., Boulter, C. "What Size is it Really? Three Teacher/Researchs Investigate How Children Deal with Representations of Scale." *Primary Science Review* 53 (1998): 11-14.

Martin, M. *Concepts of Science Education: A Philosophical Analysis*. London: Scott, Foresman, 1972.

Martin, M. "Pseudoscience, the Paranormal and Science Education." *Science & Education* 3 (1994): 357-371.

Matthews, M.R. "History, Philosophy and Science Teaching: The Present Rapproachment." *Science & Education* 1, no. 1 (1992a): 11-47.

Matthews, M.R. "Teaching about air pressure: A role for history and philosophy in science teaching." In *History and Philosophy of Science in Science Education*, edited by S. Hills, 121-133. Kingston, Ontario: Queen's University, 1992b.

Matthews, M. *Science Teaching: The Role of the History and Philosophy of Science*. New York: Routledge, 1994.

Matthews, M., ed. *Constructivism in Science Education*. Dordrecht: Kluwer, 1998.

Mayer, R. "Models for Understanding." *Review of Educational Research* 59, no. 1 (1989): 43-64.

Mayers, J.F., Jahoda, G., Nelson, I. "Patterns of visual-spatial performances and 'spatial ability': dissociation of ethnic and sex differences." *British Journal of Psychology* 79, no. 105-119 (1988).

Mayr, E. *The Growth of Biological Thought*. Cambridge, MA: Belknap Press, 1982.

McClelland, J. "Alternative frameworks: interpretation of evidence." *European Journal of Science Education* 6, no. 1 (1984): 1-6.

McCormick, R. "Issues in the Current Practice of Technology Education." Paper presented at the ITEA Conference, Salt Lake City, Utah 1991.

McCormick, R. "The Evolution of Current Practice in Technology Education." In *Teaching and Learning Technology*, edited by R. McCormick, P. Murphy and M. Harrison. Reading, MA: Addisson-Wesley and Open University, 1993.

McCormick, R., Davidson, M. "Solving and the Tyranny of Product Outcomes." *The Journal of Design and Technology Education* 1, no. 3 (1996): 230-241.

McCormick, R. "Conceptual and Procedural Knowledge." *International Journal of Technology and Design Education* 7 (1997): 141-159.

McFie, J. "Intellectual imbalance: a perceptual hypothesis." *British Journal of Social Clinical Psychology* 12 (1973): 433-434.

McGee, G.M. "Human spatial abilities: Psychometric studies and environmental, genetic, hormonal and neurological influences." *Psychological Bulletin* 86, no. 5 (1979): 889-913.

McKellar, P. *Imagination and Thinking: A Psychological Analysis*. London: Cohen and West, 1957.

Merriam, S. *Case Study Research in Education: A Qualitative Approach*. London: Jossey-Bass, 1988.

Meyer, J., Kamens, D., Benavot, A. *School Knowledge for the Masses*. Edited by Studies in Curriculum History. London: Falmer, 1992.

Middleton, J. "Student-generated analogies in biology." *The American Biology Teacher* 53, no. 1 (1991): 42-46.

Millar, R., Osborne, J. (Eds). *Beyond 2000: Science Education for the Future*. London: School of Education, King's College London, 1999.

Millar, R., Driver, R. "Beyond Processes." *Studies in Science Education* 14 (1987): 109-118.

Miller, A.I. *Insights of Genius: Imagery and Creativity in Science and Art*. New York: Springer-Verlag, 1996.

Miller, G. *Living in the Environment*. 7th edition ed. Belmont, CA: Wadsworth, 1992.

Ministry of Education. *The New Zealand Curriculum Framework*. Wellington: Learning Media, 1993.

Ministry of Education. *Technology in the New Zealand Curriculum*. Wellington: Learning Media, 1995.

Mirham, G. "The Modelling Process." *IEEE Transactions on Systems, Man and Cybernetics* SMC - 2, no. 5 (1997): 621-629.

Monk, M., Osborne, J. "Placing the History and Philosophy of Science on the Curriculum: A Model for the Development of Pedagogy." *Science Education* 81, no. 4 (1997): 405-424.

Moyles, J. *Just Playing? The Role and Statues of Play in Early Childhood Education*. Buckingham: Open University, 1989.

National Research Council, (US). *National Science Education Standards*. Washington, DC: National Academy Press, 1996.

Nersessian, N. *Faraday to Einstein: Constructing meaning in scientific theories*. Dordrecht: Martinus Nijhoff, 1984.

Nersessian, N. "A cognitive-historical approach to meaning in scientific theories." In *The Process of Science*, edited by N. Nersessian. Dordrecht: Matinus Nijhoff, 1987.

Nersessian, N. "Conceptual change in science and in science education." *Synthese* 80 (1989): 163-183.

Nersessian, N. "How do Scientists Think? Capturing the Dynamics of Conceptual Change in Science." In *Cognitive Models of Science - Vol XV*, edited by R.N. Giere, 3-44. Minnesota Studies in Philosophy of Science, Minneapolis: University of Minnesota Press, 1992a.

Nersessian, N. "Constructing and Instructing: The role of 'abstraction techniques' in creating and learning physics." In *Philosophy of Science, Cognitive Psychology and Educational Theory and Practice*, edited by R. Duschl and D. Hamilton, 48-68. New Your: State University of New York Press, 1992b.

Nersessian, N. "Should physicists preach what they practice? Constructive modelling in doing and learning physics." *Science & Education* 4 (1995): 203-226.

Newcomb, N. "Sex related differences in spatial ability: problems and gaps in current approaches." In *Spatial abilities. Development and Physiological Foundations*, edited by J. Potegal. New York: Academic Press, 1982.

Nielsen, H., Thomsen, P.V. "History and philosophy of science in physics education." *International Journal of Science Education* 12, no. 3 (1990): 308-316.

Norman, D.A. "Some observations on mental models." In *Mental Models*, edited by D. Gentner A.L. Stevens, 7-14. Hillsdale, NJ: Lawrence Erlbaum, 1983.

Nott, M., Wellington, J. "Probing Teacher's Views of the Nature of Science: How Should We Do It and Where Should We Be Looking?" In *Research in Science Education in Europe: Current Issues and Themes*, edited by G. Welford, J. Osborne and P. Scott, 283-294. London: The Falmer Press, 1996.

Novak, J. *Learning, Creating and Using Knowledge*. New Jersey: Lawrence Erlbaum, 1998.

Nussbaum, J. "Children's conceptions of the Earth as a cosmic body: a cross-age study." *Science Education* 63, no. 1 (1979): 83-93.

Nussbaum, J. "Classroom Conceptual Change: The Lesson to be learned from the History of Science." Paper presented at the Proceedings of the International Seminar on Misconceptions in Science and Mathematics 1983.

Nussbaum, J. "The Earth as a cosmic body." In *Children's Ideas in Science*, edited by E. Guesne R. Driver, A. Tiberghien, 124-144. Milton Keynes: Open University Press, 1985.

O'Reilly, T. *The Open-Source Revolution*. Edited by Release 1: http://www.edventure.com/release1/1198.html, 1998.

Ogborn, J., Millar, R. "Computational Issues in Modelling." In *Learning with Artificial Worlds*, edited by H. Mellar et al, 33-38. London: Falmer Press, 1994.

Ogborn, J., Kress, G., Martins, I., McGillicuddy, K. *Explaining Science in the Classroom*. Buckingham: Open University Press, 1996.

Ogborn, J. "Recovering Reality." *Studies in Science Education* 25 (1995): 3-38.

Olejnik, I., Farmer, B. *Biology and Industry*. Glasgow: Blackie, 1989.

Ontario Ministry of Education and Training. *The Common Curriculum Grades 1-9*. Ontario, 1993.

Ormell, C., Bentley, C., Flemming, W., Knowles, F., Malvern, D., Morgan, J., Sewell, B. *Teaching Mathematics Applicable*. London: Heinemann Educational Books, 1979.

Osborne, J. "Beyond Constructivism." *Science Education* 80, no. 1 (1996): 53-82.

Osborne, R.J., Wittrock, M.C. "The generative learning model and its implication for science education." 12 (1985): 59-87.

Pacey, A. *The Culture of Technology*. Oxford: Blackwell, 1983.

Pallrand, G.J., Seeber, J. "Spatial ability and achievement in introductory physics." *Journal of Research in Science Teaching* 21, no. 5 (1984): 507-516.

Papert, S. *Mindstorms: Children, Computers and Powerful Ideas*. Edited by Basic Books, *New York*, 1980.

Penfold, J. *Craft, Design & Technology: Past, Present & Future*. Stoke-on-Trent: Trentham, 1988.

Pevsner, N. *Pioneers of Modern Design*. Harmondsworth: Penguin, 1960.

Pfund, H., Duit, R. *Students' Alternative Frameworks and Science Education*. Kiel, Germany: Institute for Science Education, University of Kiel, 1988.

Physics, Royal Society Institute of. "The Mathematical Needs of A Level Physics Students." *Physics Education* 8 (1973): 241-246.

Physics, Royal Society/Institute of Mathematics and its applications/Institute of. *Mathematics-Physics Interface in Schools*. London: The Royal Society, 1986.

Piaget, J. *The Child's Conception of Movement and Speed*. Translated by G.E.T. Holloway, M.J. Mackenzie. New York: Ballantine Books, 1971.

Piaget, J. *The Construction of Reality in Children*. New York: Basic Books, 1954.

Piaget, J. *The Grasp of Consciousness*. Translated by S. Wedgwood. Cambridge, MA: Harvard University Press, 1976.

Piaget, J. *Judgement and Reasoning in the Child*. Totowa, New Jersey: Littlefield Adams, 1924/1969.

Piaget, J. *The Mechanisms of Perception*. London: Routledge and Kegan Paul, 1969.

Piaget, J. *Play, Dreams and Imitation*. New York: Norton, 1962.

Piaget, J, Garcia, R. *Psychogenesis and the History of Science*. New York: Columbia University Press, 1989.

Plowden, Lord. *Children and their Primary Schools*. London: HMSO, 1967.

Pope, M., Keen, T. *Personal Construct Psychology in Education*. London: Academic, 1981.

Popper, K.R. *Objective Knowledge: An Evolutionary Approach*. Revised ed. Oxford: Clarendon Press, 1979.

Posner, G., Strike, K., Hewson, P., Gertzog, W. "Accommodation of a scientific conception: towards a theory of conceptual change." *Science Education* 66, no. 2 (1982): 211-227.

Prain, V., Boulter, C., Armitage, M. "What's going to happen in the eclipse tonight: Rethinking perspectives on primary school science." *International Journal of Science Education* 20, no. 4 (1998): 487-500.

Primrose, S. *Molecular Biotechnology*. 2nd ed. Oxford: Blackwell Scientific Publications, 1991.

Psacharopoulos, G. "Are Teachers Over Paid? Some Evidence from Brazil." *Teaching & Teacher Education* 3, no. 4 (1987): 315-318.

Raymond, E. *The Cathedral and the Bazaar*: http://www.tuxedo.org/~esr/writings/cathedral-bazaar, 1998.

Resnick, M. *Turtles, Termites and Traffic Jams*. Cambridge MA: MIT Press, 1994.

Ribe, N.M. "Goethe's Critique of Newton: A Reconsideration." *Studies in History and Philosophy of Science* 16, no. 4 (1985): 315-335.

Richards, T., Richards, L. "Using Computers in Qualitative Research." In *Handbook of Qualitative Research*, edited by N. Denzin and Y. Lincoln. Newbury Park: Sage, 1994.

Richmond, G.P. "A Limited Sex Difference in Spatial Tests Scores with a Pre-adolescent Sample." *Child Development* 51, no. 2 (1980): 601-602.

Roberts, D. "Junior High School Science Transformed: Analysing a Science Curriculum Policy Change." *International Journal of Science Education* 17, no. 4 (1995): 493-504.

Roberts, P. "Of models, modelling and design: an applied philosophical enquiry." In *Modelling: the language of designing, Design: Occasional Paper No.1*, edited by B. Archer P. Roberts, K. Baynes. Loughborough: Department of Design and Technology, University of Loughborough, 1992.

Rodriguez, W.E. "A dual approach to engineering design visualisation." *Engineering Design Graphics Journal* 54, no. 3 (1990): 36-43.

Rogoff, B. *Apprenticeship in Thinking*. New York: Oxford University Press, 1990.

Ross, B., Munby, H. "Concept mapping and misconceptions: a study of high-school students' understandings of acids and bases." *International Journal of Science Education* 13, no. 1 (1991): 11-23.

Roth, W.-M. *Authentic School Science*. Dordrecht: Kluwer, 1995.

Rouse, W.B., Morris, N.M. "On looking into the black box: Prospects and limits in the search for mental models." *Psychological Bulletin* 100, no. 3 (1986): 349-363.

Russell, T., Munby, H. "Science as a Discipline, Science as seen by students and teachers' professional knowledge." In *Doing Science: Images of Science in Science Education*, edited by R. Millar. London: The Falmer Press, 1989.

Rutherford, F., Ahlgren, A. *Science for All Americans*. New York: Oxford University Press, 1990.

Rutherford, M. "Using models of colour." In *Exploring Models and Modelling in Science and Technology Education*, edited by J.K. Gilbert, 162-179. Reading: The New Bulmershe Papers, University of Reading, 1997.

Rutter, P. "The Effect of Studying A-level Mathematics on performance in A-level Physics." *Physics Education* 29, no. 1 (1994): 8-13.

Sabra, A.I. *Theories of Light: from Descartes to Newton*: Cambridge University Press, 1981.

Salomon, G. "The use of visual media in the service of enriching mental thought processes." *Instructional Science* 9 (1980): 327-339.

Sawyer, W.W. *Mathematician's Delight*. London: Pelican, 1961/1943.

SCAA, (Schools Curriculum and Assessment Authority). *Nursery Education: Desirable outcomes for children's learning on entering compulsory education*. London: QCA, 1996.

Scardamalia, M., Bereiter, C. "Higher levels of agency for children in knowledge building: a challenge for the design of new knowledge media." *The Journal of the Learning Sciences* 1, no. 1 (1991): 37-68.

Schaverien, L., Cosgrove, M. "Learning to teach generatively: Mentor-supported professional development and research in technology-and-science." *Journal of the Learning Sciences* 6, no. 3 (1997): 317-346.

Schmidt, H-J. "Applying the concept of conjugation to the Bronsted theory of acid-base reactions by senior high school students from Germany." *International Journal of Science Education* 17, no. 6 (1995): 733-742.

Schmidt, H-J. "A label as a hidden persuader: chemists' neutralisation concept." *International Journal of Science Education* 13, no. 4 (1991): 459-471.

Science for Living. The circulatory system: a prototype interactive multimedia resource. Apple Computer and Stanford University, Stanford, CA. 1990

Secondary Examinations Council, Kimbell, R. *Craft Design and Technology. GCSE: A Guide for Teachers*. Milton Keynes: Open University Press, 1986.

Shallcross, D. *Teaching Creative Behaviour*. Buffalo, New York: Prentice Hall, 1981.

Sharpe, D. "Out with the old, in with the new: transition to technology education in North America." *The Journal of Design and Technology Education* 1, no. 1 (1996).

Shoaff-Grubbs, M.M. "The effect of the graphing calculator on female students' spatial visualisation skills and levels of understanding in elementary graphing and algebra concepts." Doctoral Dissertation, College of New Rochelle, 1992.

Shulman, L. "Knowledge and teaching: foundations of the new reforms." *Harvard Educational Review* 57, no. 1 (1987): 1-22.

Shulman, L. "Those who understand: Knowledge growth in teaching." *Educational Researcher* 15 (1986a): 4-14.

Shulman, L. "Those who understand: A conception of teacher knowledge." *American Educator* 10 (1986b): 9-15.

Shulman, L. "Toward a Pedagogy of Cases." In *Case Methods in Teacher Education*, edited by J. Shulman. New York: Teachers' College Press, 1992.

Sinclair, J., Coulthard, R. *Towards an Analysis of Discourse*. London: Oxford University Press, 1975.

Smith, D.C., Cypher, A., Spohrer, J. "KidSim: Programming agents without a programming language." *Communications of the ACM 1994* (1994): 54-67.

Smith, I.M. *Spatial Ability: Its Educational and Social Significance*. London: University of London Press, 1964.

Solomon, J. "The rise and fall of constructivism." *Studies in Science Education* 23 (1994): 1-19.

Solomon, J. "Teaching the History of Science: Is nothing sacred?" In *Teaching the History of Science*, edited by M. Shortland and A. Warwick, 42-53. Oxford: Basil Blackwell, 1989.

Sorensen, R.A. *Thought Experiments*. New York: Oxford University Press, 1992.

Sparkes, J. "Modelling." In *Technology for Technology Education*, edited by C. Newey R. McCormick, J. Sparkes, 75-87. Padstow: Addison-Wesley, Open University, 1992a.

Sparkes, J. "Quality and quality assurance." In *Technology for Technology Education*, edited by C. Newey R. McCormick, J. Sparkes, 123-132. Padstow: Addision-Wesley, Open University, 1992b.

Staudenmaier, J. *Technology's Storytellers. Reweaving the Human Fabric*. Cambridge, MA: Society for the History of Technology and The Massachusetts Institute of Technology Press, 1989.

Stein, M.I. *Making the Point*. New York: The Mews Press, 1984.

Sternberg, R.J, ed. *The Nature of Creativity: Contemporary Psychological Perspectives*. Cambridge: Cambridge University Press, 1988.

Stewart, J., Hafner, B. "Extending the Conception of Problem Solving." *Science Education* 75, no. 1 (1991): 105-120.

Stork, H. "Using History and Philosophy of Science in Science Teaching." Paper presented at the Proceedings of the Third International History, Philosophy and Science Teaching Conference, Minneapolis 1995.

Strike, K., Posner, G. "A revisionist theory of conceptual change." In *Philosophy of Science, Cognitive Psychology and Educational theory and Practice*, edited by R. Duschl and R. Hamilton, 147-176. New York: State University of New York Press, 1992.

Sutton, C. "Beliefs About Science and Beliefs About Language." *International Journal of Science Education* 18, no. 1 (1995): 1-18.

Sutton, C. "Figuring Out a Scientific Understanding." *Journal of Research in Science Teaching* 30, no. 10 (1993): 1215-1227.

Swaine, M. "I Join the Rebel Alliance." *Dr. Dobb's Journal* 296 (1999): 107-109.

Swaine, M. "Learning by Building Expert Systems." In *Learning with Artificial Worlds*, edited by H. Mellar et al. London: Falmer, 1994.

Swift, D.J. "Curricular Philosophy of Science Today." In *Philosophies of Science in Science Education*, edited by D.J. Swift et al. Guilding: IET mimeograph, University of Surrey, 1982.

Talley, H.L. "The Use of Three-dimensional Visualisation as a Moderator in the Higher Cognitive Learning of Concepts in College Level Chemistry." *Journal of Research in Science Teaching* 10, no. 3 (1973): 263-269.

Teles, E.J. "Numerical and Graphical Presentation of Functions in Pre-calculus: A Doctoral Dissertation." In *Dissertation Abstracts International*. Maryland: University of Maryland College Park, 1990.

Thagard, P. "Analogy, Explanation and Education." *Journal of Research in Science Teaching* 29, no. 6 (1992): 537-544.

Thewlis, J. *Concise Dictionary of Physics and Related Subjects*. 2nd Edition ed. Oxford: Pergamon, 1979.

Thompson, N. *Thinking Like a Physicist*. Bristol: Adam Hilger, 1987.

Tiberghien, A. "Modelling as a Basis for Analysing Teaching-learning Situations." *Learning and Instruction* 4 (1994): 71-87.

Tobin, K., Tippins, D. "Constructivism as a Referent for Teaching and Learning." In *The Practice of Constructivism in Science Education*, edited by K. Tobin, 3-21. Hillsdale, N.J.: Erlbaum, 1993.

Tobin, K., Tippins, D., Gallard, A. "Research on Instructional Strategies for Teaching Science." In *Handbook of Research on Science Teaching and Learning*, edited by D. Gabel, 45-93. New York: MacMillan, 1994.

Torrance, E.P. *Guiding Creative Talent*. Englewood Cliffs, New Jersey: Prentice Hall, 1962.

Toulmin, S. *Human Understanding*. Oxford: Oxford University Press, 1972.

Tracy, D.M. "Toys, Spatial Ability and Science and Mathematics Achievement are they Related?" *Sex Roles* 17 (1987): 115-138.

Treagust, D. "Enhancing students' understanding of science using analogies." In *Teaching and Learning in Science: The Constructivist Classroom*, edited by V. Prain B. Hand, 44-61. Sydney: Harcourt Brace, 1995.

Treagust, D. "The Evolution of an Approach for using Analogies in Teaching and Learning Science." *Research in Science Education* 22 (1993) 293-301.

Treagust, D., Duit. R., Joslin, P., Lindauer, I. "Science teachers' use of analogies: observations from classroom practice." *International Journal of Science Education* 14 (1992): 413-422.

Tuckey, H.P., Selvaratnam, M., Bradley, J.D. "Identification and rectification of student difficulties concerning three dimensional structures, rotation and reflection." *Journal of Chemical Education* 68 (1991): 460-464.

Tuckey, H.P., Selvaratnam, M. "Studies involving three-dimensional visualisation skills in chemistry: A review." *Studies in Science Education* 21 (1993): 99-121.

Tufte, E.R. *Visual Explanations*. Cheshire, CT: Graphics Press, 1997.

Tversky, B. "Parts, partonomies and taxonomies." *Developmental Psychology* 25, no. 6 (1989): 983-995.

Twyman, M. "A schema for the study of graphic language." In *Processing of Visible Language*, edited by M.E. Wrolstad P.A. Kolers, H. Bouma, 117-150. New York: Plenum Press, 1979.

Twyman, M. "Using pictorial language: A discussion of the dimensions of the problem." In *Designing Usable Texts*, edited by R. Waller T.M.Duffy, 245-312: Academic Press, 1985.

UNESCO. *Technology Education as Part of General Education*. Paris: UNESCO, 1983.

Van Brunt, J. "The new biology." In *Biotechnology. The Science and the Business*, edited by V. Moses & R. Cape, 131-143. New York: Harwood Academic Publishers, 1991.

Van Driel, J. "Teachers' knowledge about the nature of models and modelling in science." Paper presented at the Paper presented at NARST (National Association for Research in Science and Technology Annual Meeting, San Diego, CA, April 19-22 1998 1998.

van Zee, E., Ministrell, J. "Reflective Discourse: Developing Shared understanding in a Physics Classroom." *International Journal of Science Education* 19, no. 2 (1996): 193-108.

Vazquez, J.L. "The effect of the calculator on student achievement in graphing linear functions: A Doctoral dissertation." In *Dissertation Abstracts International*. Florida: University of Florida, 1991.

Viennot, L. *Le raisonnement spontane en dynamique elementaire*. Paris: Hermann, 1979.

Von Laue, M. *History of Physics*. New York: Academic Press, 1950.

Vosniadou, S., Brewer, W.F. "Mental models of the earth: A study of conceptual change in childhood." *Cognitive Psychology* 24, no. 4 (1992): 535-585.

Vosniadou, S. "Universal and culture-specific properties of children's mental models of the earth." In *Mapping the mind: Domain specificity in cognition and culture*, edited by S.A. Gelman L.A. Hirschfeld. Cambridge: Cambridge University Press, 1994.

Vygotsky, L. *Thought and Language*. Cambridge, Mass.: Harvard University Press, 1978.

Vygotsky, L.S. *Mind in Society*. Cambridge: Harvard University Press, 1978.

Waber, D.P. "Sex differences in cognition: A function of maturation rate." *Science* 192, no. 572-574 (1976).

Wandersee, J.H. "Can the history of science help science educators anticipate students' misconceptions?" *Journal of Research in Science Teaching* 23, no. 7 (1985): 581-597.

Watts, D.M. Personal Communication to J.Gilbert. 1998.

Webb, M. "Learning by Building Expert Systems." In *Learning with Artificial Worlds*, edited by H. et al Mellar, 162-170. London: Falmer, 1994.

Weck, M.A. "Are today's models tomorrow's misconceptions?" Paper presented at the Proceedings of the Third International History, Philosophy and Science Teaching Conference, Minneapolis 1995.

Welch, M. "Students use of three-dimensional modelling while designing and making a solution to a technological problem." *International Journal of Technology and Design Education* 8, no. 3 (1998): 241-260.

Wertheimer, M. "The thinking that led to the theory of relativity." In *Productive Thinking*, edited by M Wertheimer. New York: Harper & Brothers, 1945.

Westfall, R. *The Life of Isaac Newton*. Cambridge: Cambridge University Press, 1993.

White, B., Frederiksen, J. "Inquiry Modelling and Metacognition: Making Science Accessible to all Students." *Cognition and Instruction* 16, no. 1 (1998): 3-118.

White, R. "Dimensions of content." In *The Content of Science: A Constructivist approach to teaching and learning*, edited by R. Gunstone P. Fensham, R. White, 225-262. London: Falmer, 1994.

Wigner, E.P. *Communications in Pure and Applied Mathematics*. Chichester: John Wiley, 1960.

Wiley, S.E. "Computer graphics and the development of visual perceptions in engineering graphics curricula." *Engineering Design Graphics Journal* 54, no. 3 (1990): 39-43.

Wilkes, K.V. *Real People: Personal identity without thought experiments*. Oxford, UK: Clarendon Press, 1989.

Willes, M. *Children into Pupils*. London: Routledge Kegan Paul, 1983.

Wilson, S.M., Shulman, L.S., Richert, A. "'150 Different Ways' of Knowing: Representations of Knowledge in Teaching." In *Exploring Teachers' Thinking*, edited by J. Calderhead, 104-124. London: Cassel, 1987.

Wingler. *The Bauhaus*. Cambridge, Mass: MIT Press, 1969.

Witkin, H.A. "Social influences in the development of cognitive style." In *Handbook of Socialisation Theory and Research*, edited by D. Gaskin. New York: Rand-McNally, 1969.

Wong, E. "Self-Generated Analogies as a Tool for Constructing and Evaluating Explanations of Scientific Phenomena." *Journal of Research in Science Teaching* 30, no. 4 (1993): 367-380.

Wubbels, T. "Taking Account of Student Teachers' Preconception." *Teaching & Teacher Education* 8, no. 2 (1992): 137-149.

Zimmermann, E., Gilbert, J.K. "Contradictory Views of the Nature of Science held by a Brazilian Secondary School Physics Teacher: Educational Value of Interviews." *Educational Research and Evaluation* 4, no. 3 (1998): 213-234.

Zimmermann, E. "The Interplay of Pedagogical and Science related issues in Physics Teachers' Classroom Activities." Unpublished Doctoral Dissertation, University of Reading, 1997a.

Zimmermann, E. *Models of Science in Science Education*. Edited by J.K. Gilbert, *Exploring Models and Modelling in Science Education: Contributions from the MISTRE Group*. Reading: New Bulmershe Papers Series, University of Reading, 1997b.

INDEX

T